GAS FLUIDIZATION TECHNOLOGY

Edited by

D. Geldart

University of Bradford, UK

A Wiley–Interscience Publication

JOHN WILEY & SONS
Chichester · New York · Brisbane · Toronto · Singapore

Library of Congress Cataloging-in-Publication Data:

Main entry under title:

Gas fluidization technology.
    'A Wiley–Interscience publication.'
    Includes index.
    1. Fluidization.   I. Geldart, D.

TP156.F65G37 1986   660.2'842   85–26532
ISBN 0 471 90806 1

British Library Cataloguing in Publication Data:

Gas fluidization technolgy.
    1. Fluidization   2. Gases
    I. Geldart, D.
    660.2'884292   TP156.F65

ISBN 0 471 90806 1

Typeset by Quadra Graphics
Printed and Bound in Great Britain

GAS FLUIDIZATION TECHNOLOGY

Contents

Contributors

DR JAN BAEYENS, *SIPEF SA, Kasteel Calesberg 2120, Schoten, Belgium*

DR J.S.M. BOTTERILL, *Department of Chemical Engineering, University of Birmingham, PO Box 363, Birmingham B15 2TT, UK*

PROFESSOR ROLAND CLIFT, *Department of Chemical and Process Engineering, University of Surrey, Guildford, Surrey GU2 5XH, UK*

DR DEREK GELDART, *Postgraduate School of Powder Technology, University of Bradford, Bradford BD7 1DP, West Yorkshire, UK*

PROFESSOR JOHN R GRACE, *Department of Chemical Engineering, University of British Columbia, Vancouver, V6T 1W5, Canada*

DR TED M. KNOWLTON, *Institute of Gas Technology, 4201 W. 36th St., Chicago, Illinois 60632, USA*

DR DAVID REAY, *Engineering Sciences Division, AERE Harwell, Oxfordshire, OX11 ORA, UK*

DR LADISLAV SVAROVSKY, *Postgraduate School of Powder Technology, University of Bradford, Bradford BD7 1DP, West Yorkshire, UK*

DR JOSEPH YERUSHALMI, *PAMA Limited, 7 Kehilat Saloniki St., Tel-Aviv 69 513, PO Box 24119, Israel*

CHAPTER 1

Introduction

D. GELDART

1.1 CHARACTERISTICS OF FLUIDIZED SYSTEMS

A fluidized bed is formed by passing a fluid, usually a gas, upwards through a bed of particles supported on a distributor. Although it is now known that even above the minimum fluidization velocity the particles are touching each other most of the time, with the exception of cohesive solids the interparticle friction is so small that the fluid/solid assembly behaves like a liquid having a density equal to the bulk density of the powder; pressure increases linearly with distance below the surface, denser objects sink, lighter ones float, and wave motion is observed. Solids can be removed from or added to the bed continuously, and this provides many processing advantages. All fine powders have a very large specific surface area — 1 m³ of 100 μm particles has a surface area of about 30,000 m² — but in a fluidized bed the stirring action of the gas bubbles continually moves the powder around, shearing it and exposing it to the gas; it is this good solids mixing which gives the high rates of heat transfer from surface to bed and gas to particle, and which is responsible for isothermal conditions radially and axially. Compared with a fixed bed of the same powder operated at the same bed depth and gas velocity, the pressure drop over a fluidized bed is much smaller, and this together with most of the characteristics described above make the fluidized bed an attractive choice as a chemical or physical processing tool.

There are, however, disadvantages which may outweigh the attractive features: for some applications the gas bubbles make scale-up more difficult and provide a means whereby the reacting gases can avoid contact with the solids; particle entrainment is almost inevitable and particle attrition and metal surface erosion occur in regions where gas velocities are high.

The basic component required for a fluidized bed are items 1 to 4 on Fig.

1

1.1 — a container, a gas distributor, a powder, and a source of gas. The provision of the other elements shown depends on the particular application contemplated. For example, fluidized bed granulators may include the spray, screw conveyor, and solids offtake pipe but not the internal cyclone or heat transfer surfaces; acrylonitrile reactors have internal cyclones and internal heat exchanger tubes but no spray nozzles, screw conveyors, or standpipes.

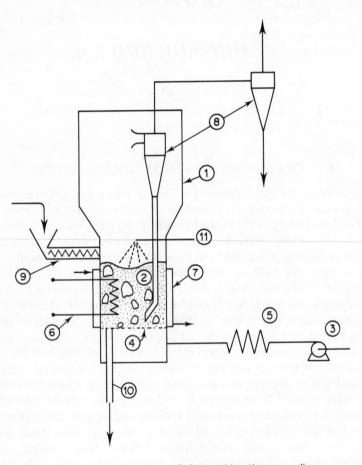

1. Shell	6. Internal heating or cooling
2. Powder	7. External heating or cooling
3. Blower	8. Cyclones
4. Gas distributor	9. Solids feeder
5. Heat exchanger	10. Solids offtake
for fluidizing gas	11. Spray feed

Figure 1.1 A conceptualized fluidized bed which could be used for a wide variety of applications.

The variety of fluidized beds encountered in commercial operation is enormous and includes powders having mean sizes as small as 15 μm and as large as 6 mm, bed diameters from 0.1 to 10 m, bed depths from a few centimetres to 10 m and gas velocities from 0.01 to 3 m/s or even as high as 10 m/s for recirculating high velocity beds. The behaviour of particulate solids in fluidized beds depends largely on a combination of their mean particle size and density (Geldart, 1973), and it has become increasingly common to discuss fluidized systems in relation to the so-called Geldart fluidization diagram. Although this is to be discussed in detail in Chapter 3 a simplified version is shown in Fig. 1.2, and this can be used to identify the 'package' of

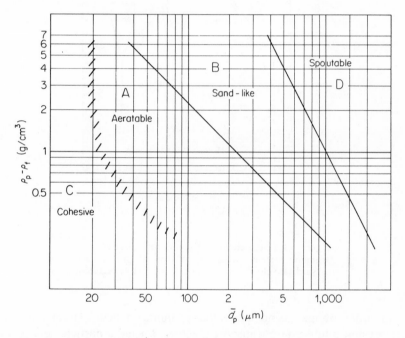

Figure 1.2 Simplified diagram for classifying powders according to their fluidization behaviour in air at ambient conditions (Geldart, 1973).

fluidization characteristics associated with fluidization of any particular powder at ambient conditions. A somewhat different 'phase diagram' has recently been proposed by Grace (1984). This is based on an earlier diagram proposed by Reh (1968) and can be used to broadly identify flow regimes appropriate to combinations of gas velocity and particle properties (Fig. 1.3). The Archimedes number, Ar (or $3/4C_D \, Re^2$, as it is sometimes written), characterizes the basic particle/gas properties and Re/C_D characterizes the gas velocity. Thus at low values of $Ar^{1/3}$ (small particles) the processing options

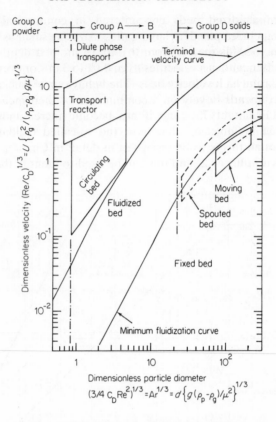

Figure 1.3 Regime/processing-mode diagram for grouping systems according to type of powder and upward gas velocity used (Grace, 1984; after Reh, 1968).

are, in order of increasing gas velocity, fluidized beds, circulating beds, transport and dilute phase transport reactors; for large particles the options are fixed, moving/spouted beds, and fluidized beds.

1.2 BRIEF HISTORY OF FLUIDIZATION

Although the technique of fluidization was in commercial use as early as 1926 for the gasification of coal, it was not until the early 1940s that its widespread use began with the construction of the first fluid bed catalytic cracker (FCC). Various accounts of the development of this and other fluidized bed processes have been given (Jahnig, Campbell, and Martin, 1980; Kunii and Levenspiel, 1969) and a comprehensive account of the industrial scene up to 1967 was given by Geldart (1967, 1968, 1969); but the most complete and carefully researched account is that given by Squires (1982), who has been associated with process invention and development involving fluid beds since 1946. Squires provides a fascinating picture of how commercial and wartime pressure, together with a compromise between daring innovation and the need to reduce catalyst losses, steered the design away from the high velocity (upflow) mode towards low velocity (downflow) operation; since that time more than 350 FCC units have been built and most are still in operation. In the late 1940s Dorr Oliver applied the technique successfully to the roasting of sulphide ores and since that time virtually all new ore roasters have been fluidized beds. Fluid bed dryers also made rapid progress and by the mid 1950s the technique was well established.

However, in all these applications the degree of conversion required was either not critical (as in FCC) or easily achievable (as in roasting and drying); a major setback occurred when the large fluidized bed Fischer–Tropsch synthesis plant at Brownsville in Texas fell far short of the conversions achieved in the pilot plant. The aim was to use natural gas to manufacture gasoline, but unlike the other processes, here bubble hydrodynamics were critical and their crucial role in scale-up was not properly appreciated. Although eventually the plant came somewhere near its designed output, with the discovery of huge quantities of oil in the Middle East, the economics became so unfavourable that the plant was shut down in 1957.

The Sohio process for making acrylonitrile in a fluidized bed was immensely successful; since 1960 virtually all new acrylo plants have been fluidized beds and 50 large units are in operation throughout the world. Undoubtedly the major success in the late 1970s and in this decade is the Union Carbide polyethylene synthesis process. The alternative high pressure liquid phase reactors were limited in scale of operation, whereas in the low pressure gas phase process single fluid bed units can be built as large as required; the better quality product and dramatic reduction in costs which are features of this fluid bed process have ensured the demise of virtually all competitors. Also in the 1970s and 1980s, fluidized bed combustion has attracted much attention largely due to its relatively low temperature operation (800 to 900°C) and its ability to absorb SO_2 through the use of limestone or dolomite. These features mean that NO_x and SO_2 emissions in the flue gases can be made acceptably low. Although such units operate at 1 to 2 m/s, an atmospheric pressure fluidized bed boiler for commercial power generation would constitute an extremely large pice of equipment. The more compact pressurized fluidized bed combustors are therefore attractive but commercially are still a long way off. A great many small atmospheric pressure fluidized combustors are in use throughout the world — probably more than 2,000 — on a variety of duties including burning plastic waste, providing hot gases for drying grass, and raising steam for process use (Highley and Kaye, 1983).

The history of fluidization contains examples of processes which were developed but were either never built on a commercial scale or were built and operated for only a short time. The Shell chlorine process in which HCl is oxidized to chlorine was systematically scaled up (de Vries *et al.*, 1972) using a combination of cold models, pilot plants, and theoretical equations; as far as is known the full-scale plants were successful, but it is believed that none is still in operation.

Use of fluid beds for ethylene oxide production and ethanol dehydration does not appear to have progressed beyond the pilot plant stage, and although the Mobil methanol-to-gasoline process has recently been demonstrated to be a technological success (Penick, Lee and Mazink, 1982), currently the economics are attractive in only a few countries.

As with all new processes, unless there is an outstanding economic or product quality advantage there is little enthusiasm for exchanging a well-known process for one which offers marginal improvement while incurring a potential scale-up risk. Where the fluidized bed will form the major part of the overall processing cost, risk-taking may be justified, but in many processes there are so many other unit operations before and after the fluid bed that its substitution makes little impact on the unit product cost. Many current industrial applications of fluidized beds are shown in Fig. 1.4., arranged in five categories according to predominating mechanisms. Some of these are treated in detail in two recent books (Hetsroni, 1982; Yates, 1983).

1.3   SOURCES OF INFORMATION

For those who wish to maintain an active interest in the field it is essential to read the general technical press, scientific journals, and proceedings of conferences which are partly or wholly concerned with fluidization. Most of the well-known chemical engineering journals regularly publish papers on fluidization, e.g. *Chemical Engineering Science, Chemical Engineering Journal, Chemical Engineering Research and Development, American Institution of Chemical Engineers Journal, Canadian Journal of Chemical Engineering, Journal of Chemical Engineering Japan,* etc. *Powder Technology* probably publishes more papers relevant to fluidization than any other single journal and at roughly two year intervals publishes a list of recent papers and ongoing research projects involving the technique (Geldart, 1982, 1983). However, conferences are undoubtedly the biggest single source of information: the Engineering Foundation (New York) has organized conferences on the subject in 1975, 1978, 1980, 1983, and 1986, and published the 60 to 80 refereed papers each time in book form; the Institute of Energy (London) has published proceedings of conferences on fluidized combustion since 1980 (e.g. Beer, Massimilla, and Sarafim, 1975), and proceedings of fluidized bed combustion conferences in the United States since 1970 are also

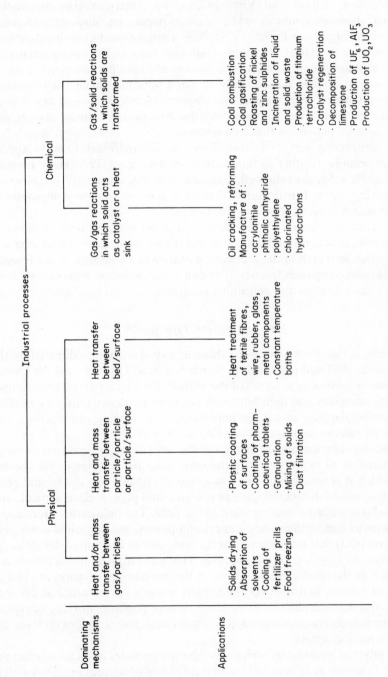

Figure 1.4 Classification of fluid bed applications according to predominating mechanisms.

available (e.g. Elliott and Virr, 1972). The American Gas Association regularly sponsors symposia which include papers on fluidized bed gasification (e.g. Elgin and Perks, 1974). The annual meeting of the American Institution of Chemical Engineers (AIChE) held in November each year usually contains four sessions on fluidization and fluid/particle systems but by no means all appear subsequently in the Chemical Engineering Progress symposium series. International symposia on chemical reaction engineering are held frequently and published by the American Chemical Society and always contain papers on fluidized systems.

The abstracting service provided by the International Energy Agency includes sections relating to fluidization, as does the HTFS (Heat Transfer and Fluid Flow Service, Harwell, United Kingdom), Chemical Abstracts, and Engineering Index. It is fortunate that these services are now computerized, making searches much less time-consuming.

The active researcher is often able to keep up with the field by personal contact with others throughout the world; those involved in fluidization and fluid/particle research in the academic world form a competitive but friendly network, and companies frequently obtain access to recent and ongoing work through short courses and consulting contracts.

1.4 SCOPE OF THE BOOK

This book is based on a series of three-to-four-day courses directed by the editor since 1968 and presented in North America, Holland, and the United Kingdom at least once a year. All the authors have lectured on these courses on many occasions and their approach has been fashioned partly by working together but largely, and more importantly, by noticing the reactions and comments of attendees and responding to their needs.

All of the contributors are experts in the field either as active researchers/consultants or as practitioners in industry. Like the lectures in the courses from which it is derived the book is aimed at engineers in design and plant operation, scientists/technologists in research and process development, and postgraduate students starting work in the field. The fluidization literature is so enormous and so many new papers and patents are published every year (Geldart, 1983) that it is difficult for the specialist to keep up — let alone for the novice to gain a good perspective. The aim throughout, therefore, has been to give the reader a clear picture of the mechanisms at work, in so far as these are known, and to cover the subject broadly yet in sufficient detail to enable practical calculations to be made. Where possible, working equations and correlations are recommended and numerical illustrations of their use are given in most chapters.

The physical parameters which describe the particles and characterize the fluidized powder at or near minimum fluidization conditions are dealt with in

Chapters 2 and 3. The hydrodynamic characteristics of bubbles, slugs, and gas distributors are discussed in Chapter 4, and their influence on mass transfer and chemical reaction between solids and gas form the subject of Chapter 11.

Solids mixing and segregation have a considerable influence on the operation of many fluid bed processes and are discussed in Chapter 5. Carryover of solids by the gas is an intrinsic feature of fluid beds, and the many factors which determine its magnitude are considered in Chapter 6, whilst beds which are designed so that all the solids pass out with the gas and are then reinjected are dealt with in Chapter 7; effective separation of solids from gas is obviously essential (Chapter 8). Fluidized beds are frequently selected as processing tools because of their excellent heat transfer properties and because they permit controlled transfer of solids into, out of, and within the system; these features are addressed in Chapters 9 and 12 respectively.

There are probably more fluid beds used as dryers than any other single application and their design is outlined in Chapter 10. In spite of the fact that much more is now known about fluidized systems it is rare for a new fluid bed process to be designed and installed without the need for cold modelling or pilot plant work, and the experimental techniques available are found in Chapter 13.

Fluidization research is an exciting field full of surprises; new applications of the technique and improvements to existing processes are continually being made, particularly in industry, and the next 20 years will undoubtedly prove to be as interesting and fruitful as the last 20 years.

1.5 REFERENCES

Beer J.M., Massimilla, L., and Sarafim, A.F. (1980) *Inst. Energy Symp. Ser.*, **1** (4), 4.
de Vries, R.J., van Swaaij, W.P.M., Mantovani, C., and Heijkoop, A. (1972). *Proc. Conf. Chem. React. Eng. Amsterdam*, **70**, 141.
Elgin, D.C., and Perks, H.R. (1974). Proc. 5th Synth. Pipeline Gas Symp., AGA Cat. No. L51173, p.145.
Elliott, D.E., and Virr, M.J. (1972). Proc. 3rd. Int. Conf. Fluidized Bed Combustion, Paper 4–1, US–EPA.
Geldart, D. (1967). *Chem. Ind.*, 1474
Geldart, D. (1968). *Chem. Ind.*, 41
Geldart, D. (1969). *Chem. Ind.*, 311
Geldart, D. (1973). *Powder Technol.*, 1, 285.
Geldart, D. (1982). *Powder Technol.*, 31, 1.
Geldart, D. (1983). *Powder Technol.*, 36, 149.
Grade, J.R. (1984). 'Gas fluidization' Course, Center for Prof. Advancement, New Jersey.
Hetsroni, G. (Ed.) (1982) *Handbook of Multiphase Systems*, Chap. 8.5, Hemisphere, Washington.
Highley, J., and Kaye, W.G. (1983). Chapter 3 in *Fluidized Beds — Combustion and Applications* (Ed. J.R. Howard), Applied Science Publishers.

Jahnig, C.E., Campbell, D.L., and Martin, H.Z. (1980). In *Fluidization* (Eds J.R. Grace and J.M. Matsen), Plenum Press, New York.

Kunii, D., and Levenspiel, O. (1969). *Fluidization Engineering*, John Wiley, New York.

Penick, J.E., Lee, W., and Mazink, J. (1982). Int. Symp. on Chem. React. Engng, ISCRE — 7, Boston.

Reh, L. (1968). *Chem-Ing-Techn.*, **40**, 509.

Squires, A.M. (1982). Proc. Joint Meeting of Chem. Ind. & Eng. Soc. of China & Am. Inst. Chem. Engrs, Beijing, Sept. 19–22, p.322.

Yates, J.G. (1983). *Fundamentals of Fluidized-bed Chemical Processes*, Butterworths. London.

CHAPTER 2

Single Particles, Fixed and Quiescent Beds

D. GELDART

2.1 INTRODUCTION

The arrival time of a space probe travelling to Saturn can be predicted more accurately than the behaviour of a fluidized bed chemical reactor! The reasons for this frustrating situation lie not in the inadequacies of chemical engineers or powder technologists but rather in the complexity of defining (and measuring) unambiguously even such fundamental parameters as the size and size distribution, shape, and density of particles. Given that these parameters influence explicitly and implicitly the behaviour of fixed and fluidized beds, it is hardly surprising that most correlations give predictions no better than ±25 per cent.

2.2 PARTICLE SIZE AND SHAPE

For a particle of any shape other than a sphere, there are many ways of defining its size; Allen (1981) lists twelve. Only four definitions are of interest for packed and fluidized beds:

d_p = *sieve size*: the width of the minimum square aperture through which the particle will pass;

d_v = *volume diameter*: the diameter of a sphere having the same volume as the particle;

d_{sv} = *surface/volume diameter*: the diameter of a sphere having the same external surface area/volume ratio as the particle;

d_s = *surface diameter*: the diameter of a sphere having the same surface as the particle.

Some of these diameters are related through Waddell's sphericity factor ψ, defined as:

$$\psi = \frac{\text{surface area of equivalent volume sphere}}{\text{surface area of the particle}}$$

$$= (d_v/d_s)^2 \qquad (2.1)$$

It can be shown that:

$$\psi = d_{sv}/d_v \qquad (2.2)$$

Because it is now well established that the most appropriate parameter for correlating the flow of fluids through packed beds is S_v (the external surface area of the powder per unit particle volume), the most relevant diameter is d_{sv}. The sphericity ψ, d_v, and d_{sv} can be calculated exactly for geometrical shapes such as cuboids, rings, and manufactured shapes (Table 2.1). Most particles, however, are irregular and their size is generally measured by

Table 2.1 Sphericities for some regular solids

Shape	Relative proportions	$\psi = d_{sv}/d_v$
Spheroid	1:1:2	0.93
	1:2:2	0.92
	1:1:4	0.78
	1:4:4	0.70
Ellipsoid	1:2:4	0.79
Cylinder	Height = diameter	0.87
	Height = 2 × diameter	0.83
	Height = 4 × diameter	0.73
	Height = ½ × diameter	0.83
	Height = ¼ × diameter	0.69
Rectangular parallelepiped	1:1:1	0.81
	1:1:2	0.77
	1:2:2	0.77
	1:1:4	0.68
	1:4:4	0.64
	1:2:4	0.68
Rectangular tetrahedron	—	0.67
Regular octahedron	—	0.83

sieving (the most common method used for powders larger than about 75 μm) or by using the Coulter counter (for particles smaller than about 75 μm). The Coulter counter gives the volume diameter, d_v, so if the sphericity is known d_{sv} can be estimated (Eq. 2.2). Unfortunately there is no simple generally accepted method for measuring the sphericity of small irregular particles, so although values have been published (Table 2.2) they should be regarded as estimates only. The tables show that ψ is between 0.64 and 1 for

Table 2.2 Sphericities of some common solids

	ψ
Crushed coal	0.75
Crushed sandstone	0.8–0.9
Round sand	0.92–0.98
Crushed glass	0.65
Mica flakes	0.28
Sillimanite	0.75
Common salt	0.84

most materials; viewing the particles through a microscope and comparison with Tables 2.1 and 2.2 will usually enable a realistic value of ψ to be estimated. A better way to find d_{sv} ($= \psi d_v$) is to use the Ergun equation (Eq. 2.17). An attempt to compare the sieve size with the volume and surface/volume diameters for crushed quartz has been described by Abrahamsen and Geldart (1980). They concluded that for materials like quartz which has a sphericity of about 0.8:

$$d_v \approx 1.13 d_p \qquad (2.3)$$

The average sphericity for regular figures in Table 2.1 is 0.773 ($s = \pm 11$ per cent). That is:

$$\frac{d_{sv}}{d_v} = 0.773 \qquad (2.4)$$

Combining Eqs. (2.3) and (2.4):

$$d_{sv} \approx 0.87 d_p \qquad (2.5)$$

Note that Eq. (2.5) is an approximation for particles which are non-spherical. For spherical or near-spherical particles:

$$d_v \approx d_{sv} \approx d_p \qquad (2.6)$$

2.3 MEAN SIZE AND SIZE DISTRIBUTION

If a powder mass M has a size range consisting of N_{p1} spherical particles of size d_1, N_{p2} of size d_2, and so on, the mean surface/volume size:

$$d_{sv} = \frac{N_{p1}d_1^3 + N_{p2}d_2^3 + \cdots}{N_{p1}d_1^2 + N_{p2}d_2^2 + \cdots}$$

$$d_{sv} = \frac{\Sigma x}{\Sigma x/d} \qquad (2.7)$$

where x is the weight fraction of particles in each size range. When sieving is used, d_1, d_2, ... are replaced by the averages of adjacent sieve apertures, d_{pi}, and the equation becomes:

$$d_p = \frac{1}{\Sigma x_i/d_{pi}} \qquad (2.8)$$

This definition of mean particle size gives proper emphasis to the important influence which small proportions of fines have. It should not be confused with another method of characterizing a powder, the *median* d_{pm}, which is the size corresponding to the 50 per cent value on the graph of cumulative percentage undersize versus size. It is, however, not directly related to the surface/volume mean.

Equation (2.8) should not be used if the powder has an unusual distribution, e.g. bi- or trimodal, or trimodal, or has an extremely wide size range. Such powders generally will not behave in an homogenous way and cannot be characterized by a single number. It is always advisable to first plot the size distribution of the powder as the weight fraction, or percentage in a size range, against the average size, that is x versus d_{pi}, because a plot of the cumulative percentage undersize can conceal peculiarities of distribution. An example of this is shown in Fig. 2.1(a). The size distribution of the powder given in Table 2.3 is rather unusual but this does not show up in Fig. 2.1(b). There is no entirely satisfactory way of comparing the width of the size distribution of two powders having different mean sizes, nor of defining how wide a distribution is. One useful way is to specify the relative spread, σ/d_{pm}. This is obtained by using the cumulative percentage undersize versus the size plot, e.g. Fig. 2.1(b), and defining the spread as:

$$\sigma = \frac{d_{84\%} - d_{16\%}}{2} \qquad (2.9)$$

In the example given, $\sigma = 105 \ \mu m$ and:

$$\frac{\sigma}{d_{pm}} = \frac{105}{270} = 0.39$$

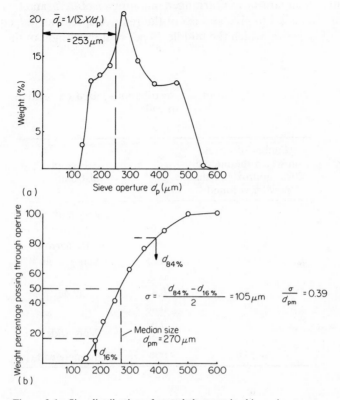

Figure 2.1   Size distribution of a sand characterized in various ways.

Table 2.3   Size distribution of a sieved sand

Sieve aperture, μm	Size d_{pi} μm	Weight percentage in range x_i	Cumulative percentage undersize
−600 + 500	550	0.5	100
−500 + 420	460	11.6	99.5
−420 + 350	385	11.25	87.9
−350 + 300	325	14.45	76.65
−300 + 250	275	20.8	62.2
−250 + 210	230	13.85	41.4
−210 + 180	195	12.5	27.55
−180 + 150	165	11.9	15.05
−150 + 125	137	3.15	3.15

$$d_p = \Sigma \frac{100}{x_i/d_{pi}} = 253 \ \mu m$$

The British Standard sieve is arranged in multiples of $^4\sqrt{2}$, and this is used as a basis in Table 2.4 to give an idea of the relative spread as judged from the number of sieves on which the middle 70 per cent. by weight of the powder is found.

Table 2.4 Width of size distributions based on relative spread

Number of sieves on which the middle 70% (approx.) of the powder is found	σ/d_{pm}	Type of distribution
1	0	Very narrow
2	0.03	Narrow
3	0.17	Fairly narrow
4	0.25	Fairly wide
5	0.33	
6	0.41	Wide
7	0.48	
9	0.6	Very wide
11	0.7	
> 13	> 0.8	Extremely wide

2.4 PARTICLE DENSITY

This is defined as (see Fig. 2.2):

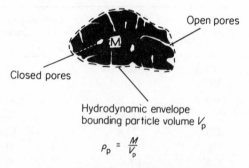

$$\rho_p = \frac{M}{V_p}$$

Figure 2.2 Definition of particle density.

$$\rho_p = \frac{\text{mass of a single particle}}{\text{volume the particle would displace if its surface were non-porous}}$$

The volume includes the voids inside the particle whether they are open or closed pores. The particle density should not be confused with the bulk density of the bed, ρ_B, which includes the voids *between* the particles; ρ_p is a 'hydrodynamic' density since it is based on the shape and volume which the flowing gas 'sees'. For nonporous solids the particle density is equal to the true, skeletal, or absolute density of the material, ρ_{ABS}, which is measured by a specific gravity bottle or air pycnometer; but for porous solids $\rho_p < \rho_{ABS}$ and cannot be measured by the usual means. A mercury porosimeter can be used to measure the particle density of coarse porous solids but is not reliable for fine powders since the mercury may not penetrate the voids between small particles.

In the petroleum industry the particle or piece density of the free-flowing cracking catalyst is estimated indirectly by measuring the pore volume. When a liquid with low viscosity and volatility, such as water, is added, the powder should remain free-flowing until the liquid has filled all the open microscopic pores. Any additional liquid then coats the external surface of each particle causing immediate caking by surface tension (the 'caking end point'). This method is not reliable for all porous powders and a simple alternative method has been developed (Abrahamsen and Geldart, 1980). This is based on the fact that the minimum packed bed voidage is virtually the same for particles of similar size and particle shape. The bed voidage is defined as:

$$\epsilon = \frac{\text{volume of bed} - \text{volume of particles}}{\text{volume of bed}}$$

That is:

$$\epsilon = 1 - \frac{M}{\rho_p V_B} \tag{2.10}$$

or:

$$\epsilon = 1 - \frac{\rho_B}{\rho_p} \tag{2.11}$$

Firstly, 0.2 to 0.25 kg of a control powder c, of known particle density ρ_{pc}, is poured into a measuring cylinder and tapped until it reaches its minimum volume, corresponding to the maximum bulk density ρ_{BTc}. The procedure is repeated with the unknown powder x. Ideally, several control powders should be used. If:

$$(\epsilon_c)_{min} = (\epsilon_x)_{min} \tag{2.12}$$

then, from Eq. (2.11):

$$\rho_{px} = k \frac{\rho_{BTx}}{\rho_{BTc}} \rho_{pc} \tag{2.13}$$

The empirical factor k is introduced because in practice it is not always possible to find control powders having the same particle shape as that of the unknown powder:

$k = 1$ if x and c are approximately the same shape;

$k \approx 0.82$ if x is rounded or spherical and c is angular;

$k \approx 1/0.82$ is x is angular and c is rounded or spherical.

2.5 PREDICTING THE BED VOIDAGE

If at all possible the voidage of a packed bed should be measured in the condition in which it will be used, but if this is impossible an estimate can often be made for group B and D solids (see Chapter 3). Obviously, the degree to which the bed is vibrated or tapped is very important and two extreme conditions are used as reference points: 'loose' packing gives the maximum possible voidage and 'dense' packing the minimum; both are based on random packings. Factors influencing the voidage are:

(a) Particle shape — the lower the sphericity ψ, the higher the voidage (see Table 2.5)

Table 2.5 Voidage versus sphericity for randomly packed beds uniformly sized particles larger than about 500 μm (Brown *et al.*, 1950)

ψ	ϵ	
	Loose packing	Dense packing
0.25	0.85	0.80
0.30	0.80	0.75
0.35	0.75	0.70
0.40	0.72	0.67
0.45	0.68	0.63
0.50	0.64	0.59
0.55	0.61	0.55
0.60	0.58	0.51
0.65	0.55	0.48
0.70	0.53	0.45
0.75	0.51	0.42
0.80	0.49	0.40
0.85	0.47	0.38
0.90	0.45	0.36
0.95	0.43	0.34
1.00	0.41	0.32

(b) Particle size — the larger the particles, the lower the loosely packed voidage (see Table 2.6); the dense packing voidage is not as dependent on size.

Table 2.6 Variation of packed bed voidage with particle size for narrow size distributions (Partridge and Lyall, 1969)

(a) Loosely packed spheres

d_p, μm	2,890	551	284	207	101	89	72	55
ϵ	0.386	0.385	0.390	0.411	0.424	0.434	0.441	0.454

(b) Loosely packed irregular sand

d_p, μm	550	460	390	330	230	140	82	72
ϵ	0.422	0.432	0.440	0.437	0.507	0.563	0.590	0.602

(c) Size distribution — the wider the size spread, the lower the voidage (see Table 2.7).

Table 2.7 Effect of size distribution on packed bed voidage

	d_p μm	σ μm	ϵ
Sand 1	195	75	0.432
Sand 2	197	7	0.469

(d) Particle and wall roughness — the rougher the surface, the higher the voidage.

2.6 FLOW AND PRESSURE DROP THROUGH PACKED BEDS

Through the work of Darcy and Poiseuille it has been known for more than 120 years that the average velocity of fluid through a packed bed, or through a pipe, is proportional to the pressure gradient:

$$U \; \alpha \; \frac{\Delta p}{H} \tag{2.14}$$

A packed bed may be thought of as a large number of smaller tortuous pipes of varying cross-section. A number of workers developed this approach, notably Carman (1937), Kozeny (1927), and Ergun (1952). The reasoning behind these equations is summarized by Allen (1981). At Reynolds numbers less than about 1 (laminar flow) the Carman-Kozeny equation applies:

$$\frac{\Delta p}{H} = \frac{K\mu}{d_{sv}^2} \frac{(1-\epsilon)^2}{\epsilon^3} U \qquad (2.15)$$

Where U is the superficial or empty tube velocity and the Reynolds number is defined as:

$$\mathrm{Re} = \frac{\rho_g U d_{sv}}{\mu} \qquad (2.16)$$

K is generally assumed to be 180 but there are indications that this may be correct only for narrow cuts, voidages between 0.4 and 0.5, and Reynolds numbers 0.1 to 1. Abrahamsen (1980) found that for fine powders of mean size 30 to 80 μm, K had average values of 263 ($s \pm 35$ per cent.) for spherical or near-spherical particles and 291 ($s \pm 26$ per cent.) for other shapes. For Reynolds numbers greater than about 1 the Ergun equation has proved satisfactory:

$$\frac{\Delta p}{H} = \frac{p_1}{\bar{p}} \left\{ 150 \frac{(1-\epsilon)^2}{\epsilon^3} \frac{\mu U_1}{d_{sv}^2} + 1.75 \frac{1-\epsilon}{\epsilon^3} \frac{\rho_{g1} U_1^2}{d_{sv}} \right\} \qquad (2.17)$$

The term $p_1/\bar{p}$ is a correction factor for compressibility, where $\bar{p}$ is the average absolute pressure in the bed. Subscript 1 denotes conditions at the inlet to the bed. Under laminar flow conditions (Re < 1) the first term on the right-hand side dominates:

$$\frac{\Delta p}{H} \alpha \frac{\mu U}{d_{sv}^2} \qquad (2.18)$$

In fully turbulent flow (Re > 1000) the second term dominates and:

$$\frac{\Delta p}{H} \alpha \frac{\rho_g U^2}{d_{sv}} \qquad (2.19)$$

Note that the surface/volume size d_{sv} is used; if only sieve sizes are available, depending on the particle shape, Eq. (2.5) or Eq. (2.6) should be used.

2.7   PRESSURE DROP ACROSS MOVING BEDS

Imagine gas molecules moving between particles in a packed bed. The average relative velocity between particles and molecules is U/ϵ when the particles are stationary. If the molecules maintain their same velocity relative to the wall and the particles now start to move with an absolute velocity v_s, then the relative velocity (interstitial) between molecules and solids becomes:

$$u_{\text{SLIP.i}} = \frac{U}{\epsilon} - v_s \qquad (2.20)$$

where v_s is positive if the particles move in the same direction as the gas or is negative if they move counter-current. The corresponding superficial slip velocity is:

$$U_{\text{SLIP}} = \epsilon \left(\frac{U}{\epsilon} - v_s \right) \qquad (2.21)$$

Substitution of U_{SLIP} for U_1 in Eq. (2.17) enables the pressure gradient to be calculated.

2.8   MINIMUM FLUIDIZATION VELOCITY U_{mf}

When gas is passed upwards through a packed bed unrestrained at its upper surface, the pressure drop increases with gas velocity according to Eq. (2.15) or Eq. (2.17) until, on the microscopic scale, the drag on an individual particle exceeds the force exerted by gravity or, on the macroscopic scale, the pressure drop across the bed equals the weight of the bed per unit area. If the bed has been compacted or is composed of interlocked, very angular or cohesive particles, then an excess pressure is required to free them (points C and C′ in Fig. 2.3) and they adopt a higher voidage configuration causing a

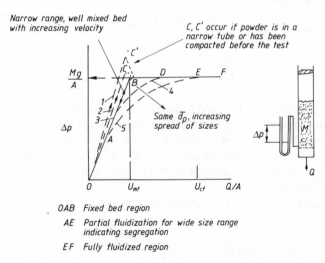

OAB   Fixed bed region

AE   Partial fluidization for wide size range
      indicating segregation

EF   Fully fluidized region

Figure 2.3   Pressure drop across fixed and fluidized beds for group B and D powders.

fall back to the theoretical pressure drop. With group B and D powders (see Chapter 3) a further increase in velocity causes the formation of small bubbles whose size increases with gas velocity. The bed pressure drop begins to fluctuate and if the bed is deep enough ($H > 2D$) the bubbles occupy a substantial proportion of the cross-sectional area. These large bubbles are called slugs and cause regular piston-like movements of the upper surface of the bed. The average bed pressure drop then increases as shown in Fig. 2.4(b). If the gas velocity is now reduced, depending on the size distribution of the powder, the pressure drop declines along curves 3, 4, or 5 in Fig. 2.3. If the powder has a narrow size range, curve 3 is followed; increasing the size distribution (but maintaining the same mean size) results in curves 4 or 5 because the larger particles settle out progressively on the distributor. Points D and E represent the minimum velocity required to fully support the solids (though not necessarily in a well-mixed state) and are called the minimum velocity of complete fluidization, U_{cf}. U_{mf} is usually defined as the intersection of the horizontal fluidized bed line EDF and the sloping packed bed line OAB, but it is clear that for solids with a very wide size range these could be drawn almost anywhere; the velocity corresponding to point E is of practical interest but relatively little work has been done to predict it (see Sec. 2.11). For the present the best recourse is to make measurements and visual observations.

The pressure drop across a fluidized bed is the only parameter which can be accurately predicted:

$$\Delta p_F = \frac{Mg}{A} \quad \text{N/m}^2 \tag{2.22}$$

or Δp_F cm water gauge (w.g.) $= \dfrac{0.1M}{A} \dfrac{\text{kg}}{\text{m}^2}$ $\tag{2.23}$

$$\frac{\Delta p_F}{H} = (\rho_p - \rho_g)(1 - \epsilon_{mf})g \tag{2.24}$$

ϵ_{mf} is the bed voidage at U_{mf} and a close approximation to it can be obtained by measuring the aerated or most packed bulk density ρ_{BLP}. This is done by pouring the powder through a vibrating sieve and allowing it to fall a fixed height into a cylindrical cup of capacity 100 cm^3. The equipment is shown later in Fig. 3.6 and for consistent results the powder should be poured through in 20 to 30 s.

If Eqs. (2.24) and (2.17) are combined (with $p_1/\bar{p} = 1$, $\Delta p/H$ eliminated, and U and ϵ set equal to U_{mf} and ϵ_{mf}, then:

$$\frac{\rho_g d_{sv}^3 (\rho_p - \rho_g)g}{\mu^2} = \frac{150(1 - \epsilon_{mf})}{\epsilon_{mf}^3} \frac{\rho_g d_{sv}}{\mu} U_{mf} + \frac{1.75}{\epsilon_{mf}^3} \frac{\rho_g^2 d_{sv}^2}{\mu^2} U_{mf}^2 \tag{2.25}$$

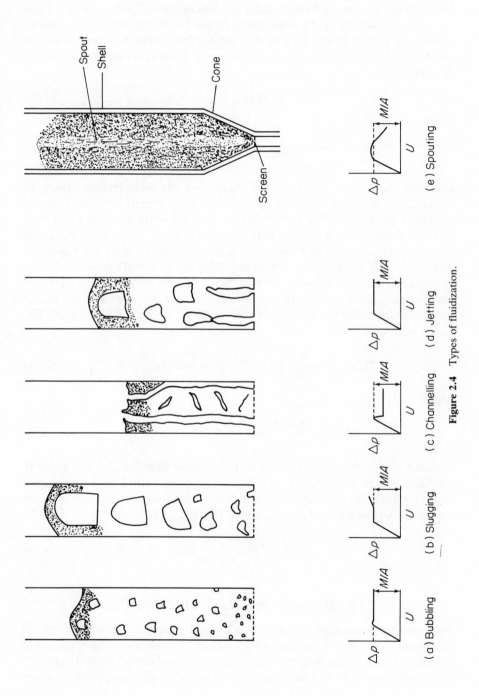

Figure 2.4 Types of fluidization.

(a) Bubbling   (b) Slugging   (c) Channelling   (d) Jetting   (e) Spouting

The group on the left-hand side is dimensionless and known as the Archimedes number or, by some workers, the Galileo number. Wen and Yu (1966) correlated many data in this form. They combined the numerical constants with the voidage terms and, using the volume diameter d_v instead of d_{sv}, proposed:

$$Ar = 1{,}650Re_{mf} + 24.5Re_{mf}^2 \qquad (2.26)$$

where

$$Re_{mf} = \frac{\rho_g U_{mf} d_v}{\mu}$$

Recently there have been several studies on the independent effects of temperature and pressure on U_{mf}. In general the qualitative effects are as predicted by Eqs. (2.27) and (2.28): in fine powders U_{mf} decreases with increasing temperature and is hardly affected by pressure (e.g. King and Harrison, 1982) whilst in coarse powders increased temperature causes an increase in U_{mf} and pressure a decrease. It should be noted that some strange effects can occur in beds of coarse particles of high temperatures (Botterill and Teoman, 1980). Also, in any powder, if softening or sintering occurs producing agglomeration, none of the equations is valid. Re-arranged, Eq. (2.26) becomes:

$$U_{mf} = \frac{\mu}{\rho_g d_v} \left\{ (1{,}135.7 + 0.0408Ar)^{1/2} - 33.7 \right\} \qquad (2.27)$$

where

$$Ar = \frac{\rho_g d_v^3 (\rho_p - \rho_g)g}{\mu^2}$$

Equation (2.27) should be used for particles larger than 100 μm (i.e. group B and D solids) in conjunction with Eqs (2.3), (2.6) and (2.8). For particles smaller than 100μm, Baeyens' equation gives the best agreement with experiments (Geldart and Abrahamsen, 1981):

$$U_{mf} = \frac{(\rho_p - \rho_g)^{0.934} g^{0.934} d_p^{1.8}}{1{,}111 \mu^{0.87} \rho_g^{0.066}} \qquad (2.28)$$

2.9   MINIMUM BUBBLING VELOCITY U_{mb}

Fine powders (cracking catalyst is a typical example) exhibit a type of behaviour not found in coarse solids, namely the ability to be fluidized at velocities beyond U_{mf} without the formation of bubbles. The bed expands, apparently smoothly and homogenously, until a velocity is reached at which small bubbles appear at the surface. These must not be confused with the continuous channels or spouts which often appear. They resemble miniature volcanoes which disappear when the side of the column is gently tapped. Further increases in velocity produce, at first, a further slight increase in bed height (see Fig. 2.6) followed by a reduction. Relatively large bubbles burst through the bed surface periodically, causing the bed to collapse rapidly; it then 'reinflates' slowly to collapse again as another swarm of bubbles bursts through. Reduction of the gas velocity produces a retracing of the height-velocity graph and finally the last bubble disappears giving once again a quiescent bed. The average of the velocities at which the bubbles appear and disappear is called the minimum bubbling velocity or bubble point and generally coincides with the maximum bed height for deep beds; premature bubbling can be caused by non-uniform distributors or protuberances in the bed, but the maximum values of U_{mb} have been correlated by Abrahamsen and Geldart (1980) and found to depend on the gas and particle properties:

$$U_{mb} = 2.07 \exp (0.716F)\frac{d_p\rho_g^{0.06}}{\mu^{0.347}} \qquad (2.29)$$

where F is the mass fraction of the powder less than 45 μm. The numerical constant is dimensional and SI units must be used. If $F \approx 0.1$ and the powder is fluidized by air at ambient conditions:

$$U_{mb} \approx 100 \, d_p \qquad (2.30)$$

Note that Eq. (2.29) implies that U_{mb} is not dependent on particle density, a finding confirmed independently by Simone and Harriott (1980). It must be stressed that Eqs (2.29) and (2.30) are valid only for powders which are fine enough (in the main, less than 100 μm mean size) to have values of U'_{mf} less than U_{mb}. If the calculations show that U_{mf} (from Eq. 2.28) is larger than U_{mb}, then the powder will start to bubble at or very slightly above incipient fluidization and Eq. (2.29) should not be used. This is discussed further in Chapter 3.

2.10 EFFECT ON U_{mb} OF TEMPERATURE, PRESSURE, AND TYPE OF GAS

This is still a relatively unknown area. Equation (2.29) indicates that U_{mb} increases with pressure: King and Harrison's (1982) data show agreement with the power on density of 0.06; Godard and Richardson (1966) indicate 0.1; Guedes de Carvalho (1981) an even higher dependency.

Piepers et al. (1984) give data which show that the power depends on the type of gas. Up to $p = 15$ bar, U_{mb} for a catalyst in H_2 is affected little, in N_2 as $\rho_g^{0.13}$, in argon as $\rho_g^{0.18}$. The reason is not clear but adsorption of the gas in the surface of the catalyst may have played a part and it may be that not all fine powders behave in the same way.

Omitting any sintering effects, increasing temperature reduces U_{mb} according to $1/\mu^{0.347}$.

2.11   VELOCITY FOR COMPLETE FLUIDIZATION U_{cf}

In materials having a wide size range, particularly those having a large mean size, segregation by size tends to occur at velocities close to U_{mf} for the mixture. Although a value of U_{mf} can be defined for the mixture based on d_p its usefulness is limited. Knowlton (1974) defined a velocity, U_{cf}, at which all the particles are fully supported (even though there may be segregation) and suggested that:

$$U_{cf} = \Sigma \, x_i U_{mfi} \tag{2.31}$$

where U_{mfi} and x_i refer to fraction of size d_{pi}. This appears to give reasonable agreement with experimental values, even at high pressures.

2.12   VOIDAGES IN FLUIDIZED BEDS

As we shall see in Chapter 4, almost all gas fluidized beds operate in the bubbling regime and consist of two phases — bubbles and the emulsion (or dense) phase. Conditions in the dense phase (gas velocity U_D and the corresponding voidage ϵ_D) are the subject of much discussion and speculation, largely because it is believed that (a) most of the chemical conversion occurs there and (b) in fine powders the equilibrium size of the bubbles may be controlled by the voidage, ϵ_D. The dense phase also figures in modes of flow in standpipes or downcomers.

U_D and ϵ_D are extremely difficult to measure directly in bubbling beds though they can be predicted for quiescent fluidization of fine powders.

(a) Voidage in non-bubbling (quiescent) beds. The earliest attempts to measure and correlate the expansion of quiescent (or particulate) fluidized beds in gas/solid systems were made by Davies and Richardson (1966) who adapted the approach used by Richardson for liquid/solid sedimentation and fluidization:

$$\frac{U}{v'_t} = \epsilon^n \tag{2.32}$$

In liquid fluidization n is a function of d_p/D and the terminal velocity Reynolds number, and varies between 4.65 and 2.4. The form of the relationship appears to hold for gas/solid systems but experimental values of n between 3.84 and 19.7 have been reported (Godard and Richardson,

1968, and Crowther and Whitehead, 1978). Recently Wong (1983) has shown that both n and v'_t/v_t increases with decreasing particle size below about 60 μm; v'_t is the intercept of the $\log\epsilon$-$\log U$ plot at $e = 1$ and v_t is the Stokes terminal velocity.

An alternative approach based on the Carmen-Kozeny equation was used by Abrahamsen and Geldart (1980). Combining Eqs (2.15) and (2.24) (with $\epsilon_{mf} = \epsilon$ and $d_p = d_{sv}$) gives:

$$\frac{(\rho_p - \rho_g)g \, d_p^2}{\mu} \frac{\epsilon^3}{1 - \epsilon} = K_f U \qquad (2.33)$$

Plots of all their data using 48 gas/solid systems gave a general expression to predict the non-bubbling expansion of a bed of fine powder, namely:

$$\frac{\epsilon^3}{1 - \epsilon} \frac{(\rho_p - \rho_g)g \, d_p^2}{\mu} = 210 \, (U - U_{mf}) + \frac{\epsilon_{mf}^3}{1 - \epsilon_{mf}} \frac{(\rho_p - \rho_g)g \, d_p^2}{\mu}$$

$$(2.34)$$

where $U_{mf} < U < U_{mb}$.

The standard deviation of the numerical constant is ± 22 per cent, and ϵ_{mf} can be found from the simple experiment to determine the aerated voidage described earlier; U_{mf} is obtained from Eq. (2.28) and U_{mb} from Eq. (2.29).

(b) Voidage in bubbling beds. The average overall voidage of a bubbling bed, i.e the bed expansion, is largely caused by the bubble hold-up, and is dealt with in Chapter 4. The voidage of the dense phase (the portion of the powder where bubbles are absent) is generally taken as being ϵ_{mf} for group B and D systems, but can be significantly higher for group A powders.

Two methods have been used to measure dense phase voidage in bubbling fluidized beds. Rowe et al. (1979) used X-rays and concluded that the dense phase voidage ϵ_D was so high that it could carry gas velocities up to $20U_{mf}$. More recent studies have given values in line with those obtained from the bed collapse technique. Rietema (1967). Abrahamsen and Geldart (1980), and Simone and Harriott (1980), using the collapse technique and certain assumptions, concluded that $\epsilon_{mf} < \epsilon_D < \epsilon_{mb}$ and hence $U_{mf} < U_D < U_{mb}$.

In collapse experiments, the powder is fluidized at a chosen velocity (say 20 cm/s) and the gas suddenly shut off. The bed height is recorded as a function of time (Fig. 2.5); it falls rapidly in the first few seconds as the bubbles rise to the surface and then much more slowly as the interstitial gas flows out and the dense phase collapses. The straight line portion of the curve is extrapolated back to time zero and it is assumed that the intercept gives the height H_D which the dense phase would occupy in the

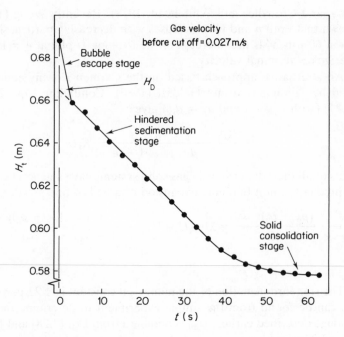

Figure 2.5 Typical collapse curve for group A powder (26 μm ballotini).

bubbling bed. This height is plotted at various superficial gas velocities on Fig. 2.6., By further assuming that Darcy's law holds within the dense phase in bubbling bed, U_D can be calculated. Abrahamsen and Geldart (1980) give:

$$\frac{H_D}{H_{mf}} = \frac{2.54 \, \rho_g^{0.016} \, \mu^{0.066} \exp (0.09F)}{d_p^{0.1} \, g^{0.118} \, (\rho_p - \rho_g)^{0.118} \, H_{mf}^{0.043}} \qquad (2.35)$$

and

$$\frac{U_D}{U_{mf}} = \frac{188 \, \rho_g^{0.089} \, \mu^{0.371} \exp (0.508F)}{d_p^{0.568} \, g^{0.663} \, (\rho_p - \rho_g)^{0.663} \, h^{0.244}} \qquad (2.36)$$

The constants are dimensional and SI units must be used. It should be noted that these empirical equations indicate that the dense phase 'opens up' as the mean particle size and particle density decrease and as gas density (pressure) and viscosity (temperature) increase. Increasing the fines fraction F also increases the dense phase voidage and gas velocity relative to conditions at incipient fluidization.

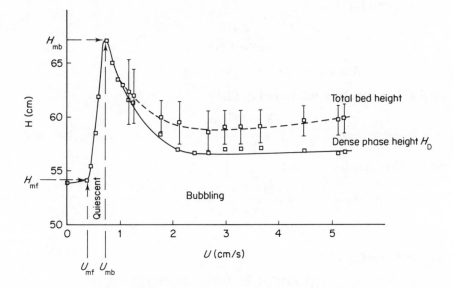

Figure 2.6 Typical expansion curve for a group A powder (60 μm ballotini)

2.13 WORKED EXAMPLE ON INCIPIENT FLUIDIZATION

A bed of angular sand of mean sieve size 778 μm is fluidized by air. The particle density is 2,540 kg/m³; μ(air) = 18.4 × 10⁻⁶ kg/m s; ρ_g = 1.2 kg/m³; and 24.75 kg of the sand are charged to a bed 0.216 m in diameter. The bed height at incipient fluidization is 0.447 m. Find:

(a) ϵ_{mf},
(b) the pressure drop across the bubbling bed,
(c) the incipient fluidization velocity.

Step 1. Bed density at U_{mf}:

$$\rho_{Bmf} = \frac{24.75}{(\pi/4)(0.216)^2 \times 0.447} = 1,511 \text{ kg/m}^3$$

Step 2. From Eq. (2.11):

$$\epsilon_{mf} = 1 - \frac{1,511}{2,540} = 0.405$$

Answer (a) ϵ_{mf} = 0.405

Step 3. From Eq. (2.23):

$$\Delta p_F = \frac{0.1M}{A} = \frac{2.475}{(\pi/4)(0.216)^2}$$

Answer (b) $\Delta p_F = 67.5$ cm w.g.

Step 4. For angular sand, from Eq. (2.3):

$$d_v = 1.13 \times 778 \times 10^{-6} \text{ m}$$
$$= 879 \times 10^{-6} \text{ m}$$

Step 5. Archimedes' number:

$$\text{Ar} = \frac{1.2 \times 2,540 \times 9.81 \times (879 \times 10^{-6})^3}{(18.4 \times 10^{-6})^2}$$
$$= 59,981$$

Step 6. From Eq. (2.26):

$$\text{Re}_{mf} = (1,135.7 + (0.0408 \times 59,981))^{\frac{1}{2}} - 33.7$$
$$= 26.15$$

Therefore:

$$U_{mf} = \frac{26.15 \times 18.4 \times 10^{-6}}{1.2 \times 879 \times 10^{-6}}$$

Answer (c) $U_{mf} = 0.456$ m/s (cf. experimental value 0.504 m/s)

2.14   NOMENCLATURE

A	cross-sectional area of bed $(= (\pi/4) D^2)$	m^2
Ar	Archimedes number $\rho_g (\rho_p - \rho) g d_v^3/\mu^2$	—
d_{pi}	arithmetic mean of adjacent sieve apertures	μm or m
d_p	mean sieve size of a powder	μm or m
d_{sv}	diameter of a sphere having the same surface/volume ratio as the particle	μm or m
d_v	diameter of a sphere having the same volume as the particle	μm or m
D	diameter of bed	m
F	mass fraction of powder less than 45 μm	—
g	acceleration due to gravity	9.81 m/s^2

H	height of packed bed	m
H_D	height of dense phase in a bubbling bed	m
H_{mf}	height of incipient fluidized bed	m
M	mass of powder in a bed	kg
p	absolute pressure at bed exit	N/m^2
$\bar{p}$	absolute mean pressure in the bed $(= p + \Delta p/2)$	N/m^2
Δp	pressure drop across a packed bed	N/m^2
Δp_F	pressure drop across a fluidized bed	N/m^2
Re	Reynolds number	—
Re_{mf}	Reynolds number at incipient fluidization velocity $(= \rho_g U_{mf} d_v/\rho)$	—
s	standard deviation	—
S_p	surface area of one particle	m^2
S_v, S_B	surface area of powder per unit volume of powder and per unit volume of bed, respectively	m^2/m^3
U, U_{mf}	superficial gas velocity at bed exit and at incipient fluidization velocity, respectively	m/s
U_{cf}	velocity at which all particles are fully supported (Eq. 2.31)	m/s
U_D	superficial velocity of gas in dense phase in a bubbling bed	m/s
U_{mb}	superficial incipient bubbling velocity	m/s
V_B	volume of bed	m^3
V_p	volume of one particle	m^3
v_t, v_t'	particle terminal velocity, intercept at $\epsilon = 1$ on log ϵ – log U plot	
x_i	weight fraction of powder of size d_{pi}	—
$\epsilon, \epsilon_{mf}, \epsilon_{mb}$	voidages of packed, incipiently fluidized, and incipiently bubbling beds	
μ	fluid viscosity	kg/m s
ρ_{ABS}	absolute density of material comprising a powder	kg/m^3
ρ_B	bulk density	kg/m^3
ρ_{BLP}	aerated or most loosely packed bulk density	kg/m^3
ρ_{BT}	tapped or maximum bulk density	kg/m^3
ρ_g	gas density	kg/m^3
ρ_p	particle density	kg/m^3
σ	spread of particle size (Eq. 2.9)	μm or m
ψ	sphericity of particle	—

2.15 REFERENCES

Abrahamsen, A.R. (1980). M.Sc. Dissertation, University of Bradford.

Abrahamsen, A.R., and Geldart, D. (1980). *Powder Technol.*, **26**, 35.

Allen, T. (1981). *Particle Size Measurement*, 3rd ed., Chapman and Hall, London.

Botterill, J.S.M., and Teoman, Y. (1980). In *Fluidization* (Eds J.R. Grace and J.M. Matsen) Plenum Press, p. 93).

Brown, G.G., *et al.* (1950). *Unit Operations*, John Wiley and Company, New York, p. 214.

Carman P.C. (1937). *Trans. Instn. Chem. Engrs (Lond.)*, **15**, 150.

Crowther, M.E., and Whitehead, J.C. (1978). In *Fluidization* (Eds. J.F. Davidson and D.L. Keairns), Cambridge University Press, p.65.

Davies, L., and Richardson, J.F. (1966). *Trans. Instn Chem. Engrs*, **44**, 293.

Ergun S. (1952). *Chem. Engng Prog.*, **48**, 89.

Geldart, D., and Abrahamsen, A.R. (1981). *Chem. Eng. Prog. Symp. Ser.* **77** (205), 160.

Godard, K., and Richardson, J.F. (1968). In *Fluidization*, Instn. Chem. Engrs Symp. Ser., p. 126.

Guedes de Carvalho, J.R.F. (1981). *Chem. Eng. Sci.,* **36**, 1349.

King, D.F., and Harrison, D.L. (1982). *Trans. Instn Chem. Engrs,* **60**, 26.

Knowlton, T.M. (1974). Paper 9b, 67th Ann. meeting of A.I.Ch.E., Washington, D.C., Dec. 1–5.

Kozeny, J. (1927). *S.B. Akad. Wiss. Wien. Abt. IIa,* **136**, 271.

Partridge, B.A., and Lyall, E. (1969). AERE Rep. M.2152.

Piepers, H.W., Cotaar, E.J.E., Verkooijen, A.H.M., and Rietema, K. (1984). EFCE Conf. on Role of Particle Interaction in Powder Mechanics, *Powder Technol.* **37**, 55.

Rietema, K. (1967). In Proc. Int. Conf. on Fluidization (Ed. A.A.H. Drinkenburg), Neth. Univ. Press. Amsterdam, p.154.

Rowe, P.N., MacGillivray, H.J., and Cheesman, D.J. (1979). *Trans. Instn. Chem. Eng,* **57**, 194.

Simone, S., and Harriott, P. (1980). *Powder Technol.* **26**, 161.

Wen, C.Y., and Yu, Y.H. (1966). *A.I.Ch.E.J.,* **12**, 610.

Wong, A.C.Y. (1983). Ph.D. Thesis, University of Bradford.

CHAPTER 3

Characterization of Fluidized Powders

D. GELDART

3.1 INTRODUCTION

Much of the early experimental work on fluidization related to cracking catalyst, a finely divided, porous, low density powder having excellent fluidization properties. Based on the practical experience gained from operating full-scale catalytic crackers and experimental units for other organic processes, practitioners of the art formulated rules of thumb concerning the conditions needed for the successful operation of fluidized beds. As the technique was applied to processes far removed from oil cracking, different solids had to be used, but there was a tendency to assume that published conclusions drawn from using cracking catalyst were also applicable to these other powders which often had quite different particle sizes and densities.

Experience with catalyst suggested that a powder with a wide size distribution fluidized more satisfactorily than a powder having a narrow size range. The term 'more satisfactorily' is used loosely, but workers in the field understand it to imply smaller wind box pressure fluctuations, less vibration of the bed, and less tendency to slug, and this is attributed to smaller bubble sizes promoted by the wide size distribution.

However, my own experimental work on bubble sizes in sands having mean sizes 75 to 470 μm (Geldart, 1972) showed no effect due to size distribution, and it was the study of these and data from the literature which eventually led to the idea of the powder groups which form the subject of this chapter.

Various attempts (Wilhelm and Kwauk, 1948; Jackson, 1963; Molerus, 1967; Simpson and Rodger, 1961; Verloop and Heertjes, 1970; Oltrogge, 1972) have been made to formulate criteria which can predict whether a fluid/solid system would fluidize in a 'particulate' or 'aggregative' (bubbling) manner, and a more detailed discussion is given by Geldart (1973). Most liquid fludized systems are 'particulate' in the sense that as the superficial velocity is increased the particles move further apart in a more or less uniform way until they are carried out of the tube; most gas fluidized systems are 'aggregative' in that when gas additional to that required for minimum fludization is supplied, it passes through the bed as bubbles, leaving the dense phase at much the same voidage as it is at the minimum fluidization velocity U_{mf}. However, some liquid systems, like lead shot/water, can behave in a bubbling mode, and some gas systems, e.g. a fine catalyst in a high pressure gas, in a 'particulate' or non-bubbling mode; some of the criteria predict correctly that these extremes of behaviour should occur, but they are not capable of distinguishing the other differences of behaviour found in gas/solid systems which are discussed below.

3.2   DESCRIPTION OF POWDER GROUPS

Before discussing the numerical criteria used to discriminate between the various groups (see Section 3.3) qualitative descriptions of their fluidization behaviour are given. The groups are dealt with in order of increasing particle size.

3.2.1   Group C

Powders which are in any way *cohesive* belong in this category. 'Normal' fluidization of such powders is extremely difficult; the powder lifts as a plug in small diameter tubes, or channels badly, i.e. the gas passes up interconnected vertical and inclined cracks extending from the distributor to the bed surface (Fig. 3.1b). The difficulty arises because the interparticle forces are greater than those which the fluid can exert on the particle, and these are the result of very small particle size, generally < 20 μm, strong electrostatic charges, wet, sticky particle surfaces, soft solids, or particles having a very irregular shape. The pressure drop across the bed is, on the whole, lower than the theoretical value (bed weight per unit cross-sectional area) and can be as little as half. Particle mixing and, consequently, heat transfer between a surface and the bed is much poorer than with powders of groups A or B. The hardness of the particles has a strong influence, soft materials being more cohesive since they deform readily and give a larger area for interparticle contacts.

Fluidization can usually be made possible or improved by the use of mechanical stirrers or vibrators which break up the stable channels, or, in the case of some powders, usually plastic materials, by the addition of a fumed silica of sub-micrometre size. Porous particles tend to be non-conductive and agglomeration may occur due to excessive electrostatic charging. Improvement can generally be effected by humidifying the incoming gas or by making the equipment wall conducting, e.g. by coating glass with a very thin layer of

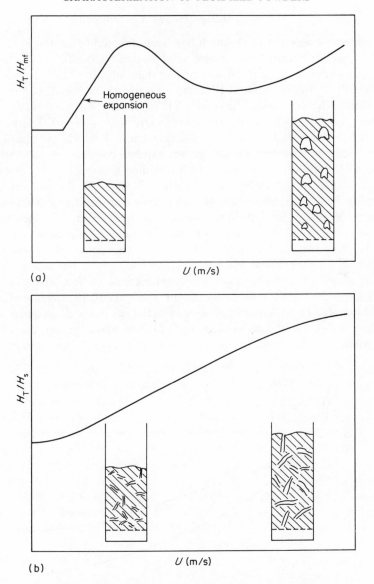

Figure 3.1 Bed expansion of (a) group A (b) group C powders.

tin oxide. However, with non-porous particles humidifying the gas beyond 65 per cent. can increase cohesiveness due to deposition of a liquid film.

If powders in this group do become fluidized or aerated, e.g. by being transported pneumatically or on an air slide, they may remain aerated for many minutes after transfer into storage hoppers.

3.2.2   Group A

A considerable amount of research has been devoted to aeratable group A powders, largely because most commercial fluidized bed catalytic reactors use them; our understanding of the structure of these powders has been advanced significantly by the research of Rietema and coworkers (e.g. Rietema, 1984) and Donsi, Moser, and Massimilla (1975) and their coworkers. Their extensive researches have shown convincingly that interparticle forces are present even in powders such as cracking catalyst which exemplifies this group. Beds of powders in this group expand considerably at velocities between U_{mf} and the velocity at which bubbling commences, U_{mb} (see Fig. 3.1a), because such powders are slightly cohesive. As the gas velocity is increased above U_{mb} the passage of each bubble disrupts the weak metastable structure of the expanded dense phase; the bed height becomes smaller because the dense phase voidage is reduced more quickly with increasing gas velocity than the bubble hold-up increases. The dense phase eventually assumes a stable voidage between ϵ_{mf} and ϵ_{mb}, and increasing the gas velocity above about 6 cm/s produces a net increase in bed expansion.

When the gas supply is suddenly cut off, the bed collapses slowly at a rate, U_c, comparable to the superficial velocity of the gas in the dense phase of the bubbling bed (0.1 to 0.6 cm/s) (See Fig. 3.2). This makes group A solids easy to circulate around fluidized and pneumatic conveying loops; however, the

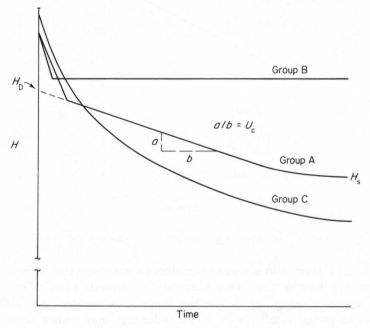

Figure 3.2 Typical deaeration curves for powders in groups A, B, and C.

ease with which they become aerated also makes them liable to flood on discharge from hoppers (Geldart and Williams, 1985).

Gross circulation of the powder (akin to convection currents in liquids) occurs even when few bubbles are present, producing rapid mixing and making the surface resemble a boiling liquid. Bubbles in a two-dimensional bed appear to split and recoalesce very frequently, resulting in a restricted bubble size. The bubble size is affected by the mean particle size, the mass fraction less than 45 μm, pressure, and temperature. Considerable back-mixing of gas in the dense phase occurs, and gas exchange between bubbles and the dense phase is high, probably due to splitting and recoalescence.

All bubbles rise more rapidly than the interstitial gas velocity, but in freely bubbling beds the velocity of small bubbles (< 4 cm) appears to be about 30 to 40 cm/s regardless of bubble size, suggesting that it is the gross circulation currents which control the rise velocity.

When the superficial gas velocity is sufficiently high and the bed diameter sufficiently small to cause the formation of slugging conditions, the slugs produced are axisymmetric; at superficial gas velocities between 0.5 and 1.5 m/s bubble and slug flow breaks down as entrainment becomes enormous and a transition to transport occurs. Further increase in velocity and the recirculation of elutriated solids results in fast fluidization (see Chapter 7).

3.2.3 Group B

Sand typifies powders in this group, which contains most solids in the mean size and density ranges:

$$60 \ \mu\text{m} < d_\text{p} < 500 \ \mu\text{m} \text{ when } \rho_\text{p} = 4 \text{ g/cm}^3$$

$$250 \ \mu\text{m} < d_\text{p} < 100 \ \mu\text{m} \text{ when } \rho_\text{p} = 1 \text{ g/cm}^3$$

In contrast with group A powders, interparticle forces are negligible and bubbles start to form in this type of powder at or only slightly above minimum fluidization velocity. Bed expansion is small and the bed collapses very rapidly when the gas supply is cut off (see Fig. 3.2).

There is little or no powder circulation in the absence of bubbles and bubbles burst at the surface of the bed as discrete entities. Most bubbles rise more quickly than the interstitial gas velocity and bubble size increases with both bed height and excess gas velocity ($U - U_\text{mf}$); coalescence is the predominant phenomenon and there is no evidence of a maximum bubble size. When comparisons are made at equal values of bed height and $U - U_\text{mf}$, bubble sizes are independent of both mean particle size and size distribution. Backmixing of dense phase gas is relatively low, as is gas exchange between bubbles and dense phase. When the gas velocity is so high that slugging commences, the slugs are initially axisymmetric, but with a further increase in gas velocity an increasing proportion becomes asymmetric, moving up the bed wall with an enhanced velocity rather than up the tube axis. Transition to transport occurs between 1 and 3 m/s.

3.2.4   Group D

Large and/or dense particles belong to this group. All but the largest bubbles rise more slowly than the interstitial fluidizing gas, so that gas flows into the base of the bubble and out of the top, providing a mode of gas exchange and by-passing which is different from that observed with group A or B powders (see fig. 3.7). The gas velocity in the dense phase is high and solids mixing is relatively poor; consequently, backmixing of the dense phase gas is small. Segregation by size is likely when the size distribution is broad, even at high gas velocities. The flow regime around particles in this group may be turbulent, that is $\rho_g U_{mf} d_p/\mu > 1,000$, causing some particle attrition with rapid elutriation of the fines produced. Relatively sticky materials can be fluidized since the high particle momentum and fewer particle–particle contacts minimize agglomeration.

Horizontal voids appear close to the distributor: in a narrow column these may extend across the tube giving bridging and piston-like slugs of solids which slide up the tube and collapse into wall slugs near the surface; in a large column these voids often split into bubbles which appear to grow without coalescence, as through draining gas from the surrounding dense phase. Slug breakdown and the onset of turbulent fluidization occur above about 3 m/s.

Bubble sizes are similar to those in group B powders at equal values of bed height and $U - U_{mf}$. If gas is admitted only though a centrally positioned hole, group D powders can be made to spout even when the bed depth is appreciable.

A qualitative summary of these properties is given in Table 3.1.

3.3   NUMERICAL CRITERIA FOR GROUPS

3.3.1   The group A/B Boundary

The most easily observed difference between powders in groups A and B is whether or not the bed bubbles at or very close to minimum fluidization. If there is appreciable bed expansion before bubbling commences, then the powder belongs to group A and is likely to have the other properties associated with that group.

The equations for the incipient fluidization and bubbling velocities, Eqs (2.28) and (2.29), are plotted on Fig. 3.3 for particles of two densities fluidized by air at ambient conditions. The < 45 μm fines fraction F is assumed to be 0.1.

Table 3.1 Summary of group properties

Increasing size and density
→

Group	C	A	B	D
Most obvious characteristic	Cohesive, difficult to fluidize	Bubble-free range of fluidization	Starts bubbling at U_{mf}	Coarse solids
Property — Typical solids	Flour, cement	Cracking catalyst	Building sand, table salt	Crushed limestone coffee beans
1. Bed expansion	Low when bed channels, can be high when fluidized	High	Moderate	Low
2. Deaeration rate	Initially fast, exponential	Slow, linear	Fast	Fast
3. Bubble properties	No bubbles. Channels, and cracks	Splitting/recoalescence predominate; maximum size exists; large wake	No limit on size	No known upper size; small wake
4. Solids mixing[a]	Very low	High	Moderate	Low
5. Gas backmixing[a]	Very low	High	Moderate	Low
6. Slug properties	Solids slugs	Axisymmetric	Axisymmetric, asymmetric	Horizontal voids, solids slugs, wall slugs
7. Spouting	No	No, except in very shallow beds	Shallow beds only	Yes, even in deep beds
Effect on properties 1 to 7 of: Mean particle size within group	Cohesiveness increases as d_p decreases	Properties improve as size decreases	Properties improve as size decreases	Not known
Particle size distribution[b]	Not known	Increasing <45 μm fraction improves properties	None	Increases segregation
Increasing pressure, temperature, viscosity, density of gas	Probably improves	Definitely improves	Uncertain, some possibly	Uncertain, some possibly

[a] At equal $U - U_{mf}$.
[b] At equal d_p.

Consider particles of density difference 1,000 kg/m^3 and mean particle size 100 μm. Entering the graph from the left, the minimum fluidization velocity line is encountered first with a U_{mf} of 0.4 cm/s and then the minimum bubbling line at 1 cm/s, giving a value of $U_{mb}/U_{mf} = 2.5$, indicating a bubble-free region and group A powder. By contrast, 100 μm alumina ($\rho_p = 4,000$ kg/m^3) has a theoretical U_{mb} of 1 cm/s and a U_{mf} of 1.5 cm/s. If will therefore bubble at minimum fluidization and fall into group B.

Thus for a powder to belong to group A or C:

$$\frac{U_{mb}}{U_{mf}} > 1 \qquad (3.1)$$

Substituting from Eqs (2.28) and (2.29), a powder will be in groups A or C if:

$$\frac{2{,}300\, \rho_g^{0.126}\, \mu^{0.523}\, \exp\,(0.716F)}{d_p^{0.8}\, g^{0.934}\, (\rho_p - \rho_g)^{0.934}} > 1 \qquad (3.2)$$

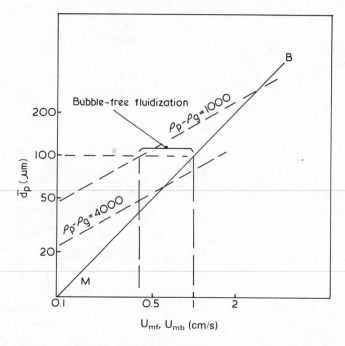

Figure 3.3 Minimum fluidization velocity and minimum bubbling velocity for air versus particle size.

If the physical properties of air at ambient conditions are substituted into Eq. (3.2), F is put equal to 0.1 and ρ_g is assumed to be negligible compared with ρ_p, then:

$$\rho_p^{0.934}\, d_p^{0.8} < 1 \qquad (3.3)$$

Equation (3.3) is shown in Fig. 3.4 as line XX. Powders with size/density combinations to the left of the line belong in groups A or C when fluidized by air at ambient conditions.

The ratio of the incipient bubbling and fluidization velocities — the velocity ratio — is a useful guide in characterising the fluidization behaviour of fine powders. From Eq. (3.2) we see that U_{mb}/U_{mf} increases as particle size and density decrease and from Fig. 3.5 that the bed expansion ratio in bubbling and non-bubbling beds increases with an increase in the velocity ratio. However, this trend does not continue indefinitely for if the powder

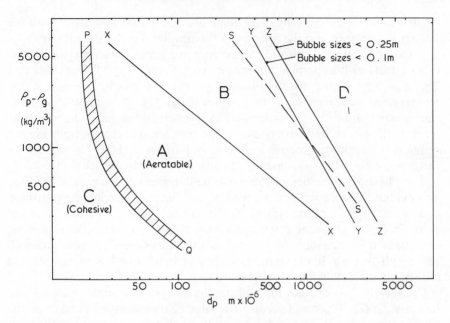

Figure 3.4 Diagram for classifying powders into groups having broadly similar fluidization characteristics in air at atmospheric temperature and pressure (after Geldart, 1973)

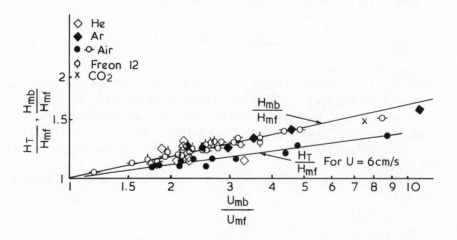

Figure 3.5 Expansion of group A powders at minimum bubbling and at 6 cm/s.

becomes too fine (typically, less than about 20 μm) cohesivity assumes greater importance and the 'quality of fluidization' rapidly deteriorates.

If the physical properties of the powder are maintained constant and no other effects such as particle sintering come into play, Eq. (3.2) predicts that U_{mb}/U_{mf} (and hence the fluidization properties) should improve if the temperature and/or pressure increase, or if the gas composition changes in such a way that $\mu^{0.523} \rho_g^{0.126}$ increases. Conversely, if the gas is changed from air to hydrogen at ambient conditions the powder may change from smooth group A behaviour to group B with its large bubbles or slugs. The reasons for this are not clear but may involve gas adsorption on to particles.

The physical properties of the gas usually remain constant in a fluidized process but those of the powder may not. If fines are produced by attrition and are not lost from the system, F will increase and d_p decrease. This will increase U_{mb}/U_{mf} and at first improve the fluidization, but if it continues, eventually it may change the powder into the excessively cohesive group C, and result in poor fluidization; this kind of behaviour has been observed industrially as well as in the laboratory.

Conversely, loss of fines through inefficient cyclones would reduce F and increase d_p; U_{mb}/U_{mf} could decline to a value near or less than 1 and move the powder into group B, causing large bubbles or slugging and loss of conversion in a chemical reactor. This has also been observed experimentally.

3.3.2   Group A/C Boundary

Molerus (1982) has attempted to distinguish between groups A and C by considering the balance between hydrodynamic and cohesive forces. He gives the criterion that a powder belongs to group C if:

$$\frac{10 \, (\rho_p - \rho_g) \, d_p^3 \, g}{F_H} < 10^{-2} \qquad (3.4)$$

where F_H is the adhesion force transmitted in a single contact between two adjacent particles. Its magnitude depends upon the surface geometry of the contacting particles, the hardness of the particulate material, and on the type of forces (van der Waals, capillary, gas adsorption, etc.). Rietema (1984) has also addressed the problem and produced a *qualitative* relationship involving F_H, d_p, ρ_p, and μ.

Wong (1983), in a parallel approach, gives the criterion that for a group C powder:

$$\frac{\rho_g^{0.06} \, d_p^2 \, \mu^{0.653}}{F_H} < 2.72 \times 10^{-6} \qquad (3.5)$$

At present it is not possible to resolve which of these equations is correct. Both predict that as F_H increases, e.g. by the solids becoming softer, group C will include larger and larger particles. The area PQ (Fig. 3.4) drawn in by Geldart (1973) on the basis of fludization properties reported in the literature implicitly recognizes this, since low density particles (e.g plastics) tend to be soft and high density materials (e.g. metals) hard. A transition category designated group AC by Dry, Shingles, and Judd (1983) having properties of the two others has recently been proposed.

On a strictly empirical basis, a simple test procedure can be used to ascertain to which group any particular powder belongs. Geldart, Harnby, and Wong (1984) have shown that standardized measurements of bulk densities can be used. The bulk densities of the powders are determined by a standardized technique in the apparatus which is shown in Fig. 3.6. These features are available on the Nauta-Hosakawa powder tester. Particular care is taken to ensure that the centre of the stationary chute is in alignment with the centre of the pre-weighed 100 cm^3 cup. The most loosely packed bulk density, ρ_{BLP}, is obtained by pouring the powder through a vibrating sieve and allowing it to fall a fixed height (about 25 cm) into a cylindrical cup. The amplitude of the vibration is set so that the powder will fill the cup in 20 to 30 s. The excess powder is removed from the cup by drawing the flat edge of a ruler over the top. Care should be taken to ensure the cup is not jarred as this causes the powder to compact and gives a false reading. The powder and cup are weighed and an extension piece is then fixed to the cup. The cup is tapped 480 times (with a fall from a height of about 3 cm) whilst extra powder is added to ensure the powder level does not drop below the surface of the cup. When tapping ceases the extension is carefully removed, the excess powder scraped off, and the powder and the cup reweighed. The tapped bulk density, ρ_{BT}, can then be calculated.

Strong interparticle forces prevent the particles from rolling over each other when poured into the cup, giving an open powder structure and low values of ρ_{BLP} for cohesive powders. However, this open structure can easily be disrupted when work is done on the system, so that the powder can be compacted to produce a closer matrix under mechanical vibration. This phenomenon gives a high ratio ρ_{BT} to ρ_{BLP} in cohesive solids and a low ratio in free-flowing powders. Thus the ratio ρ_{BLP} is a reflection of the cohesivity of a powder. ρ_{BT}/ρ_{BLP} is called the Hausner ratio, HR (Grey and Beddow, 1968/69). If the ratio HR > 1.4 the powder is in group C; if $1.25 <$ HR < 1.4 the powder is in the transition AC group; and if HR < 1.25 the powder is in group A, B, or D.

3.3.3 Group B/D Boundary

There are two criteria which can be used to determine the B/D boundary. The first is based on the mode of gas by-passing, discussed in Chapter 11. In small

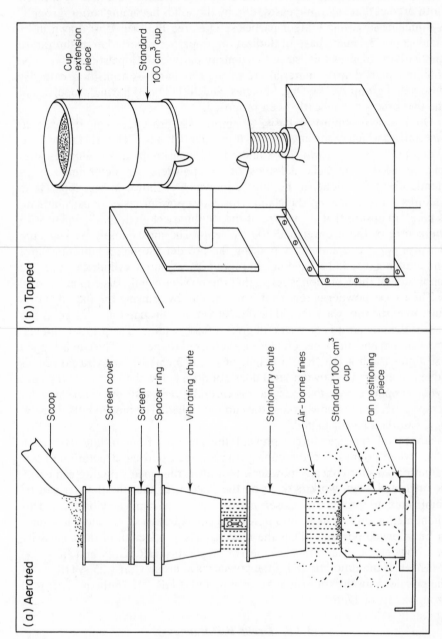

Figure 3.6 Equipment for standardized measurement of most loosely packed, and tapped, densitities (based on Nanta-Hosakawa powder tester).

particle systems all but the smallest bubbles travel much faster than the
interstitial gas, and gas tends to stay in the bubble except during coalescence
and splitting (see Fig. 3.7a).

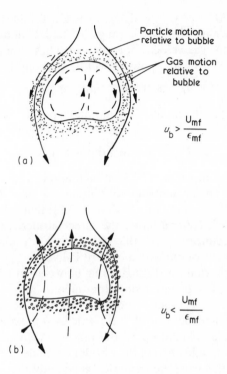

Particle motion
relative to bubble

Gas motion
relative to
bubble

$$u_b > \frac{U_{mf}}{\epsilon_{mf}}$$

(a)

$$u_b < \frac{U_{mf}}{\epsilon_{mf}}$$

(b)

Figure 3.7   Modes of gas by-passing.

In large particle systems, all but the largest bubbles travel slower than the
interstitial gas and gas short-circuits the bed by entering the bottom of a
bubble and leaving the top. It is clear that when considering groups B and D
we are really looking at a continuum because the form of by-passing which
occurs depends not only on U_{mf} but also on the velocities of bubbles and
hence their sizes. The bubble size depends on the excess gas velocity $U–U_{mf}$
and the height above the distributor; so within a large, deep fluidized bed
near the distributor there would be small bubbles travelling more slowly than
U_{mf}/ϵ_{mf} and higher up the bed, large bubbles travelling more quickly.

In spite of this complexity the situation can be simplified considerably by
choosing a bubble size and calculating the density/particle size combinations
of powders in which bubbles smaller than this would rise faster than U_{mf}/ϵ_{mf},
i.e. in the throughflow mode (Fig. 3.7b).

Thus a powder is in group D if:

$$u_b < \frac{U_{mf}}{\epsilon_{mf}} \qquad (3.6)$$

For large particles, $0.5 > \epsilon_{mf} > 0.4$, say 0.45. Bubbles in commercial fluidized beds usually have sizes in the range 0.1 to 0.25 m due to limitations imposed by bed depth, gas velocity, or internal tubes. For bubbles smaller than 0.25 m, $u_b < 1.1$ m/s and a powder will be in group D if:

$$U_{mf} > 0.5 \text{ m/s (approx.)} \qquad (3.7)$$

For bubbles smaller than 0.1 m, $u_b < 0.7$ m/s and a powder will be in group D if:

$$U_{mf} > 0.3 \text{ m/s (approx.)} \qquad (3.8)$$

The minimum fluidization velocity U_{mf} can be found from Eq. (2.27) and this has been done for air at ambient conditions to give the size/density combinations which must satisfy Eqs (3.7) and (3.8), that is lines ZZ and YY respectively in Fig. 3.4. By and large, the effect of increasing temperature and/or pressure is to reduce U_{mf} and this would cause YY and ZZ to move to the right; i.e. the system would become more B-like.

There is another criterion for distinguishing between groups B and D; this is entirely empirical, being based on experimental measurements of the spoutability of powders. Whether or not a powder can be spouted depends not only on the size and density of the particles — nozzle diameter, bed depth, and bed diameter all play a part. Baeyens and Geldart (1973) adopted the arbitrary criterion that if a bed of the powder 0.3 m deep can be spouted it belongs to group D; this gave the numerical inequality that if:

$$(\rho_p - \rho_g) \, d_p^{1.24} > 0.23 \qquad (3.9)$$

the powder is spoutable and is therefore in group D. This equation is plotted as line SS in Fig. 3.4 and gives results similar to those based on the mode of gas by-passing, King and Harrison (1980) found that the minimum spouting velocity decreases with increasing pressure which suggests that, in common with YY and ZZ, SS should move to the right as pressure increases.

3.4 APPLICATION TO LIQUID FLUIDIZATION

It has long been known that in some liquid fluidized systems instabilities which resemble bubbles can be seen, for example 770 μm lead shot in water (see Davidson and Harrison, 1963). The instabilities develop and grow more rapidly as the bed expands (i.e. as U/U_{mf} increases), as the density and size of the particles increase, and as the liquid viscosity decreases (Jackson, 1971). All these parameters are in the correct qualitative relationship to one another in Eqs (3.1) and (3.2), which were developed for gas fluidized systems. Replacing subscript g by l to denote liquid fluidization,

$$\frac{U_{mb}}{U_{mf}} = \frac{2,300 \, \rho_l^{0.126} \, \mu_l^{0.523} \, \exp(0.716F)}{d_p^{0.8} \, g^{0.934} \, (\rho_p - \rho_l)^{0.934}} \tag{3.10}$$

Since there are rarely any particles less than 45 μm in a liquid fluidized system $F = 0$ and the exponential term in the numerator becomes equal to unity.

In a series of photographs, Harrison, Davidson, and de Kock (1961) show the appearance of liquid fluidized beds of various solids. Lead shot was fluidized by various concentrations of glycerol solutions in order to vary the liquid viscosity, and paraffin was used to fluidize particles having approximately the same size but different densities.

In Table 3.2 the basic data relating to the experiments of Harrison, Davidson and de Kock (1961) are listed, together with values of U_{mb}/U_{mf} calculated from Eq. (3.10) and the qualitative appearance of the beds as judged from the photographs reproduced in Davidson and Harrison (1963). It can be seen that at values of U_{mb}/U_{mf} less than about 2 the beds exhibited slightly or appreciably disturbed fluidization, which may be construed as equivalent to bubbling in gas fluidized beds; at higher values of U_{mb}/U_{mf} the beds appeared to be quiescent. This qualitative agreement is encouraging and warrants further investigation. For the present the inequality below may be used with caution to identify liquid fluidized systems likely to exhibit non-quiescent behaviour at voidages up to 0.75. Thus 'bubbling' will occur if

$$\frac{2,300 \, \rho_l^{0.126} \, \mu_l^{0.523}}{d_p^{0.8} \, g^{0.934} \, (\rho_p - \rho_l)^{0.934}} < 2 \tag{3.11}$$

3.5 CONCLUSIONS

The concept of four broad groups of powders each of which has distinctive fluidization characteristics is now widely accepted. Their representation on a 'phase' diagram (Fig. 3.4) when air is the gaseous medium is useful but shows only part of the picture. As we move away from groups D and B towards groups A and C, parameters other than particle density and mean size assume increasing importance, and a multi-dimensional diagram is required. There remains much to be understood, but two lessons emerge from the ideas presented above.

Firstly, it is unwise to assume that theories deduced on the basis of data gathered from a powder in one group are necessarily applicable to a powder in another. Secondly, although cold modelling is useful and often essential when studying scale-up, changing the temperature and pressure of the gas may affect crucially the behaviour of the powder, particularly if it happens to have a size/density which puts it near the boundary of groups A and C where interparticle forces play an important role.

Table 3.2 Comparison of experimental data on liquid fluidization with Eq. (3.10). Data from Harrison, Davidson, and de Kock (1961).

(a)

Weight (%) of glycerol	ρ_1 kg/m^3	μ_1 N s/m^2	Appearance of bed of voidages of:		U_{mb}/U_{mf}
			0.6	0.75	(Eq. 3.10)
75	1,190	36.3×10^{-3}	Q	Q	6.57
66	1,170	16.6×10^{-3}	Q	Q	4.36
39	1,110	3.58×10^{-3}	SB	B	1.94
0	1,000	$1 \quad \times 10^{-3}$	SB	B	0.98

Particles : lead shot $\rho_p = 11,320$ kg/m^3, $d_p = 770$ μm

(b)

ρ_p, kg/m^3	d_p, μm	Appearance of bed at voidages of:		U_{mb}/U_{mf}
		0.65	0.75	(Eq. 3.10)
1,500	600	Q	Q	19.8
2,900	775	Q	Q	5.9
7,430	770	SB	SB	2.03
11,320	770	B	B	1.32

Liquid : $\rho_1 = 780$ kg/m^3
$\mu_1 = 2 \times 10^{-3}$ Ns/m^2

Q = quiescent, SB = slightly bubbling, B = bubbling

3.6 EXAMPLE

A new catalyst to be used in a hydrogenation process has the following properties:

Skeletal (absolute) density $= 3000$ kg/m^3
Particle porosity $= 0.4$ cm^3/gm

Table 3.3 Size Distribution

Weight (%)	7	12	10	16	18	12	10	8	7
$d_p(\mu m)$	-30 $+20$	-45 $+30$	-53 $+45$	-63 $+53$	-75 $+63$	-90 $+75$	-105 $+90$	-125 $+105$	-150 $+125$

Cold model tests using air show the powder to have ideal fluidization properties.

Conditions in the full-scale plant are such that the gas density and viscosity will be 1 kg/m^3 and 1.3×10^{-5} N s/m^2, respectively.

Over a period of several months the conversion efficiency decreases and a sample of powder from the bed is found to have a particle porosity of 0.1 cm^3/gm and only 5 per cent by weight less than 45 μm.

Step 1. Calculate particle densities and mean sizes:

$$\text{Initial particle density } \rho_{p1} = \frac{1}{\dfrac{\text{particle}}{\text{porosity}} + \dfrac{1/\text{skeletal}}{\text{density}}}$$

$$= \frac{1}{0.4 + \frac{1}{3}} = 1.364 \text{ gm/cm}^3$$

$$\text{Final particle density } \rho_{p2} = \frac{1}{0.1 + \frac{1}{3}} = 2.31 \text{ gm/cm}^3$$

From the size distribution table and $d_p = 1/\Sigma x_i/d_{pi}$) (Eq. 2.8):

Initial mean particle size $d_{p1} = 58$ μm, fines fraction $F_1 = 0.19$

Final mean particle size $d_{p2} = 70$ μm, $F_2 = 0.05$

Step 2. Calculate U_{mb}/U_{mf} for cold model, initial, and final reactor conditions, using Eq. (3.2):

$$\text{Cold model } \frac{U_{mb}}{U_{mf}} = \frac{2{,}300 \times 1.2^{0.126} (1.84 \times 10^{-5})^{0.523} \exp (0.716 \times 0.19)}{(58 \times 10^{-6})^{0.8} [9.81 \times (1364 - 1.2)]^{0.934}}$$

$$= 3.09$$

Initial reactor conditions $\dfrac{U_{mb}}{U_{mf}} = 2.51$

Final reactor conditions $\dfrac{U_{mb}}{U_{mf}} = 1.115$

The powder/gas system is clearly in group A under both cold model and initial reactor conditions, which would give ideal fluidization behaviour. Under the final reactor conditions the powder is close to group B with consequent reduction in bed expansion and catalyst fluidity.

Step 3. Estimate the maximum stable bubble sizes for initial and final reactor using Eq. (4.13)

(a) Initial conditions. Calculate terminal velocity for particles:
    2.7 d_p, that is 2.7×58 μm $= 157$ μm and $\rho_p = 1{,}364$ kg/m^3
    From Eq. (6.12), $C_D Re^2 = 408$; from Fig. 6.7, $Re_t = 9.6$; therefore:
    $v_t = 0.795$ m/s

    From Eq. (4.12): $(d_{eq})_{max} = \dfrac{2\,v_t^2}{g} = 0.127$ m

(b) Final conditions. Use the calculation procedure as in (a):
    $(d_{eq})_{max} = 0.467$ m

(c) Comments
    The much lower U_{mb}/U_{mf} value for the final reaction conditions is a clear predictor of deteriorating fluidization conditions, and the larger maximum stable bubble size confirms this. Coupled with a lower particle (internal) porosity it is not surprising that the conversion efficiency suffered.

3.7 NOMENCLATURE

d_p	mean sieve size of a powder	m
F	mass fraction of powder less than 45 μm	—
F_H	adhesive force transmitted in a single contact between two particles	N
g	acceleration of gravity	9.81 m/s^2
U	superficial velocity of gas	m/s
u_b	rise velocity of an isolated bubble	m/s
U_{mf}, U_{mb}	incipient fluidization and bubbling velocities	m/s
v_t	terminal velocity of particle	m/s

ϵ_{mf}, ϵ_{mb}	bed voidage and incipient fluidization and bubbling velocities	—
μ	viscosity of gas	kg/ms
μ_l	viscosity of liquid	kg/ms
ρ_{BT}	tapped (maximum) bulk density of powder	kg/m^3
ρ_{BLP}	most loosely packed (minimum) bulk density of powder	kg/m^3
ρ_g	density of gas	kg/m^3
ρ_l	density of liquid	kg/m^3
ρ_p	particle or piece density	kg/m^3

3.8 REFERENCES

Baeyens, J., and Geldart, D. (1973). In *Fluidization and Its Applications*, Toulouse, p. 263.
Davidson, J.F., and Harrison, D. (1963). *Fluidized Particles*, Cambridge University Press.
Donsi, G., Moser, S., and Massimilla, L. (1975). *Chem. Eng. Sci.*, **30**, 1533.
Dry, R.J. Shingles, T., and Judd, M.R. (1983). *Powder Techno.*, **34**, 213.
Geldart, D. (1972). *Powder Technol.*, **6**, 201.
Geldart, D. (1973). *Powder Technol.*, **7**, 285.
Geldart, D., Harnby, N., and Wong. A.C.Y. (1984). *Powder Technol.*, **37**, 25.
Geldart, D., and Williams, J.C. (1985). *Powder Technol.*, **43**, 181.
Grey, R.O., and Beddow, J.K. (1968/69). *Powder Technol.*, **2**, 323.
Harrison, D., Davidson, J.F., and de Kock, J.W. (1961). *Trans. Instn Chem. Engrs, Lond.*, **37**, 328.
Jackson, R. (1963). *Trans. Instn Chem. Engrs*, **41**, 13.
Jackson, R. (1971). In *Fluidization* (Eds. J.F. Davidson and D. Harrison), Academic Press.
King, D., and Harrison, D. (1980). *Powder Technol.*, **26**, 103.
Molerus, O. (1967). In *Proc. Intern Symp on Fluidization*, Eindhoven, Netherlands University Press, p. 134.
Molerus, O. (1982). *Powder Technol.*, **33**, 81.
Oltrogge, R.D. (1972). Ph. D. Thesis, University of Michigan.
Rietema, K. (1984). *Powder Technol.*, **37**, 5.
Simpson, H.C., and Rodger, B.W. (1961). *Chem. Eng. Sci.*, **16**, 153.
Verloop, J., and Heertjes, P.M. (1970). *Chem. Eng. Sci.*, **25**, 825.
Wilhelm, R.H., and Kwauk, M. (1948). *Chem. Eng. Prog.*, **44**, 201.
Wong, A.C.Y. (1983). Ph. D. Dissertation, University of Bradford.

Gas Fluidization Technology
Edited by D. Geldart
Copyright © 1986 John Wiley & Sons Ltd.

CHAPTER 4

Hydrodynamics of Bubbling Fluidized Beds

R. CLIFT

4.1 INTRODUCTION

The behaviour of most gas fluidized beds is dominated by the rising gas voids, conveniently termed 'bubbles', which characterize these systems. In analysing the behaviour of bubbling fluidized beds, it is essential to distinguish between the *bubble phase* (or 'lean phase'), i.e. the gas voids containing virtually no bed particles, and the *particulate phase* (also known as 'dense phase' or 'emulsion phase') consisting of particles fluidized by interstitial gas. A bubbling bed can conveniently be defined as a bed in which the bubble phase is dispersed and the particulate phase is continuous. The flow regimes observed at high velocity, discussed in Chapter 7, correspond to beds in which the volume occupied by the lean phase is so high that the particulate phase no longer forms a continuous medium between discrete bubbles.

The rising bubbles cause motion of the particulate phase which is the main source of solids mixing in bubbling beds, discussed in Chapter 5. This particle motion in turn causes the temperature uniformity and high bed/surface heat transfer coefficients characteristic of fluidized beds and reviewed in Chapter 9. Since the gas in a bubble is not in direct contact with the bed particles, it cannot take part in any reaction between gas and solids. Thus interchange of gas between bubbles and interstitial gas in the particulate phase can determine the performance of a fluidized bed reactor. The influence of bubble characteristics on fluid bed reactors is explained in Chapter 11. Thus an understanding of the behaviour of the bubble phase in fluidized beds is essential for understanding the applications of these devices. Section 4.2 of the present chapter considers the behaviour of individual bubbles and of

53

slugs, i.e. bubbles so large that their shape and rise velocity are determined by the dimensions of the column containing the fluidized bed rather than by their own volume.

The gas which fluidizes the bed and forms bubbles must normally be distributed uniformly over the cross-section of the bed; this is achieved by introducing it through a *gas distributor*. Design requirements for gas distributors are discussed in Section 4.3, and the flow patterns observed in the vicinity of the distributor are reviewed in Section 4.4. Finally, the behaviour of continuously slugging and freely bubbling beds is discussed in Sections 4.5 and 4.6, concentrating on prediction of the most important characteristics: bubble size and flowrate, and bed expansion.

In many respects, bubbles in gas fluidized beds are analogous to gas bubbles in viscous liquids (see, for example, Davidson, Harrison, and Guedes de Carvalho, 1977). This analogy has been of immense value in interpreting the behaviour of bubbling fluidized beds, since bubbles are more readily observable in transparent liquids, and the analogy is used throughout this chapter. There is, however, an important difference between liquids and fluidized beds. A gas bubble in a liquid is bounded by a distinct interface; material can be transferred across this interface by diffusion but not by bulk flow. On the other hand, the surface of a bubble in a fluidized bed is a boundary between a particle-lean region and a region of high solids concentration which is permeable to gas. Therefore bulk flow can occur between a bubble and the surrounding particulate phase, so that material can be transferred across the boundary by convection as well as by diffusion. These processes of interphase transfer are considered in detail in Chapter 11.

4.2   BUBBLES AND SLUGS

4.2.1   Shapes of Single Bubbles

Figure 4.1 shows an idealized bubble, the general shape being well known from photographic studies of bubbles in liquids (see, for example, Clift, Grace and Weber, 1978) and X-ray studies of bubbles in fluidized beds (see, for example, Rowe, 1971). The upper surface of the bubble is approximately spherical, and its radius of curvature will be denoted by r. The base is typically slightly indented. Since r is not readily determinable, it is usually more convenient to express the bubble size as its 'volume-equivalent diameter', i.e. the diameter of the sphere whose volume is equal to that of the bubble:

$$d_{eq} = (6V_B/\pi)^{1/3} \qquad (4.1)$$

where V_b is the bubble volume. A bubble in a viscous liquid shows the form of Fig. 4.1 if it is sufficiently large for surface tension forces to be negligible (Clift, Grace, and Weber, 1978), i.e. if

$$Eo = \frac{\rho_l \, d_{eq}^2 \, g}{\sigma} > 40 \qquad (4.2)$$

where ρ_l and σ are the liquid density and interfacial tension, and the dimensionless group Eo is known as the Eötvös number. Since fluidized beds lack a phenomenon equivalent to interfacial tension, σ is effectively zero. Thus bubbles in fluidized beds satisfy Eq. (4.2), and the analogy is reasonable.

The shape of such a bubble can conveniently be described by the 'wake angle' between the nose and the lower rim, θ_w in Fig. 4.1. Grace (1970) showed that θ_w is a function of the bubble Reynolds number:

$$Re_b = \frac{\rho_l \, u_b \, d_{eq}}{\mu_l} \qquad (4.3)$$

where u_b is the bubble rise velocity and μ_l the liquid viscosity. Observed shapes of bubbles in liquids are represented (Clift, Grace, and Weber, 1978) by:

$$\theta_w = 50 + 190 \exp(-0.62 \, Re_b^{0.4}) \qquad (Eo > 40, Re_b > 1.2) \qquad (4.4)$$

where θ_w is expressed in degrees. This relationship is shown in Fig. 4.2, with a theoretical relationship derived by Davidson et al. (1977):

$$Re_b = \left(\frac{4}{2 - 3 \cos \theta_w + \cos^3 \theta_w} \right)^{2/3} \qquad (5 < Re_b < 100) \qquad (4.5)$$

which is in qualitative agreement with the empirical result. It is worth noting that, for $Re_b < 100$, θ_w is constant at $50°$, as predicted by Rippin and Davidson (1967); this is the shape observed for spherical cap bubbles in low viscosity liquids such as water. Other parameters of interest are the frontal diameter of the bubble d_b and the volume of the circumscribing sphere not occupied by the bubble V_w. For reasons explained below, V_w is termed the 'wake volume', and the ratio of wake/sphere volume, that is $3V_w/4\pi r^3$, is known as the 'wake fraction' f_w. Figure 4.2 shows values of d_{eq}/r, d_b/r, and the wake fraction, calculated assuming that the wake angle is given by Eq. (4.4) and the base of the bubble is flat.

---

*Note that the wake fraction is sometimes defined as the wake volume V_w divided by the bubble volume V_b, in which case it is given the symbol β_w. The relationship between the two is $\beta_w = f_w/(1 - f_w)$.

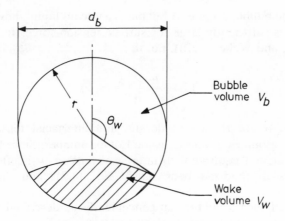

Figure 4.1   Spherical cap bubble.

Quantitative application of these results to fluidized beds is limited by lack of sufficient information on the effective viscosity of the particulate phase, or even on how realistic is the assumption that the particulate phase behaves as a Newtonian medium. However, the effect of viscosity is a convenient starting point for discussing the effect of particle characteristics on bubble behaviour in fluidized beds. The essential feature of Fig. 4.2 is that decreasing the viscosity of the medium surrounding the bubble increases Re_b and therefore causes the bubble to become flatter (lower θ_w) and the wake fraction to increase. Grace (1970) used this effect to infer the effective viscosity of fluidized beds from bubble shapes reported by Rowe and Partridge (1965). Values of θ_w ranged from 90 to 134°, corresponding to Re_b from 2 to 10, implying effective viscosities from 0.4 to 1.3 N s/m² (i.e. similar to the viscosity of engine oil). Apparent viscosities have also been measured by more conventional techniques (see Schügerl, 1971), involving immersing rotating bodies or surfaces in the bed. Since these techniques disturb the local structure of the bed, the significance of the results is questionable. However, the values obtained are generally consistent with those inferred by Grace.

Hetzler and Williams (1969) have given a semiempirical expression for the effective low-shear viscosity of a fluidized medium. It was derived mainly from data for liquid/fluidized systems and its general applicability to gas/fluidized beds is untested, but it seems to explain some of the differences in behaviour between particle groups. The effective viscosity is predicted to increase with particle diameter for group A powders, to increase weakly with d_p in group B powders, and to be high but insensitive to d_p in group D powder. The viscosity estimates of Grace (1970) and Schügerl (1971) show this general trend. The results given by Rowe (1971) show the corresponding effect on bubble shape : on increasing d_p from group A to group B, bubbles become

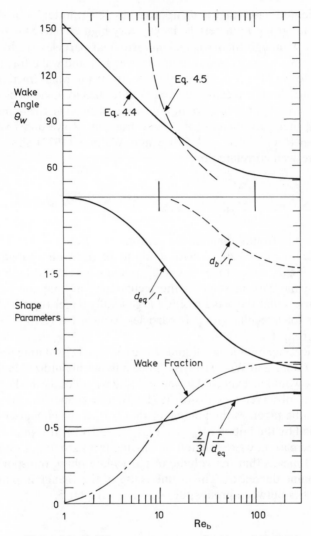

Figure 4.2 Bubble parameters as a function of Reynolds number.

more rounded and the wake fraction decreases, from 0.4 for 50 μm cracking catalyst to 0.28 to 0.3 in group B ballotini. More recently, Rowe and Yacono (1976), working with freely bubbling beds of groups A and B, reported that particle size had no detectable effect on average bubble shape. However, it is likely that the changes of shape resulting from bubble interaction and coalescence (see Section 4.6) masked any effect of particle characteristics. Cranfield and Geldart (1974) report that bubbles in group D beds lack a deeply indented base (that is θ_w approaches 180°); in terms of Fig. 4.2, this confirms that group D materials display very high effective viscosity, so that Re_b is low. Although theory is lacking, irregular particles are likely to show higher effective viscosity and therefore lower Re_b and wake fraction. Results quoted by Rowe (1971) show exactly this effect: the wake fraction in beds of irregular particles is typically 0.2 compared to 0.28 to 0.3 in beds of spherical material. Increasing bubble size in a medium of fixed viscosity increases Re_b and should therefore increase the wake fraction. Rowe and Everett (1972) observed such an effect, and Rowe and Widmer (1973) showed that the results were well correlated by:

$$\frac{V_b}{V_{sph}} = 1 - \frac{V_w}{V_{sph}} = \exp{(- 0.057 d_b)} \qquad (4.6)$$

where d_b is the frontal diameter in centimetres. Equation (4.6) was derived from observations of bubbles with d_b up to 16 cm, with a range of group B materials; for d_b larger than 16 cm, it predicts unrealistically low values of V_b/V_{sph}. In the current stage of the technology, all that can be said about other groups is that the wake fraction is generally larger (up to 0.4) for group A, smaller for irregular group B particles (typically 0.17 to 0.22), and very small for group D.

A spherical cap bubble in a liquid carries a closed circulating wake if $Re_b <$ 110 (Clift, Grace and Weber, 1978). Since bubbles in fluidized beds fall in this range, they would be expected to carry closed wakes of particulate phase, and this has been observed (see Rowe, 1971). To a good approximation, the wake volume can be taken as the volume of that part of the circumscribing sphere not occupied by the bubble, V_w (see Figs 4.1 and 4.2). It is important to note that the decrease in wake fraction on moving from group A through group B to group D means that the volume of particulate phase transported per unit bubble volume decreases. This is one cause of the change in particle mixing characteristics between groups, discussed in Chapter 5.

4.2.2. Rise Velocity of an Isolated Bubble

Davies and Taylor (1950) showed that the rise velocity u_b of a spherical cap bubble is related to its radius of curvature by:

$$u_b = 2/3 \ \sqrt{(gr)} \qquad (4.7)$$

Experimental results show that Eq. (4.7) is reliable for bubbles in liquids if Re_b is greater than about 40; below this value, departures from the spherical cap shape have a significant effect on u_b (Clift, Grace and Weber, 1978). Bubbles in fluidized beds typically have Reynolds numbers of order 10 or less, below the range for which Eq. (4.7) is strictly valid in liquids. Even so, Eq. (4.7) has been widely used for fluidized beds, and inaccuracies are no doubt masked by the erratic velocity variations typically observed (Rowe and Yacono, 1976). In terms of the volume-equivalent diameter, Eq. (4.7) can be written:

$$u_b = \left\{ \frac{2}{3} \sqrt{\left(\frac{r}{d_{eq}} \right)} \right\} \sqrt{(gd_{eq})} \qquad (4.8)$$

where the braced term is a weak function of Re_b, as shown in Fig. 4.2. For $Re_b > 100$ this term is constant at 0.71, and the resulting expression:

$$u_b = 0.71 \sqrt{(gd_{eq})} \qquad (4.9)$$

has been widely used to predict the rise velocity of isolated bubbles in fluidized beds (Davidson and Harrison, 1963). However, observed rise velocities are generally smaller than predicted by Eq. (4.9), values of $u_b/\sqrt{(gd_{eq})}$ from 0.5 to 0.66 being typical for groups A and B (see Davidson, Harrison, and Guedes de Carvalho, 1977). From Fig. 4.2, these values correspond to $Re_b < 60$, broadly consistent with the range inferred from the bubble shape.

For groups A and B, it is possible to estimate bubble rise velocities from the shape, as expressed by the wake fraction V_w/V_{sph}, and this approach is at least self-consistent. From Eq. (4.8), it follows that:

$$u_b = \frac{\sqrt{2}}{3} \left(\frac{1}{1 - V_w/V_{sph}} \right)^{1/6} \sqrt{(gd_{eq})} \qquad (4.10)$$

For rounded group B particles, Eq. (4.6) can be used to estimate V_w/V_{sph}. For angular group B particles, the wake fraction is typically 0.2, so that $u_b/\sqrt{(gd_{eq})}$ should be roughly 0.5. The shapes of bubbles in group A particles suggest that Re_b is about 10, so that $u_b/\sqrt{(gd_{eq})}$ should be from 0.5 to 0.6.

For group D particles, Cranfield and Geldart's (1974) results give:

$$u_b = 0.71 \sqrt{(gd_b)} \qquad (4.11)$$

Since d_b and d_{eq} are virtually the same in group D beds, Eqs (4.9) and (4.11) are equivalent. This is unexpected, since observed bubble shapes suggest that group D prticles should be treated as low Re_b systems (see Section 4.2.1). In the present state of the technology, all that can be concluded is that treating the dense phase as a Newtonian medium does not explain all features of bubble behaviour and that Eqs (4.9) and (4.11) can be used in group D beds.

4.2.3 Bubble Break-Up

Bubbles in viscous liquids and in fluidized beds break up by the process shown schematically in Fig. 4.3 (Clift and Grace, 1972). An indentation forms on the upper surface of the bubble and grows as it is swept around the periphery by the motion of the particles relative to the bubble. If the curtain grows sufficiently to reach the base of the bubble before being swept away, the bubble divides. When the surface tension effects are negligible, as in fluidized beds, the growth rate of such a disturbance decreases with increasing kinematic viscosity of the medium surrounding the bubble, so that the maximum stable size a bubble can attain before splitting increases with increasing kinematic viscosity (Clift, Grace and Weber, 1974). Analysis of this process leads to realistic predictions of the maximum stable size of bubbles and drops of conventional fluids (Clift, Grace, and Weber, 1978). Application to fluidized beds is again limited by lack of sufficient data on effective viscosities, but it does explain the differences between powder groups. In group A particles the effective viscosity and hence maximum bubble size are relatively small, and both increase with particle diameter. For group B particles the effective viscosity and bubble size are larger and less dependent on particle size, while for Group D particles the maximum bubble size is too large to be realized.

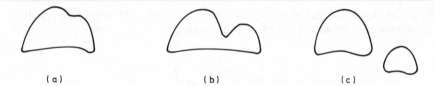

(a)   (b)   (c)

Figure 4.3 Bubble splitting.

For quantitative predictions of maximum stable size, a method developed by Harrison *et al.* (see Davidson and Harrison, 1963) must be applied. It was proposed that bubbles split when particles are carried up from the base of the bubble by the circulation of gas within the bubble. The gas was assumed to circulate with velocity of order u_b, so that a bubble breaks up if u_b exceeds the terminal velocity of a bed particle. Since the predictions are at best approximate, Eq. (4.9) may be used to estimate u_b. The resulting expression for the maximum stable bubble size is:

$$(d_{eq})_{max} = \frac{2v_t^2}{g} \tag{4.12}$$

where v_t is the particle terminal velocity. Although the physical process on which Eq. (4.12) is based is now known not to occur, there is some evidence that Eq. (4.12) predicts the correct order of magnitude for the maximum bubble size. Matsen (1973) found that Eq. (4.12) underpredicts the maximum bubble size, typically by a factor of 4, the difference being larger for very fine particles. Geldart (1977) suggests that a better fit with experimental data for group A powders is obtained if the terminal velocity in Eq. (4.12) is calculated for particles of size:

$$d_p' = 2.7d_p \qquad (4.13)$$

where d_p is the mean size of the powder (Eq. 2.8). Figure 4.4 shows $(d_{eq})_{max}$, for particles of density 2,500 kg/m^3, fluidized by air at 1 and 15 bar, and 300 and 1,000 K, calculated from Eq. (4.12) with v_t estimated from the correlations given by Clift, Grace and Weber (1978) (see Fig. 6.7) and using Eq. (4.13). For group D and large group B particles, the maximum stable bubble size is so large that in practice it would not impose a limitation on the size of bubbles present in a bubbling bed. For both group A and group B particles, increasing temperature and, especially, increasing pressure reduces the maximum bubble size, thus tending to make fluidization 'smoother'. This is consistent with observations (see Chapter 3).

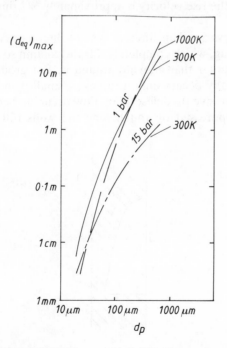

Figure 4.4 Maximum stable bubble size as a function of mean size of powder, and gas pressure or temperature (calculated using Eqs 4.12 and 4.13).

4.2.4 Wall Effects : Slugs

Provided that d_{eq} is less than about 0.125 times the bed diameter D, the shape and rise velocity of a bubble are unaffected by the walls. For d_{eq}/D greater than 0.125, wall effects reduce the rise velocity and cause the bubble to be more rounded, so that the wake fraction is reduced (Clift, Grace, and Weber, 1978). For d_{eq}/D between 0.125 and 0.6, the rise velocity can be estimated roughly (Wallis, 1969) by:

$$u_b = 1.13u_{b\infty} \exp -\left(\frac{d_{eq}}{D}\right) \qquad (4.14)$$

where $u_{b\infty}$ is the rise velocity in the absence of wall effects. For d_{eq}/D greater than 0.6, the bubble velocity becomes completely controlled by the bed diameter, and the bubble is then termed a 'slug'. Figure 4.5(a) shows schematically the shape of an ideal slug. The rise velocity of such a slug is given (see Clift, Grace, and Weber, 1978) by:

$$u_{s1} = 0.35 \; \surd(gD) \qquad (4.15)$$

Slugs sometimes adhere to the wall of the tube, as shown schematically in Fig. 4.5(b); in this case the rise velocity is approximately $\surd2$ times the value from Eq. (4.15).

Figure 4.5(c) shows a rather different type of slug, more accurately termed a 'plug', or solids slug, which completely fills the column so that particles rain through the void rather than moving around it. For groups A and B, this phenomenon normally occurs only in tubes of small diameter. For coarse, very angular, or cohesive particles, 'plug' flow occurs in beds of much larger diameter and is suppressed by roughening the walls (Geldart, Hurt, and Wadia, 1978).

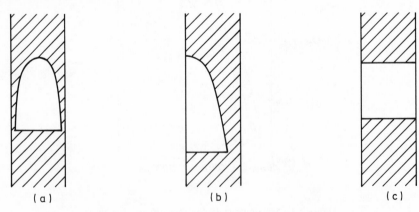

   (a)   (b)   (c)

Figure 4.5 (a) Axisymmetric slugs, (b) asymmetric slugs, (c) solids, square-nosed or 'plug' slugs.

Clearly slugs can only exist if the diameter of the largest bubble is greater than $0.6D$. The question of whether a continuously fluidized bed contains bubbles or slugs is considered in Section 4.6.

Example 4.1: Calculation of the properties of single bubbles

In the following example, bubble properties are estimated for a fluidized bed roaster for zinc sulphide ore. Typical conditions for such an operation are given by Avedesian (1974):

Bed conditions : 1,260 K, 1 bar
Off-gas composition (% molar)

  O_2 : 4.0
  N_2 : 77.3
  H_2O : 7.8
  SO_2 : 10.9

The bed is operated at velocities many times minimum fluidization, so that it will be assumed to be fully mixed with the off-gas composition also representative of the gases within the bed. The bed particles, primarily ZnO, have density 5,300 kg/m^3. The calculations are carried through for 30 μm particles (group A) and 100 μm particles (group B). The latter is typical for the surface-volume mean diameter of bed particles in such an operation. Two possible beds are considered (Fig. 4.6):

(a) A production plant of large diameter.
(b) A pilot plant 0.2 m diameter.

Solution

(a) *Maximum stable bubble size.* The properties of the above gas are taken as:

Density: 0.32 kg/m^3
Viscosity: 5×10^{-5} N s/m^2

Particle diameter: d_p	30 μm	100 μm
Particle type:	Group A	Group B
d_p' (from Eq. 4.13)	81 μm	270 μm
$C_D Re_t^2$	4.72	174.6
Re_t	0.2	5
v_t (m/s)	0.42	3.18
(d_{eq}) from Eq. (4.12)	3.7 cm	2.06 m
Frontal diameter $(d_b)_{max}$	4.3 cm	2.46 m

Notes

Equation (4.12) predicts reasonable values for $(d_{eq})_{max}$ for both group A and group B particles. In practice it is unlikely that the bubbles would reach 2.06 m in diameter in the group B particles because the normal operating bed depth would be insufficient. Frontal diameter in each case is calculated assuming that the wake fraction is 0.4; this value is reasonable for the group A particles, and is probably reasonable also for group B given that the particles are not smooth and spherical (see Section 4.2.1).

b)   *Isolated bubble rise velocity.*   Bubble rise velocity is calculated from Eq. (4.10). Assuming that the wake fraction, V_w/V_{sph}, is constant at 0.4, Eq. (4.10) gives:

$$u_{b\infty} = 0.51 \ \sqrt{(gd_{eq})}$$
$$= 1.61 \ \sqrt{d_{eq}} \ \text{m/s with } d_{eq} \text{ in m}$$

Resulting values, appropriate to a bed of large diameter, are shown as curve (a) in Fig. 4.6.

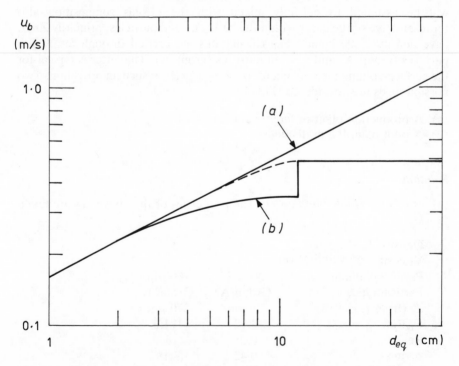

Figure 4.6 Bubble rise velocities calculated in worked Example 4.1: (a) in large production plant, (b) in 0.2 m diameter pilot unit.

For a bed of diameter 0.2m, wall effects become significant for $d_{eq} >$ 2.5 cm (see Section 4.2.4); they are therefore slightly significant for bubbles in the 30 μm particles. In a bed of 100 μm particles, wall effects must be considered. Curve (b) in Fig. 4.6 shows values calculated from Eq. (4.14) for 2.5 cm $< d_{eq} <$ 12 cm and from Eq. (4.15) for $d_{eq} >$ 12 cm. Clearly the two equations do not match at $d_{eq}/D = 0.6$. The discontinuity occurs because Eq. (4.14) was developed for bubbles with high Re_b, for which $u_{b\infty}$ is given by Eq. (4.9). Bubbles in fluidized beds are more 'rounded' (that is θ_w is larger), so that for given d_{eq} they rise more slowly, have smaller d_b, and are less sensitive to wall effects. Equation (4.15) is well tested for fluidized beds, and may therefore be applied with more confidence. Best estimates are therefore obtained by replacing Eq. (4.14) by an interpolation of the form shown by the broken curve in Fig. 4.6.

4.3   GAS DISTRIBUTION

4.3.1   Types of Gas Distributor

For satisfactory operation of a fluidized bed, the gas must normally be distributed uniformly across the bed area. In addition, the distributor must prevent the bed solids from falling through into the 'windbox' and must be able to support forces due to the pressure drop associated with gas flow during operation and the weight of the bed solids during shutdown.

For laboratory studies, it is frequently convenient to use a distributor of some porous material (see Chapter 13), but this is not practical for industrial fluid beds. One of the simplest types of distributor, also much used in laboratory studies, is a single perforated plate; i.e. a metal sheet drilled or punched with a regular array of orifices or slots. Provided that the diameter or width of the perforations is less than about ten particle diameters, little 'weeping' of solids occurs. Devices to reduce weeping with this type of distributor include the use of two or more staggered perforated plates and a 'sandwich' of coarse particles contained between two perforated plates or meshes (see Kunii and Levenspiel, 1969). If the fluidizing gas contains dust or liquid droplets, these particles may deposit in the perforations to block the distributor. Such deposition is reduced or eliminated by using tapered orifices or holes as shown in Fig. 4.7(a) (Doganoglu et al., 1978).

Figures 4.7(b), (c), and (d) show types of gas distributors which are particularly suitable for high temperature operation. In each case, the gas passes up an array of vertical pipes or 'risers', through the metal load-bearing plate of the distributor. The plate is covered by a layer of refractory, normally laid or cast between the risers. The insulation provided by the refractory layer ensures that the plate is at a temperature close to that of the incoming gases, although the bed itself may be at a much higher temperature. In the type of distributor usually known as a 'tuyere', shown in Fig. 4.7(b), the riser has a single hole whose axis is vertical. Alternatively, the riser may be capped and the cap drilled with radial holes, usually four, six or eight in number, as shown in Fig. 4.7(c). With this type of distributor, sometimes termed 'multi-hole tuyere', the gas enters the bed horizontally. The operating advantage is that weeping and dumping of solids through the distributor are much reduced if the radial passages are sufficiently long. A similar advantage is achieved by the type of distributor shown in Fig. 4.7(d), known as a 'quonset' ('Quonset' being the American term for what is known in the British Commonwealth as a 'Nissen hut'). Here the riser is capped by a horizontal length of pipe cut on a plane through its axis to form a half-cylinder whose open side is level with the top of the refractory. The risers may also be capped by bubble caps similar to the devices used in distillation (see Kunii and Levenspiel, 1969).

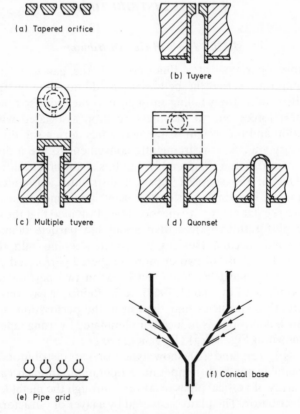

(a) Tapered orifice

(b) Tuyere

(c) Multiple tuyere

(d) Quonset

(e) Pipe grid

(f) Conical base

Figure 4.7   Types of gas distributor.

If it is desirable to separate the load-bearing and gas distribution requirements, the gas can be introduced into the bed by a pipe grid, shown schematically in Fig. 4.7(e). This arrangement is also suitable if more than one gas stream is to be introduced to the bed without premixing. The pipe grids are typically horizontal pipes, each with an array of branching smaller pipes. These small pipes may point upwards, as in Fig. 4.7(e), or be inclined downwards to reduce passage of bed solids into the pipe grid.

If it is necessary to remove solids continuously from the bed, e.g. ash in coal processing, it may be convenient to provide a conical base to the bed as shown in Fig. 4.7(f). Ash is removed from the apex of the cone and fluidizing gas introduced through pipes in the tapered section. If these pipes are inclined downwards, as shown in Fig. 4.7(f), solids weeping is reduced.

4.3.2   Design Criteria

From the point of view of gas distribution, the essential requirement is to design the distributor so that the gases passing through it experience a sufficiently high pressure drop, Δp_D. The critical requirement is generally held to be the ratio of the distributor pressure drop to that across the fluidized bed, $\Delta p_D/\Delta p_B$. Agarwal, Davis, and King, (1962) suggest that Δp_D should be 0.1 to 0.3 times Δp_B or 3.4×10^3 N/m^2 (0.5 lb/in^2), whichever is the greater, and this is broadly consistent with the recommendations of Hanway (1970), Whitehead (1971) and Qureshi and Creasy (1979) point out that some fluidized beds operate successfully at much lower values of $\Delta p_D/\Delta p_B$. Apart from the 'ease of fluidization' of the bed solids, a property which is related to cohesiveness or stickiness but is impossible to define or quantify, the value of $\Delta p_D/\Delta p_B$ required depends on the ratio of bed depth H to diameter D. Qureshi and Creasy (1979) concluded, from a review of published information, that the distributor pressure drop required for satisfactory operation was given by:

$$\frac{\Delta p_D}{\Delta p_B} = 0.01 + 0.2 \left\{ 1 - \exp \left(\frac{-D}{2H} \right) \right\} \qquad (4.16)$$

so that a pressure drop ratio of 0.21 is required for large diameter beds. If the bed solids contain tars or other 'sticky' components, it is advisable to increase this value to at least 0.3.

Since the value of Δp_B is fixed by the weight of solids per unit bed area (see Chapter 2), selection of an appropriate value for $\Delta p_D/\Delta p_B$ fixes the required Δp_D. The number of orifices, nozzles, etc., to achieve this pressure drop must now be selected. We consider first the simplest case of a perforated plate distributor. If the superficial gas velocity in the windbox is U_w and the fractional open area of the distributor (i.e. the fraction of the total area open to gas flow) is f_{or}, then the average gas velocity through the orifices is:

$$U_{or} = \frac{U_w}{f_{or}} \qquad (4.17)$$

It should be noted that U_w differs from the superficial gas velocity in the bed if the entering gas is at a different temperature from the bed, or if the processes occurring in the bed cause a change in molar flowrate, or if the distributor pressure drop is a significant fraction of the absolute operating pressure.

The pressure drop through the distributor then follows from Bernoulli's equation:

$$\Delta p_D = \frac{\rho_{gw}}{2} \left\{ \left(\frac{U_{or}}{C_d} \right)^2 - U_w^2 \right\} \qquad (4.18)$$

where ρ_{gw} is the density of gas in the windbox and C_d is the orifice discharge coefficient, which accounts for the fact that the gas does not pass uniformly through the whole orifice area so that the true average velocity is greater than U_{or}. Normally $f_{or} \ll 1$ so that $U_{or} \gg U_w$ and Eqs (4.17) and (4.18) can be simplified to give:

$$f_{or} = \frac{U_w}{C_d} \sqrt{\left(\frac{\rho_{gw}}{2\Delta p_D} \right)} \qquad (4.19)$$

The discharge coefficient depends on the shape of the orifice and is also a weak function of f_{or} (Perry and Chilton, 1973). There is also a possibility that the gas passing through an orifice into a fluidized bed experiences less pressure drop than in the absence of particles (see Qureshi and Creasy, 1979). For square-edged circular orifices with diameter d_{or} much larger than the plate thickness t, C_d can be taken as 0.6. For $t/d_{or} > 0.09$, C_d can be estimated by a correlation given by Qureshi and Creasy:

$$C_d = 0.82 \left(\frac{t}{d_{or}} \right)^{0.13} \qquad (4.20)$$

Fixing f_{or} by Eq. (4.19) fixes the ratio of centre-to-centre hole spacing s to d_{or}. For orifices on equilateral triangle pitch:

$$\frac{s}{d_{or}} = \left(\frac{\pi}{2f_{or}\sqrt{3}} \right)^{1/2} = \frac{0.9523}{\sqrt{f_{or}}} \qquad (4.21)$$

while for square pitch:

$$\frac{s}{d_{or}} = \left(\frac{\pi}{4f_{or}} \right)^{1/2} = \frac{0.8862}{\sqrt{f_{or}}} \qquad (4.22)$$

Ideally, d_{or} should be as small as is feasible: reducing d_{or} decreases the size of bubbles formed at the distributor, and therefore improves gas/solid contacting in the bed (see Chapter 11). It is also advisable to check that all orifices will be equally 'active' in passing gas into the bed, by a criterion developed by Fakhimi and Harrison (1970) which can be written:

$$\left(\frac{U_w}{U_{mfw}} \right)^2 > 1 + \left(1 - \frac{2}{\pi} \right) \frac{2\rho_p \, g(1 - \epsilon_{mf}) \, L_j}{\rho_{gw}} \left(\frac{C_d \, f_{or}}{U_{mfw}} \right)^2 \tag{4.23}$$

where ρ_p is the particle density, ϵ_{mf} is the void fraction in the bed at minimum fluidization, L_j is the length of the jet above the orifice (see Section 4.4.1), and U_{mfw} is the minimum fluidization velocity corresponding to gas at windbox conditions. If Eq. (4.23) is not satisfied, the orifice diameter should be reduced to reduce L_j.

With distributors of the tuyere type, the form and dimensions of each tuyere are normally fixed in advance by other considerations. The number of holes or orifices is then selected to give the required Δp_D. For an initial estimate, Eq. (4.18) can be used, rearranged as:

$$U_{or,1} = C_d \left(\frac{2\Delta p_D}{\rho_{gw}} + U_w^2 \right)^{1/2} \tag{4.24}$$

$$\approx C_d \sqrt{\left(\frac{2\Delta p_D}{\rho_{gw}} \right)} \tag{4.25}$$

where $U_{or,1}$ is the first estimate for the mean gas velocity issuing from each hole into the bed and C_d is estimated for the exit holes; i.e. the upper orifice for a tuyere (Fig. 4.7b) and the radial holes for a multi-hole tuyere (Fig. 4.7c). If the exit area per hole is a_{or}, the number of holes required per unit bed area is:

$$N_1 = \frac{U_w}{a_{or} \, U_{or,1}} \tag{4.26}$$

and the volumetric gas flowrate per nozzle is:

$$Q_{or,1} = a_{or} \, U_{or,1} \tag{4.27}$$

The initial estimate N_1 is now improved by estimating the pressure drop $\Delta p_{D,1}$ experienced by the gas in passing through each hole at rate $Q_{or,1}$ using standard results for flow through pipe fittings (Perry and Chilton, 1973). Thus entry into the riser is treated as a sudden contraction, associated with pressure drop:

$$\frac{\rho_{gw}}{2} (U_R^2 - U_w^2) + \rho_{gw} \frac{U_R^2}{4} = \rho_{gw} \left(\frac{3U_R^2}{4} - \frac{U_w^2}{2} \right) \tag{4.28}$$

where U_R is the mean gas velocity in the riser. The first term on the left of Eq. (4.28) results from application of the Bernoulli equation, while the second represents the mechanical energy loss resulting from formation of a *vena contracta* at the sharp contraction.

Pressure drop due to wall friction in the riser is usually negligible. The upper orifice of a tuyere is treated as a contraction, sharp or rounded depending on the profile. Entry to the outlet holes of a multi-hole tuyere is treated as a 'tee' followed by a contraction, while a 'quonset' is equivalent to a 'tee' with flow entering the branch. The pressure drop $\Delta p_{D,1}$ is evaluated as the sum of the contributions from each change in gas velocity or direction. The final estimate N for the number of holes per unit area is obtained using the fact that the pressure drop is proportional to U_{or}^2:

$$N = N_1 \sqrt{\left(\frac{\Delta p_{D,1}}{\Delta p_D}\right)} \qquad (4.29)$$

The average gas velocity entering the bed is then

$$U_{or} = \frac{U_w}{a_{or} N} \qquad (4.30)$$

For complicated tuyere designs it is advisable to determine the pressure drop in a pilot plant or 'cold model', so that U_{or} can be selected to give the required pressure drop under operating conditions.

Normally design calculations of the type summarized here are based on flowsheet gas rates. It should be noted that Δp_D is proportional to U_w^2; therefore operation below design rates may give insufficient Δp_D and hence poor fluidization. If operation with a range of gas rates is anticipated, the distributor should be designed to give sufficient pressure drop at the lowest gas rate foreseen. Designing for 'turndown' in this way will inevitably lead to higher Δp_D and hence higher operating costs at the higher gas velocities. As a further check, it is also advisable to ensure that Δp_D is much larger than the dynamic pressure of gases entering the windbox, i.e.:

$$\Delta p_D \gg \rho_{gw} U_{in}^2 \qquad (4.31)$$

where U_{in} is the mean gas velocity in the duct entering the windbox. Normally this requirement imposes no further restriction on Δp_D.

Example 4.2: Calculation of tuyere and multi-hole tuyere distributors

This example considers possible distributor designs for the zinc sulphide roaster introduced in Example 4.1. Since the solids may have a tendency to 'stickiness', the distributor will be designed following the criterion of Agarwal, Davis, and King (1962) (see Section 4.3.2) to give Δp_D as the greater of $0.3\Delta p_B$ and 3.4×10^3 N/m^2. Avedesian (1974) gives the total pressure drop across the distributor and bed as 1.54×10^4 N/m^2 and estimates the distributor pressure drop as approximately 0.53×10^4 N/m^2. Hence:

Bed pressure drop $= 1.54 \times 10^4 - 0.53 \times 10^4$
$$= 1.01 \times 10^4 \quad \text{N/m}^2$$

Distributor pressure drop is the greater of

$$3.4 \times 10^3 \text{ N/m}^2 \text{ and } 0.3 \times 1.01 \times 10^4$$
$$= 3.0 \times 10^3 \text{ N/m}^2$$
$$\text{i.e. } \Delta p_D = 3.4 \times 10^3 \text{ N/m}^2$$

(which is only 64 per cent of that estimated by Avedesian for a specific operating unit).
Calculations are based on gas at the windbox conditions reported by Avedesian:

Air at 298 K and 1.154 bar
Density $\rho_{gw} = 1.35$ kg/m^3
Corresponding superficial velocity $U_w = 0.159$ m/s

Solution

(a) *Single-hole tuyeres*   Figure 4.8(a) shows the cast steel tuyeres used in the installation described by Avedesian (1974). The pressure drop across each tuyere is made up of three components:
 (i) Contraction loss of entering the 'riser' tube of 22 mm i.d. If the gas velocity in the 'riser' is U_R metres per second then the contraction loss is estimated from Eq. (4.28) as:

$$\rho_{gw} \left(\frac{3U_R^2}{4} - \frac{U_w^2}{2} \right) = 1.0125 \, U_R^2 - 0.01706 \text{ N/m}^2$$

 (ii) Frictional loss in riser. It may readily be demonstrated that this component is negligible.
(iii) Loss on passing through 6 mm hole, Since the length of the 6 mm opening is several times its diameter, this hole is treated as a sharp contraction rather than an orifice.

$$\text{Mean velocity in nozzle } = U_R \times \left(\frac{22}{6} \right)^2$$

$$= 13.44 \, U_R$$

$$\text{Area ratio of contraction } = \left(\frac{6}{22} \right)^2 = 0.0744$$

which is sufficiently low that the appropriate form of Eq. (4.28) can be applied (Perry and Chilton, 1973). The contribution to the pressure drop is then estimated as:

$$\rho_{gw} \left[\frac{3}{4} (13.44U_R)^2 - \frac{U_R^2}{2} \right] = 182.2 \ U_R^2 \ N/m^2$$

which is clearly the dominant contribution.

Hence:
Total pressure drop

$$= 183.2 \ U_R^2 - 0.01706 \ N/m^2$$

Required pressure drop $= 3.4 \times 10^3 \ N/m^2$

$$\therefore U_R^2 = 18.66 \ (m/s)^2$$

$$U_R = 4.32 \ m/s$$

Hence, if N holes per square metre are required:

$$N \times \frac{\pi}{4} \times (0.022)^2 \ U_R = \text{superficial velocity} = 0.159 \ m/s$$

so that $N = 96.8$ holes/m^2. Since there is only one hole per tuyere 96·8 tuyeres/m^2 are needed.

Area associated with one tuyere $= 1.03 \times 10^{-2} \ m^2$

i.e. if the tuyeres are arranged on square pitch, the centre to centre spacing is $\sqrt{(1.03 \times 10^{-2})} = 0.102$ m. This is precisely the value reported by Avedesian (1974).

(b) Multi-hole tuyeres Figure 4.8(b) shows a possible alternative design utilizing a similar 'riser', but with four radial holes each of 3 mm internal diameter. The total hole area is the same as for the single-hole tuyere in case (a). The pressure drop components considered in case (a) are still present, with an additional term arising from the deflection of the gas entering the holes. Treating the configuration as a 'tee', the extra pressure drop is estimated (Perry and Chilton, 1973) as $\rho_{gw} \ U_R^2/2 = 7.95 \times 10^{-2} \ U_R^2 \ N/m^2$. Hence the total pressure drop is $183.3U_R^2 - 0.01706 \ N/m^2$. This is essentially indistinguishable from the value for case (a), because the pressure drop is dominated by the velocity in the exit holes which is the same in the two cases. Thus the spacing for these multi-hole tuyeres should be the same as for the single-hole tuyeres.

In each case, the gas enters the bed at pressure $1.154 - 0.034 = 1.12$ bar at density $1.35 \times 1.21/1.154 = 1.31 \ kg/m^3$. Correcting to these conditions, the gas velocity entering the bed, U_{or}, is $13.44 \ U_R \times 1.154/1.12 = 59.8 \ m/s$.

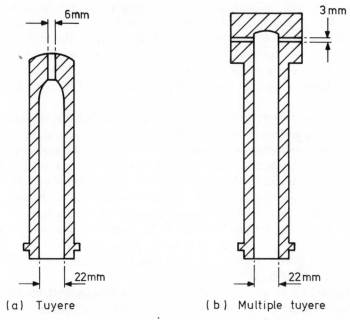

Figure 4.8   Alternative tuyere designs for example 4.2.

4.4 BUBBLE FORMATION AND JETTING

4.4.1 Gas Jets at Distributors

Considerable academic argument has been engendered over whether gas entering a fluidized bed from an orifice or hole forms a series of bubbles or a permanent gas jet with low solids concentration. Rowe, MacGillivray, and Cheeseman (1979) showed that, when gas is introduced through a vertical hole into a bed held close to minimum fluidization by a background gas flow, a stream of bubbles forms immediately at the hole. However, when there is no background gas flow — the case of common practical interest — a gas jet forms. As in all multi-phase systems, the jet is unsteady, and when the gas flow through the hole is more than about three times the flow needed to fluidize the bed, it may break down into a stream of fast-moving bubbles close to the hole. Whether or not there is a permanent gas void in the region of the hole, there is ample evidence that the average gas and particle velocities are very much higher and the average solids concentration is very much lower in this region than in the rest of the bed. There seems little harm in calling these regions simply 'jets'. Massimilla and his colleagues have modelled these regions successfully by a modification of conventional turbulent jet theory (see, for example, Donsi, Massimilla, and Colantuoni, 1980), so that the analogy between fluidized beds and conventional liquids holds qualitatively. In group D particles, rather different phenomena occur (Cranfield and Geldart, 1974). Voids break away periodically from the distributor, spread horizontally to form lenticular cavities, and then break up to form a number of bubbles on a common level. The process is roughly analogous to formation and break-up of toroidal gas bubbles in liquids (see Clift, Grace and Weber, 1978).

The practical interest in gas jets arises from the fact that they are regions of very high gas and solids velocity. Therefore solid surfaces in the jet region are subject to rapid erosion. Distributors such as multi-hole tuyeres which introduce gas horizontally into the bed must be arranged so that the jets do not impinge on neighbouring tuyeres or on the wall of the bed. When the gas enters the bed vertically, the jet penetration determines the lowest level at which surfaces such as heat exchange tubes should be located, and imposes a lower limit on operating bed depth. In addition, mass and heat transfer between gas and particles is rapid in the jet region (Donsi, Massimilla, and Colantuoni, 1980). The implications for the performance of fluid bed reactors are discussed in Chapter 11.

Yang and Keairns (1979) reviewed the correlations available for predicting the length of the 'jet' established when gas enters a fluidized bed vertically, concluded that none was sufficiently general, and proposed their own:

$$\frac{L_j}{d_{or}} = 15 \left\{ \left(\frac{\rho_g}{\rho_p - \rho_g} \right) \frac{U_{or}^2}{g d_{or}} \right\}^{0.187} \tag{4.32}$$

where L_j is the jet length and ρ_g and ρ_p are the densities of the gas and particles. While Eq. (4.32) appears to be the most reliable correlation for temperatures close to ambient, the scant available evidence (Ghadiri and Clift, 1980) suggests that only the correlation due to Merry (1975) gives reasonable predictions for elevated temperatures:

$$\frac{L_j}{d_{or}} = 5.2 \left(\frac{\rho_g d_{or}}{\rho_p d_p} \right)^{0.3} \left\{ 1.3 \left(\frac{U_{or}^2}{g d_{or}} \right)^{0.2} - 1 \right\} \tag{4.33}$$

where d_p is the mean bed particle diameter. Equations (4.32) and (4.33) were developed for particles in groups A and B, but there is insufficient evidence to assess their validity for group D. For horizontal jets, another correlation due to Merry (1971) may be used:

$$\frac{L_j}{d_{or}} = 5.25 \left[\frac{\rho_g d_p}{\rho_p d_{or}} \left\{ \frac{\rho_{gw} U_{0r}^2}{\rho_p d_p g (1 - \epsilon_{mf})} \right\}^2 \right]^{0.2} - 4.5 \tag{4.34}$$

where ρ_g and ρ_{gw} are the gas densities in the bed and windbox. The appropriate density for use in Eqs (4.32) and (4.33) is not clear; it is probably more realistic to use ρ_{gw} where this differs from ρ_g.

4.4.2   Bubble Formation

Davidson and Schüler (see Davidson and Harrison, 1963) have given an analysis of bubble formation at a single upward-facing orifice in a fluidized bed, in which the particulate phase is treated as an incompressible fluid of zero viscosity. The initial bubble volume V_o is predicted to be a function of the volumetric gas flow rate through the orifice Q_{or}, but not of the orifice diameter:

$$V_o = 1.38 \, Q_{or}^{1.2} g^{-0.6} \qquad (4.35)$$

The coefficient in Eq. (4.35) is appropriate for an orifice in a flat plate; for a protruding tuyere it should be replaced by 1.14. The frequency of bubble formation follows as:

$$f_{b,o} = \frac{Q_{or}}{V_o} \qquad (4.36)$$

The treatment of Davidson and Schüler assumes that all gas entering the bed forms bubbles; in fact the bubble flowrate is usually less than Q_{or}, so that V_o is overestimated. The error becomes particularly significant for Q_{or} less than about 20 cm^3/s; in this case, more complicated semiempirical equations given by Clift, Grace, and Weber, (1978) should be applied. Equations (4.35) and (4.36) are consistent with the observation that bubble frequency is only weakly sensitive to Q_{or} but at high Q_{or} the predicted frequency tends to be too low. Bubble frequencies around 20 s^{-1} (Rowe, MacGillivray, and Cheesman, 1979) have been reported for high Q_{or}. Therefore, if Eq. (4.36) predicts $f_{b,o}$ less than about 10 s^{-1}, it is probably more realistic to assume that $f_{b,o}$ is 10 s^{-1} and estimate V_o as $Q_{or}/f_{b,o}$.

Semiempirical correlations to predict the size of bubbles formed at a distributor containing a number of gas injection points are summarized in Table 4.1. In these expressions, $A_o (U - U_{mf})$ is the gas flow in excess of that required for minimum fluidization passing through each orifice (since $A_o = 1/N$), and is thus analogous to Q_{or} in Eq. (4.35) (see Section 4.5.2). The form of the expressions derives directly from Eq. (4.35), which can be written in terms of the volume – equivalent diameter, $d_{eq,o}$, as:

$$d_{eq,o} = 1.38 \, Q_{or}^{0.4} g^{-0.2} \qquad (4.37)$$

and the result given by Mori and Wen (1975) is essentially Eq. (4.37) in c.g.s. units. Geldart (1972) and Darton et al. (1977) give slightly larger values, but the differences are small and there is no clear reason for preferring any one correlation. The results all refer to particles in groups A and B; their accuracy for group D is untested.

Table 4.1   Correlations for bubble size in fludized beds   In the following expressions, z indicates distance above distributor, D the bed diameter, and A_o the distributor area per orifice for multiple orifice or tuyere type distributors. The forms given here for the correlations of Geldart (1972) and Mori and Wen (1975) require dimensions in centimetres and velocities in centimetres per second. Any consistent set of units can be used in the correlations of Darton *et al.* (1977) and Rowe (1976). The parameter $d_{eq,m}$ in the Mori and Wen correlation represents the diameter which would be attained if all the fluidizing gas above that required for minimum fluidization were to form a train of bubbles rising on the bed axis. The parameter z_o in Rowe's correlation is a characteristic of the distributor, effectively zero for a porous plate.

Author	Expression for bubble size	Initial bubble size (at $z = 0$)
Darton *et al.* (1977)	$d_{eq} = 0.54(U-U_{mf})^{0.4}(z + 4\sqrt{A_o})^{0.8}g^{-0.2}$	$d_{eq,o} = 1.63\,\{A_o(U-U_{mf})\}^{0.4}g^{-0.2}$ $[\,= 0.411\,\{A_o(U-U_{mf})\}^{0.4}$ in centimetre units]
Geldart (1972)	$d_b = 1.43\,\{A_o(U-U_{mf})\}^{0.4}g^{-0.2}$ $+ 0.027z(U-U_{mf})^{0.94}$	$d_{b,o} = 1.43\,\{A_o(U-U_{mf})\}^{0.4}g^{-0.2}$ $[\,= 0.361\,\{A_o(U-U_{mf})\}^{0.4}$ in centimetre units]
Mori and Wen (1975)	$d_{eq} = d_{eq.m} - (d_{eq.m} - d_{eq.o})e^{-0.3z/D}$ where $d_{eq.m} = 0.374\,\{\pi D^2(U-U_{mf})\}^{0.4}$	$d_{eq,o} = 0.347\,\{A_o(U-U_{mf})\}^{0.4}$ for perforated plates $= 3.76 \times 10^{-3}(U-U_{mf})^2$ for porous plates
Rowe (1976)	$d_b = (U-U_{mf})^{0.5}(z + z_o)^{0.75}g^{-0.25}$	$d_{b,o} = 0$ for porous plates; no general results given for other types of distributor

Example 4.3: Jet penetration and initial bubble size for distributor of Example 4.2

(a) Jet penetration

For each of the distributors in Example 4.2, the gas enters the bed with velocity $U_{or} = 59.8$ m/s and density $\rho_{gw} = 1.31$ kg/m^3. The bed particles, described in Example 4.1, have density $\rho_p = 5{,}300$ kg/m^3 and diameter $d_p = 10^{-4}$ m.

Single-hole tuyere.   For the first distributor considered in Example 4.2, the gas enters the bed vertically through holes of diameter $d_{or} = 6 \times 10^{-3}$ m. Two estimates for the jet length are then:
  (i) Yang and Keairns (Eq. 4.32):

$$L_j = 15d_{or}\left\{\left(\frac{\rho_{gw}}{\rho_p - \rho_{gw}}\right)\frac{U_{or}^2}{gd_{or}}\right\}^{0.187}$$
$$= 0.15 \text{ m}$$

(ii) Merry (Eq. 4.33):

$$L_j = 5.2d_{or} \left(\frac{\rho_g d_{or}}{\rho_p d_p} \right)^{0.3} \left[1.3 \left(\frac{U_{or}^2}{g d_{or}} \right)^{0.2} - 1 \right]$$

$$= 0.095 \text{ m}$$

It is typical that the estimate from Merry's correlation is smaller than that from Yang and Keairns'. To ensure that excessive erosion of surfaces such as heat exchange tubes is avoided, such surfaces would be located no less that 0.15 m above the ends of the tuyeres.

Multi-hole tuyere. For the second distributor of Example 4.2, the gas enters the bed horizontally through nozzles of diameter $d_{or} = 3 \times 10^{-3}$ m. From Merry's correlation, Eq. (4.34):

$$L_j = 5.25d_{or} \left[\frac{\rho_g d_p}{\rho_p d_{or}} \left\{ \frac{\rho_{gw} U_{or}^2}{\rho_p d_p \, g \, (1 - \epsilon_{mf})} \right\}^2 \right]^{0.2} - 4.5 \, d_{or}$$

$$= 7.8 \times 10^{-3} \text{ m}$$

where ϵ_{mf} has been assumed to be 0.4 and ρ_g is 0.32 kg/m^3 (see Example 4.1). If this estimate, that the horizontal jets persist for less than 1 cm from the holes, is reliable, then no especial precautions need to be taken in the orientation of the holes. If the jet penetration is comparable to the tuyere spacing, then the holes should be directed so that a jet cannot reach neighbouring tuyeres and cause excessive erosion.

(b) *Initial Bubble Size*

Gas entering the bed reaches the bed temperature rapidly. Avedesian (1974) reports that the superficial velocity in the bed, corresponding to conditions for which the distributor has been designed, is $U = 0.78$ m/s, while the minimum fluidizing velocity U_{mf} is approximately 0.005 m/s.

For the single-hole tuyeres, the bed area per hole is $A_o = 1.03 \times 10^{-2}$ m^2 (see Example 4.2). Hence the excess gas flow per hole is:

$$Q_{or} = A_o (U - U_{mf}) = 1.03 \times 10^{-2} (0.78 - 0.005)$$
$$= 8.0 \times 10^{-3} \text{ m}^3/\text{s}$$

The Davidson and Schüler result (Eq. 4.37) then gives the initial bubble size as

$$d_{eq,o} = 1.38 \times (8.0 \times 10^{-3})^{0.4} (9.81)^{-0.2}$$
$$= 0.127 \text{ m}$$

Since the spaces between tuyeres are filled with castable refractory to the level of the nozzles, the form appropriate to orifices in a flat plate has been used.

This result may be compared with the expression in Table 4.1. In c.g.s. units:

$$\left\{ A_o \left(U - U_{mf} \right) \right\}^{0.4} = \left(8.0 \times 10^3 \right)^{0.4} = 36.4$$

Hence we obtain the following expressions:

Darton *et al*: $d_{eq,o} = 15.0$ cm $= 0.15$ m

Geldart: $d_{b,o} = 13.1$ cm $= 0.131$ m

                    $(d_{eq,o} = 0.6^{1/3} \times 0.131 = 0.110$ m$)$

Mori and Wen: $d_{eq,o} = 12.6$ cm $= 0.126$ m

Thus the estimates for $d_{eq,o}$ are comparable and all are larger than the tuyere spacing (0.102 m; see Example 4.2). It thus seems unlikely that the tuyeres act independently. However, in the current state of the art, little further can be said.

For multi-hole tuyeres, the area and gas flow per injection point is one quarter of the value for the single-hole tuyeres. Assuming that the same relationships can be used (questionable for horizontal gas entry), we obtain the following estimates for $d_{eq,o}$:

Davidson and Schuler: $d_{eq,o} = 0.073$ m   (Eq. 4.37)

Darton *et al*: $d_{eq,o} = 0.086$ m

Geldart: $d_{b,o} = 0.075$ m $(d_{eq,o} = 0.063$ m$)$

Mori and Wen: $d_{eq,o} = 0.073$ m

Again, the values for $d_{eq,o}$ are sufficiently large to suggest that the tuyeres are not generating bubbles independently.

4.5 FREELY BUBBLING BEDS

4.5.1 Overall Flow Patterns

When a bed of particles is fluidized at a gas velocity above the minimum bubbling point, bubbles form continuously and rise through the bed, which is then said to be 'freely bubbling'. As they rise, bubbles interact and coalesce, so that the average bubble size increases with distance above the distributor. Growth by coalescence proceeds until the bubbles approach their maximum stable size, when splitting becomes frequent. Thereafter, splitting and recoalescence cause the average bubble size to equilibrate at a value close to the maximum stable size. In group A particles, for which the maximum stable diameter is relatively small and is therefore attained close to the distributor, the average bubble size is therefore typically constant over much of the depth of the bed. For group B particles, the maximum stable diameter is larger and typically only attained in the upper levels of deep beds; generally, the average bubble size then increases steadily with distance from the distributor. Measurements of bubble size reported by Werther (1975) give an excellent illustration of this difference between groups A and B. Group D particles show a behaviour which is again different (Cranfield and Geldart, 1974): bubbles rise as horizontally associated swarms, and grow by absorption of neighbouring bubbles rather than conventional coalescence.

In addition to controlling the vertical profile of bubble size, coalescence also sets up horizontal non-uniformity in the bubble distribution in beds of groups A and B. This effect is shown schematically in Fig. 4.9(a), with typical coalescence processes shown in Fig. 4.9(b). Bubbles generally coalesce by overtaking a bubble in front, shown by (i) in Fig. 4.9(b), and may move sideways into the track of such a bubble before coalescing, shown by (ii) in Fig. 4.9(b). Thus coalescence can cause lateral motion of bubbles. Bubbles formed close to the wall of the bed can only move inwards as a result of this process, since the bubbles surrounding them are only on the side away from the wall, whereas bubbles nearer to the axis have equal probability of migrating in any horizontal direction. As a result of this preferential movement of bubbles away from the wall, an 'active' zone of enhanced bubble flowrate forms a small distance in from the wall. The higher concentration of bubbles leads to more frequent coalescence and hence larger bubbles in this zone. Since the region between the wall and the annular 'active' zone is lean in bubbles, coalescence continues to cause preferential migration towards the bed axis. Thus the 'active' zone intensifies and moves closer to the axis with increasing height above the distributor. At heights of the order of the bed diameter, virtually the whole of the bubble flow occurs close the the axis. This effect was first observed and explained by Grace and Harrison (1968), but has since been confirmed by many independent studies. Large beds essentially divide into cells, to form several bubble tracks rather than a single axial track (Whitehead, 1971). Quantitative prediction of the development of bubble profiles is possible (Farrokhalaee and Clift, 1980), but generally requires extensive computer calculations. Qualitative recognition of this effect helps to interpret the behaviour of freely bubbling beds. An important consequence of the non-uniformity in bubble flow is gross solids circulation (see Chapter 5). Generally, solids move up in regions of high bubble activity and downwards elsewhere; the resulting circulation pattern (observed, for example, by Masson, 1978) is shown schematically in Fig. 4.9(a). In the upper levels of the bed, particle motion is commonly described as 'gulf-streaming', up on the axis (or bubble tracks) and downwards near the walls. At lower levels, the particle motion is down near the axis. Although coalescence sets up this kind of circulation even if gas enters the bed uniformly across the distributor, the effect is amplified by the motion of the dense phase, which reduces bubble formation near the axis and enhances it near the walls (Werther, 1975; Farrokhalaee and Clift, 1980).

Because of the importance of the walls in initiating the flow pattern shown in Fig. 4.9(a), the details of the bubble flow pattern depend on the scale of the equipment as well as on the characteristics of the particles and distributor. The resulting difficulty in predicting the behaviour of a large bed from tests on a smaller pilot unit lies behind the problem of scale-up of fluidized bed reactors discussed in Chapter 11. Moreover, since coalescence effects differ quantitatively in two and three dimensions, one may not simply assume that a bubble size distribution measured in a two dimensional bed is the same as that existing in a three-dimensional bed of the same particles (see Chapter 13).

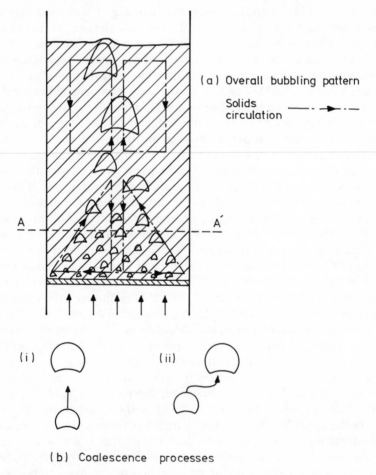

(a) Overall bubbling pattern

Solids circulation — · → · —

(i)   (ii)

(b) Coalescence processes

Figure 4.9 (a) Bubble and solids flow pattern and (b) coalescence modes.

4.5.2   Bubble Flowrate

The 'visible flowrate' in a fluidized bed Q_b is defined as the rate at which bubble volume crosses any level in the bed. For example, the visible bubble flowrate across section AA' in Fig. 4.9(a) could, in principle, be determined by noting the volume of each bubble crossing the section in some period of time and dividing the sum of these volumes by the time of observation. It is important to note that, because of the bulk 'throughflow' of gas into and out of each bubble (see Section 4.1), the flow of gas in the bubble phase is greater than the visible bubble flowrate; the importance of this effect for chemical reactors is discussed in Chapter 11.

A first estimate for Q_b is given by the 'two-phase theory of fluidization', proposed by Toomey and Johnstone (1952) and developed by Davidson and Harrison (1963). As defined by the latter authors, the theory models a fluidized bed as consisting of:

(a) A particulate phase in which the flow rate is equal to the flowrate for incipient fluidization, i.e. the voidage is essentially constant at ϵ_{mf}; and
(b) A bubble phase which carries the additional flow of fluidizing fluid.

This model is then commonly applied by estimating the visible bubble flowrate as the excess gas flow above that required for minimum fluidization, i.e.:

$$\frac{Q_b}{A} = U - U_{mf} \qquad (4.38)$$

where A is the cross-sectional area of the bed.

In practice, Q_b is rather smaller than the value given in Eq. (4.38): reported values for the deficit are typically 10 per cent for group A particles, 20 to 30 per cent for group B, and 40 to 50 per cent for group D. The deficit must result in part from the gas 'throughflow' in bubbles, but this effect alone does not account satisfactorily for the measured values of Q_b (Grace and Clift, 1974). The gas flow through the particulate phase appears to exceed the value implied by assumption (a) of the two-phase theory, but the extra interstitial flow cannot at present be estimated. Simply replacing U_{mf} by the minimum bubbling velocity U_{mb} does not markedly improve agreement with experiment. Yacono, Rowe and Angelino (1979) have proposed an analysis which accounts for both 'throughflow' and increased interstitial flow, but the result is complex and its general applicability remains to be tested. Equation (4.38) thus remains the best initial estimate for bubble flowrate, and has been the basis for much work in fluidization. At present, there is no ready alternative to using Eq. (4.38) and perhaps reducing the value for Q_b obtained by an empirical factor appropriate to the type of particle being fluidized (see Chapter 5, Section 5.4, Eq. 5.2 and Fig. 5.7).

4.5.3   Bubble Size and Velocity

Correlations abound in the literature for the mean bubble size across horizontal sections in freely bubbling beds. Some of the more recent and more successful are summarized in Table 4.1. That due to Geldart (1972) is based primarily on data for relatively small beds, and generally gives values for frontal diameter (d_b) which are too high for large-scale beds; its value lies primarily in estimating the bubble diameter formed at the distributor ($d_{b,o}$). The remaining three correlations generally give comparable estimates, which is perhaps not surprising since they are derived by similar semiempirical arguments and are based on much the same data. The value of $d_{eq,m}$ in the correlation of Mori and Wen (1975) represents the maximum bubble size attainable by coalescence, and is distinct from the maximum stable size $(d_{eq})_{max}$ discussed in Section 4.2.3. The correlation of Rowe (1976) contains a parameter z_o characterizing the distributor: essentially, z_o is the height above a porous distributor at which bubbles would have attained the size formed at the distributor actually used. Rowe gives some typical values for z_0, but no general results from which it can be estimated. It is therefore necessary to estimate $d_{b,o}$ from, for example, the correlation of Geldart (1972) and then estimate z_o as:

$$ z_o = \left\{ \frac{d_{b,o}^4 \, g}{(U - U_{mf})^2} \right\}^{1/3} \tag{4.39} $$

Werther (1975) has given a correlation for beds with porous distributors which could perhaps be adapted to other types of distributors in the same way.

There is no strong reason for preferring any one of the correlations by Darton *et al.*, Mori and Wen, and Rowe; probably the safest approach is to use all three and average their results. The correlations should be used up to the height at which either the bubbles attain their maximum stable diameter, $(d_{eq})_{max}$, or become large enough to be treated as slugs ($d_{eq} = 0.6D$). In the former case, the bubble size should be assumed constant at higher levels; in the latter case, the upper levels of the bed should be treated as slugging. These results were derived for groups A and B. For group D particles, Cranfield and Geldart (1974) give a correlation for estimating the frontal diameters of bubbles:

$$ d_b = 0.0326 \, (U - U_{mb})^{1.11} \, z^{0.81} \tag{4.40} $$

in c.g.s. units, where U_{mb} is the minimum gas velocity at which bubbles form. Since bubbles do not originate at the distributor in group D particles, Eq. (4.40) contains no term describing the distributor and should only be used for $h > 5$ cm.

Davidson and Harrison (1963) suggested that the average velocity of a bubble in a freely bubbling bed is:

$$u_A = (U - U_{mf}) + u_b \qquad (4.41)$$

where u_b is the rise velocity of an isolated bubble of the same size (see Section 4.2). Equation (4.41) is developed by analogy with slug flow, and the applicability to freely bubbling beds is questionable (see Section 4.6 and Turner, 1966). In fact, $u_A > u_b$, primarily as a consequence of the increase in bubble velocity caused by coalescence, and there is some evidence (Farrokhalaee and Clift, 1980) that Eq. (4.41) accounts approximately for the average effect of coalescence in a freely bubbling bed. Thus Eq. (4.41) can be applied to a bed overall, although instantaneous bubble velocities vary widely about this result and the mean velocity for a given bubble diameter varies with position in the bed, being highest in the 'active' zones where coalescence is frequent.

4.5.4 Bed Expansion

Consider a section such as AA′ in Fig. 4.9(a), across which the visible bubble flowrate is Q_b and the average bubble velocity is u_A. The average fraction of the bed area occupied by bubbles is then:

$$\epsilon_b = \frac{Q_b}{Au_A} \qquad (4.42)$$

where A is the cross-sectional area of the bed. The volume of particulate phase in an elementary height dz at this section in a bed of height H is then:

$$AH_D = A \cdot \int_0^H (1 - \epsilon_b)\, dz = A \int_0^H \left(1 - \frac{Q_b}{Au_A}\right) dz \qquad (4.43)$$

from Eq. (4.42). Following assumption (a) of the two-phase theory (see Section 4.5.2), if the particulate phase remains at the void fraction corresponding to minimum fluidization, then AH_D is equal to the total bed volume at minimum fluidization, AH_{mf}. Equation (4.43) then becomes:

$$H_{mf} = \int_0^H \left(1 - \frac{Q_b}{Au_A}\right) dz$$

that is:

$$H - H_{mf} = \int_0^H \frac{Q_b}{Au_A}\, dz \qquad (4.44)$$

which gives the expansion of the bed above its depth at minimum fluidization. This general result can be simplified if Q_b/A is approximated by the two-phase theory (Eq. 4.38) and u_A is estimated from Eq. (4.41):

$$H - H_{mf} = \int_0^H \frac{(U - U_{mf}) \, dz}{(U - U_{mf}) + u_b} \tag{4.45}$$

In general, u_b is a function of z because bubble diameter is a function of z. For detailed calculations of bed expansion, it is therefore advisable to evaluate the integral in Eq. (4.45) analytically or numerically, using values of u_b calculated from the profile of bubble diameters obtained from the correlations in Table 4.1. If u_b varies little or if a single mean bubble diameter and rise velocity are to be used for the purpose of estimate, then Eq. (4.45) simplifies to:

$$\frac{H - H_{mf}}{H} = \frac{U - U_{mf}}{U - U_{mf} + u_b} \tag{4.46}$$

which may be rearranged more conveniently as:

$$\frac{H - H_{mf}}{H_{mf}} = \frac{U - U_{mf}}{u_b} \tag{4.47}$$

Equation (4.47), derived by Davidson and Harrison (1963) for the case where u_b is constant in the bed, is equivalent to a result for slug flow given in Section 4.6. In addition to design calculations of bed expansion, it is also frequently used to infer mean values of u_b and hence d_{eq} or d_b from measurements of bed expansion.

Example 4.4:   Bubble size distribution and bed expansion

In the following example, bubble properties and mean bed density are estimated for the fluidized bed roaster considered in Examples 4.1 to 4.3. The bed in question, as described by Avedesian (1974), has diameter $D = 6.38$ m and expanded depth 1 metre. The distributor is of the single-hole tuyere type, designed in Example 4.2 (a), with one such tuyere per 1.03×10^{-3} m^2 of bed area. The superficial velocity at bed conditions is taken as 0.78 m/s (see Example 4.3b), so that

$$U - U_{mf} = 0.775 \text{ m/s}$$

Solution

(a) *Estimation of bubble size*

Initial bubble sizes are calculated in Example 4.3(b):

Darton *et al*: $d_{eq,0} = 0.150$m

Geldart: $d_{b,o} = 0.131$ m $(d_{eq,0} = 0.110$ m)

Mori and Wen: $d_{eq,0} = 0.126$ m

For the correlation of Rowe, we characterize the distributor using the Geldart value for $d_{b,0}$, using Eq. (4.39):

$$z_o = \left(\frac{0.131^4 \times 9.81}{0.78^2} \right)^{1/3} = 0.168\text{m}$$

suggesting that the initial bubble size is equal to that present 0.168 m above a porous distributor. The correlation of Mori and Wen also requires:

$$d_{eq,m} = 0.374 \ (\pi \times 638^2 \times 77.5)^{0.4} \text{ cm}$$
$$= 590 \text{ cm} = 5.9 \text{ m}$$

This is much larger than the maximum stable bubble size, estimated as:

$$(d_{eq})_{max} = 2.06 \text{ m}$$

in a bed of 100 μm particles (see Example 4.1).

Table 4.2 summarizes resulting values for mean bubble size, which are also shown in Fig. 4.10. Clearly Geldart's correlation predicts unrealistically rapid bubble growth, but the other three correlations are in broad agreement. Reasonable estimates may therefore be obtained by averaging the three correlations; resulting values are also shown in Table 4.2 and Fig. 4.10. It is predicted that the maximum stable bubble size is not reached.

(b) *Bubble velocity and bed expansion*

Using the values of d_{eq} given in column 5 in Table 4.2, the isolated rise velocity is estimated as in Example 4.1. The average rise velocity at any level is then calculated from Eq. (4.41), and the average local bubble fraction is then estimated from Eq. (4.42), in the simplified form derived from the two-phase theory:

$$\epsilon_b = \frac{U - U_{mf}}{u_A}$$

Table 4.2 Estimates for bubble properties in fluidized roaster (values in SI units)

	1. Darton *et al.*	2. Geldart		3. Mori and Wen	4. Rowe		5. Average			
z	d_{eq}	d_b	d_{eq}	d_{eq}	d_b	d_{eq}	d_{eq}	u_b	u_A	ϵ_b
0^a	0.150	0.131	0.110	0.126	$(0.13)^b$	(0.110)	0.129	0.578	1.35	0.573
0.25	0.220	0.534	0.450	0.193	0.256	0.218	0.210	0.738	1.51	0.512
0.5	0.285	0.937	0.790	0.260	0.367	0.310	0.285	0.86	1.635	0.474
0.75	0.347	1.340	1.130	0.326	0.467	0.393	0.355	0.96	1.734	0.447
1.0	0.406	1.740	1.470	0.391	0.559	0.471	0.422	1.046	1.82	0.425

a The initial bubble size is taken to apply at the level of the distributor, although it could be argued that these values apply at the ends of the distributor jets (see Example 4.3a).
b Values for use in the Rowe correlation deduced from Geldart's correlation.

Resulting values are given in Table 4.2 and Fig. 4.10. The local average bubble velocity increases, and the average local bubble fraction decreases with height above the distributor.

Bed expansion is estimated by applying Eq. (4.45) to estimate H_{mf}. In terms of ϵ_b, Eq. (4.45) is written:

$$H - H_{mf} = \int_0^H \epsilon_b \, dz$$

From the values in Table 4.2 and Fig. 4.10, ϵ_b can conveniently be approximated by a piecewise linear form shown by the broken lines in Fig. 4.10:

$$\epsilon_b = 0.573 - 0.212z \quad (z \leq 0.4)$$
$$= 0.53 - 0.105z \quad (z \geq 0.4)$$

Hence:

$$H - H_{mf} = \int_0^{0.4} (0.573 - 0.212z) \, dz + \int_{0.4}^H (0.53 - 0.105z) dz$$

$$= 0.53H - 0.052H^2 + 0.008$$

Using Avedesian's value for the expanded hed height, $H = 1.0$ m:

$$H - H_{mf} = 0.486 \text{ m}$$

that is:

$$H_{mf} = 1.0 - 0.486 = 0.514$$

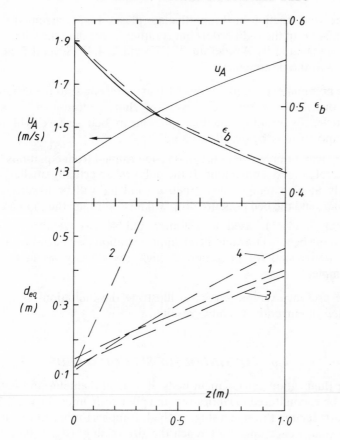

Figure 4.10 Bubble sizes, rise velocities, and volume fractions as a function of level in bed.

The corresponding average bubble fraction in the bed is:

$$\epsilon_b = \frac{H - H_{mf}}{H} = 0.486$$

The value of H_{mf} corresponds to the equivalent depth of particulate phase at minimum fluidization conditions. It therefore indicates the weight of particles per unit cross-sectional area supported by the fluidizing gas, i.e. the bed pressure drop. Using the above value, the bed pressure drop is estimated as:

$$\rho_p (1 - \epsilon_{mf}) \, g \, H_{mf} = 5{,}300 \times 0.5 \times 9.81 \times 0.514$$
$$= 1.34 \times 10^4 \text{ N/m}^2$$

where the void fraction in the particulate phase, ϵ_{mf}, is assumed to be 0.5. This estimate is of the right order, but is rather larger than the value of 1.01×10^4 N/m^2 estimated by Avedesian (see Example 4.2). Several factors could contribute to the discrepancy:

(i) The nominal expanded depth of 1.0 m corresponds to the distance from the distributor to the solids overflow chute. Because of 'splashing' of particles by erupting bubbles, the mean bed surface will be some distance below the overflow chute.

(ii) The mean bubble size in the bed is determined from equations based on data relating to ambient air. If the bubbles are actually smaller, as seems likely at high temperature, then u_b and u_A will be lower, ϵ_b will be higher, and the bed pressure drop will be lower than the estimated value.

(iii) Equation (4.41), used to estimate bubble rise velocity in a freely bubbling bed, is known to be an approximation (see Section 4.5.3) and is not well tested for operation at high gas velocity, as in the present example.

Thus the present example serves to illustrate the limitations as well as the application of currently available results.

4.6   CONTINUOUSLY SLUGGING BEDS

Slugging fluidization can occur in beds of a small diameter and is therefore likely to be encountered in laboratory and pilot-scale units. Figure 4.11 shows a schematic section through a slugging bed. Bubbles formed at the distributor grow by coalescence until they reach the size of slugs ($d_{eq} > 0.6D$). Above that height, the bed is filled by a succession of slugs, and the bed surface oscillates widely. Each time a slug breaks through the surface, the level drops sharply, then rises steadily again until the next slug breaks through.

A fluidized bed will show slug flow if three conditions are met:

(a) The maximum stable bubble size, $(d_{eq})_{max}$, is greater than $0.6D$, where D is the tube diameter.
(b) The gas velocity is sufficiently high.
(c) The bed is sufficiently deep.

Condition (a) depends on the properties of gas and particles (see Section 4.2.3). Conditions (b) and (c) are combined in a criterion for slug flow developed by Baeyens and Geldart (1974). The minimum gas superficial velocity for slugging, U_{ms}, is related to the velocity U_{mf} and bed depth H_{mf} at minimum fluidization by:

$$U_{ms} = U_{mf} + 1.6 \times 10^{-3} (60D^{0.175} - H_{mf})^2 + 0.07 \sqrt{(gD)} \quad (4.48)$$

with c.g.s. units used. The second term on the right is omitted if $H_{mf} > 60D^{0.175}$, and Eq. (4.48) then becomes identical with a result derived for deep beds by Stewart (see Hovmand and Davidson, 1971).

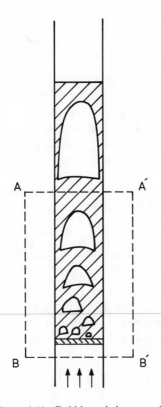

Figure 4.11 Bubble and slug growth.

The simple two-phase theory (see Section 4.5.2) generally appears to be a better approximation for slugging beds than for freely bubbling beds. Therefore the average 'visible flowrate' of slugs can be estimated as $A(U - U_{mf})$, and the void fraction in the dense phase can be assumed to be constant at ϵ_{mf}. Consider now the control volume bounded by AA'B'B in Fig. 4.11, where the section AA' is deliberately chosen to pass through the dense, or particulate, phase between slugs. By continuity, the volume flowrate of the dense phase across AA', Q_d, must be equal to the rate at which the fluidizing gas forms slugs within AA'B'B:

$$Q_d = A \ (U - U_{mf}) \tag{4.49}$$

Since the particulate phase moves essentially as a plug (another respect in which fluidized particles differ from Newtonian liquids), the particulate phase velocity across AA' is Q_d/A, that is $(U - U_{mf})$. The slug approaching AA' moves with its isolated velocity, u_{sl}, relative to these particles so that its absolute rise velocity is

$$u_{sA} = (U - U_{mf}) + u_{sl} \qquad (4.50)$$

$$= (U - U_{mf}) + 0.35 \sqrt{(gD)} \qquad (4.51)$$

using the expression for u_{sl} for an ideal slug from Eq. (4.15). Equation (4.50) has been carried over by analogy to freely bubbling beds (see Davidson and Harrison, 1963) and is the origin of Eq. (4.41). The derivation of Eq. (4.50) shows one of the reasons why the theoretical basis of Eq. (4.41) is uncertain: in a freely bubbling bed, it is not normally possible to define a section like AA' which passes through no bubbles and across which the particulate phase velocity is constant.

Matsen, Hovmand, and Davidson (1969) showed that the maximum height of a slugging bed, H_{max}, is related to its depth at minimum fluidization, H_{mf}, by:

$$\frac{H_{max} - H_{mf}}{H_{mf}} = \frac{U - U_{mf}}{u_{sl}} \qquad (4.52)$$

which is analogous to Eq. (4.47) for freely bubbling beds. In order to design the depth of vessel needed for a bed in which slug flow is anticipated, the overall maximum bed depth is required. Equation (4.52) should therefore be used with the minimum value for u_{sl}, i.e. the value for an ideal slug (Eq. 4.15), giving

$$\frac{H_{max} - H_{mf}}{H_{mf}} = \frac{U - U_{mf}}{0.35 \sqrt{(gD)}} \qquad (4.53)$$

If bubbles require a significant distance above the distributor to grow to slugs, it is best to treat the lowermost region as a bubbling bed, with Eq. (4.55) applied above the level at which d_{eq} reaches $0.6D$.

Example 4.5: Slug flow in a pilot-scale bed

In the following example, we consider fluidization of the particles considered in Examples 4.1 to 4.4 in a column of diameter 0.2 m (that is 20 cm). For purposes of illustration, the gas velocity is taken as that in Example 4.4:

$$U = 0.78 \text{ m/s} = 78 \text{ cm/s}$$

that is:

$$U - U_{mf} = 0.775 \text{ m/s} = 77.5 \text{ cm/s}$$

and the bed depth corresponding to minimum fluidization conditions is taken as the value calculated in Example 4.4, that is:

$$H_{mf} = 0.514 \text{ m} = 51.4 \text{ cm}$$

(a) *Flow regime*

At such high fluidizing velocity in a column of diameter less than $(d_{eq})_{max}$, slug flow can clearly be anticipated. Applying the criterion developed by Baeyens and Geldart, Eq. (4.48), slug flow will occur for gas velocities in excess of:

$$U_{ms} = 0.5 + 1.6 \times 10^{-3} (60 \times 20^{0.175} - 49.6)^2 + 0.07 \sqrt{(981 \times 20)}$$
$$= 0.5 + 4.0 + 9.3 = 14.3 \text{ cm/s}$$
$$= 0.143 \text{ m/s}$$

For a very deep bed, the simple form of Eq. (4.48) developed by Stewart gives:

$$U_{ms} = 0.5 + 0.07 \sqrt{(981 \times 20)} = 10.3 \text{ cm/s} = 0.103 \text{ m/s}$$

Since the operating velocity is so much larger than U_{ms}, it is likely that slug flow will obtain throughout most of the depth of the bed. From Section 4.2.4, flow may be considered slugging if:

$$d_{eq} > 0.6D$$
$$= 0.12 \text{ m in the present case}$$

The bed depth required to attain this bubble size depends on the distributor used. For the tuyere distributor considered in Example 4.2(a), the initial bubble size was estimated in Example 4.4 as 0.129 m, so that slug flow could be assumed to occur throughout the bed; this seems intuitively correct, since the required tuyere spacing would give the bed only three nozzles. For a porous distributor, the initial bubble size would be small, so that a finite bed depth would be required for the bubbles to attain $d_{eq} = 0.12$ m. The correlations in Table 4.1 can be used to give a rough estimate for this depth. Using the correlation of Darton *et al.*, with $A_o = 0$ for a porous distributor:

$$d_{eq} = 0.54 \times 0.775^{0.4} \times 9.81^{-0.2} \times z^{0.8}$$
$$= 0.309 z^{0.8}$$

whence:

$$z = 0.31 \text{ m} \quad \text{when } d_{eq} = 0.12 \text{ m}$$

This estimate is likely to be generous, because the constraining effect of the walls causes more coalescence and hence more rapid bubble growth than in a large bubbling bed.

(b) *Bed expansion*

Taking slug flow to occur throughout the bed, Eq. (4.52) may be applied to predict the maximum depth:

$$H_{max} = H_{mf} \left(1 + \frac{U - U_{mf}}{u_{sl}} \right)$$

$$= 0.514 \left(1 + \frac{0.775}{0.490} \right)$$

where the rise velocity of a single slug, u_{sl}, is taken as 0.490 m/s, the value calculated in Example 4.1. From the above:

$$H_{max} = 1.33 \text{ m}$$

4.7 NOMENCLATURE

a_{or}	cross-sectional area of one hole or orifice	m^2
A	cross-sectional area of bed	m^2
A_o	area of distributor per orifice	m^2
C_d	discharge coefficient of orifice	—
C_D	drag coefficient	—
d_b	frontal diameter of bubble	m
$d_{b,0}$	initial frontal diameter of bubble	m
d_{eq}	equivalent volume diameter of bubble $(= (6V_b/\pi)^{1/3})$	m
$d_{eq,m}$	maximum equivalent bubble size attainable by coalescence	m
$d_{eq,o}$	initial equivalent bubble size	m
d_{or}	diameter of hole or orifice	m
d_p	mean sieve size of particles $(= 1/\Sigma \, (x_i/d_{pi}))$	m
d_{pi}	average of adjacent sieve apertures	m
d_p'	particle size used to calculate v_t in Eq. (4.12) $(= 2.7d_p)$	m
D	diameter of fluidizing column	m
Eo	Eötvös number $(= \rho_1 \, d_{eq}^2 \, g/\sigma)$	—
$f_{b,o}$	initial bubble frequency	s^{-1}
f_{or}	fraction of distributor open to gas $(= Na_{or})$	—
f_w	ratio of bubble wake volume to sphere volume $(= 6V_w/\pi d_b^3)$	—

g	acceleration of gravity	9.81 m/s
H	height of fluidized bed	m
H_d	height of dense phase in a bubbling bed	m
H_{mf}	height of bed at minimum fluidization velocity	m
H_{max}	maximum bed height	m
L_j	length of jet	m
N	number of orifices per unit cross-sectional area	m^{-2}
Q_b	visible bubble flowrate	m^3/s
Q_{or}	volumetric gas flowrate per hole	m^3/s
r	bubble radius	m
Re_b	bubble Reynolds number (Eq. 4.3)	—
Re_t	particle Reynolds number at terminal velocity	—
s	hole spacing in a drilled distributor	m
t	thickness of distributor plate	m
u_A	rise velocity of a bubble in a bubbling bed	m/s
u_b	rise velocity of an isolated bubble	m/s
u_{sl}	rise velocity of an isolated slug	m/s
u_{sA}	rise velocity of a slug in a slugging bed	m/s
U	superficial velocity of gas	m/s
U_{in}	mean gas velocity in duct entering windbox	m/s
U_{mb}, U_{mf}	minimum bubbling and fluidization velocities, respectively	m/s
U_{mfw}	minimum fluidization velocity related to windbox	m/s
U_{ms}	minimum slugging velocity conditions	m/s
U_{or}	velocity of gas in hole in distributor	m/s
U_R	velocity of gas in riser of tuyere	m/s
U_w	superficial velocity of gas in windbox	m/s
v_t	particle terminal or free-fall velocity	m/s
V_b	bubble volume	m^3
V_0	initial bubble volume	m^3
V_{sph}	volume of bubble and wake	m^3
V_w	volume of wake	m^3
z	distance above distributor	m/s
z_0	parameter in correlation of Rowe (1976)	m
β_w	volume of solids in wake as a fraction of bubble volume	—

$\Delta p_B, \Delta p_D$	pressure drop across bed and distributor, respectively	N/m^2
ϵ_b	fraction of bed occupied by bubbles	—
ϵ_{mf}	voidage of bed at minimum fluidization	—
θ_w	wake angle	—
μ	viscosity of gas	$N\,s/m^2$
μ_l	viscosity of liquid	$N\,s/m^2$
ρ_g	gas density	kg/m^3
ρ_{gw}	gas density in windbox	kg/m^3
ρ_l	liquid density	kg/m^3
ρ_p	particle density	kg/m^3
σ	interfacial tension	N/m

4.8 REFERENCES

Agarwal, J.C., Davis, W.L., and King, D.T. (1962). *Chem. Eng. Prog., 58* (Nov.), 85.
Avedesian, M.M. (1974). Paper 5d, 24th Annual Conference of C.S.Ch.E.
Baeyens, J., and Geldart, D. (1974). *Chem. Eng. Sci., 29*, 255.
Clift, R., and Grace, J.R. (1972). *Chem. Eng. Sci., 27*, 2309.
Clift, R., Grace, J.R., and Weber, M.E. (1974). *Ind. Eng. Chem. Fund., 13*, 45.
Clift, R., Grace, J.R., and Weber, M.E. (1978). *Bubbles, Drops and Particles*, Academic Press.
Cranfield, R.R., and Geldart, D. (1974) *Chem. Eng. Sci., 29*, 935.
Darton, R.C., LaNauze, R.D., Davidson, J.F., and Harrison, D. (1977). *Trans. Instn Chem. Engrs, 55*, 274.
Davidson, J.F., and Harrison, D. (1963). *Fluidised Particles*, Cambridge University Press.
Davidson, J.F., Harrison, D., Darton, R.C., and LaNauze, R.D. (1977). Chapter 10 in *Chemical Reactor Theory, A Review* (Eds L. Lapidus and N.R. Amundson), Prentice Hall.
Davidson, J.F., Harrison, D., and Guedes de Carvalho, J.R.F. (1977). *Ann. Rev. Fluid Mech., 9*, 55.
Davies, R.M., and Taylor, G.I. (1950). *Proc. Roy. Soc., Ser A, 200*, 375.

Doganoglu, Y., Jog, V., Thambimuthu, K.V., and Clift, R. (1978). *Trans. Instn Chem. Engrs,* **56**, 239.
Donsì, G., Massimilla, L., and Colantuoni, L. (1980). *3rd Eng. Found. Conf. on Fluidization* (Eds J.M. Matsen and J.R. Grace), Plenum Press, p.297.
Fakhimi, S., and Harrison, D. (1970). *Chemeca '70,* Instn Chem. Engrs, p.29.
Farrokhalaee, T., and Clift R. (1980). *3rd Eng. Found. Conf. on Fluidization* (Eds J.M. Matsen and J.R. Grace), Plenum Press.
Geldart, D. (1972). *Powder Technol.,* **6**, 201.
Geldart, D. (1977). 'Gas Fluidization', Short Course, University of Bradford, p.114.
Geldart, D., Hurt, J.M., and Wadia, P.W. (1978). *A.I.Ch.E. Symp. Ser.,* **74** (176), 60.
Ghadiri, M., and Clift, R. (1980). *Ind. Eng. Chem. Fund.,* **19**, 440.
Grace, J.R. (1970). *Can. J. Chem. Eng.,* **48**, 30.
Grace, J.R., and Clift, R. (1974). *Chem. Eng. Sci.,* **29**, 327.
Grace, J.R., and Harrison, D. (1968). *Instn Chem. Engrs Symp. Ser.,* **30**, 105.
Hanway, J.E. (1970). *Chem. Eng. Prog. Symp. Ser.,* **66** (105), 253.
Hetzler, R., and Williams, M.C. (1969). *Ind. Eng. Chem. Fund.,* **8**, 668.
Hovmand, S., and Davidson, J.F. (1971). Chapter 5 in *Fluidization* (Eds J.F. Davidson and D. Harrison), Academic Press.
Kunii, D., and Levenspiel, O. (1969). *Fluidization Engineering*, John Wiley and Sons Inc.
Masson, H. (1978). *Chem. Eng. Sci.,* **33**, 621.
Matsen, J.M. (1973). *A.I.Ch.E. Symp. Ser.,* **69** (128), 30.
Matsen, J.M., Hovmand, S., and Davidson, J.F. (1969). *Chem. Eng. Sci.,* **24**, 1743.
Merry, J.M.D. (1971). *Trans. Instn Chem. Engrs,* **49**, 189.
Merry, J.M.D. (1975). *A.I.Ch.E.J.,* **21**, 507.
Mori, S., and Wen, C.Y. (1975). *A.I.Ch.E.J.,* **21**, 109.
Perry, R.H., and Chilton, C.H. (1973). *Chemical Engineers' Handbook*, 5th ed., McGraw-Hill, Kogakusha.
Qureshi, A.E., and Creasy, D.E. (1979). *Powder Technol.,* **22**, 113.
Rippin, D.W.T., and Davidson, J.F. (1967). *Chem. Eng. Sci.,* **22**, 217.
Rowe, P.N. (1971). Chapter 4 in *Fluidization* (Eds J.F. Davidson and D. Harrison), Academic Press.
Rowe, P.N., (1976). *Chem. Eng. Sci.,* **31**, 285.
Rowe, P.N., and Everett, D.J. (1972). *Trans. Instn Chem. Engrs,* **50**, 55.
Rowe, P.N., MacGillivray, H.J., and Cheesman, D.J. (1979). *Trans. Instn Chem. Engrs,* **57**, 194.
Rowe, P.N., and Partridge, B.A. (1965). *Trans. Instn Chem. Engrs,* **43**, T157.
Rowe, P.N., and Widmer, A.J. (1973). *Chem. Eng. Sci.,* **28**, 980.
Rowe, P.N., and Yacono, C.X.R. (1976). *Chem. Eng. Sci.,* **31**, 1179.
Schügerl, K. (1971). Chapter 6 in *Fluidization* (Eds J.F. Davidson and D. Harrison)., Academic Press.
Toomey, R.D., and Johnstone, H.F. (1952). *Chem. Eng. Prog.,* **48**, 220.
Turner, J.C.R. (1966). *Chem. Eng. Sci.,* **21**, 471.
Wallis, G.B. (1969). *One-Dimensional Two-Phase Flow*, McGraw-Hill.
Werther, J. (1975). *Fluidization Technology* (Ed. D.L. Keairns), Vol. 1, Hemisphere Publishing Co., p. 215.
Whitehead, A.B. (1971). Chapter 19 in *Fluidization* (Eds J.F. Davidson and D. Harrison), Academic Press.
Yacono, C.X.R., Rowe, P.N., and Angelino, H. (1979). *Chem. Eng. Sci.,* **34**, 789.
Yang, W.–C., and Keairns, D.L. (1979). *Ind. Eng. Chem. Fund.,* **18**, 317.

CHAPTER 5

Solids Mixing

J. BAEYENS AND D. GELDART

5.1 INTRODUCTION

Particle mixing is of importance in the design of both batch and continuous fluidized beds. For example, the axial and radial transport of solids within the bed can influence:

(a) Gas–solid contacting
(b) Thermal gradients between a reaction zone and the zone in which heat transfer surfaces are located
(c) Heat transfer coefficients
(d) The position and number of solids feed and withdrawal points
(e) The presence and extent of dead zones at distributor level.

Problems of mixing in fluid beds fall into four categories:

1. Particles have uniform size and density.
2. Particles have uniform size but variable density.
3. Particles are all of same density but vary in size.
4. Particles vary in both size and density, e.g. small/heavy — large/light; small/light — large/heavy.

In category 1 the main problem, generally, is to ensure that the overall solids transport or circulation flux is high enough to eliminate temperature gradients; in categories 2, 3, and 4 an additional concern is whether the local composition of the powder — however defined — is everywhere equal to the overall average. The amount of solids circulation and degree of local mixing or segregation is primarily determined by the gas velocity, but particle shape,

97

size, density, stickiness, and size distribution all play a part. Bed and column geometry can also be important.

5.2 MECHANISMS OF MIXING AND SOLIDS CIRCULATION

Although mechanisms of mixing are now better understood, there is still much to be learnt and experimentation is essential.

5.2.1 Experimental Methods

A selection of the most frequently used methods is listed in Table 13.5 of Chapter 13. None is entirely satisfactory; they either make use of indirect measurements (e.g. heat transfer) from which circulation has to be inferred or involve the use of tracers in batch experiments or in artificial systems such as two-dimensional beds. There is a great need for a reliable on-line technique so that mixing can be monitored continuously. 'Freezing' the bed is the most tedious but simplest technique and is often used. In one version a layer of tracer is positioned in the bed, the gas flow is turned on for a given time, then suddenly shut off, and the bed removed a section at a time. In another, a tracer is released into an already fluidized bed and the bed shut down after a specified elapsed time.

Where the process demands that solids are fed and removed continuously it is often important to know the residence time distribution (RTD); experimental techniques to determine RTD are generally more straightforward and depend on taking continuous samples of magnetic, chemically active, or radioactive tracers.

5.2.2 Bubble-Induced Particle Mixing and Circulation

It is well known that vertical mixing in fluidized beds is many times faster than lateral mixing. Indeed it seems likely that solids move in a net horizontal direction largely by being (a) carried up to the surface where they are dispersed sideways by bursting bubbles and (b) carried down to, and across, the distributor by bubble-free flows of dense phase material.

Much of the early work on mixing mechanisms was done by Rowe and Partridge (1962, 1965). Layers of coloured and colourless particles were disturbed by injected bubbles and photographed in two-dimensional beds. These studies showed that in an idealized case, particles are carried upwards in the *wake* of the bubble (defined as the solids occupying the bottom of the completed sphere) and in the *drift* (defined as the region behind the completed sphere) (see Figs 5.1 and 5.2). Particles appear to be pulled into the wake or drift, carried up the bed for a distance, and then shed. Thus, particles travel upwards where there are bubbles and, by continuity, downwards where there are not.

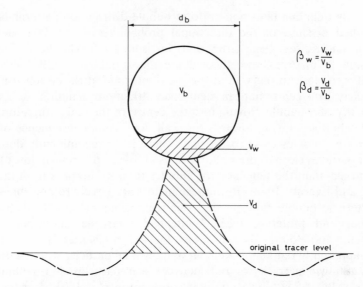

Figure 5.1 Wake and drift for a bubble injected into a two-layer bed.

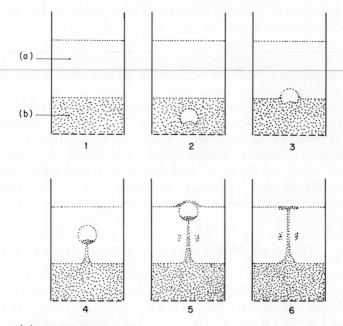

(a) Colourless solid
(b) Coloured solid

Figure 5.2 Displacement and mixing of solids by a rising bubble
(Rowe and Partridge, 1962).

In freely bubbling beds non-uniform bubble distribution patterns become established because of the directional probability of bubble coalescence (Geldart and Kelsey, 1968; Grace and Harrison, 1969; Whitehead, Gartside, and Dent, 1976). Areas near the wall are almost bubble free. Bubbles which are initiated close to the wall at the distributor are attracted towards other bubbles near the centre but can themselves attract only a limited number; the probability of a bubble moving into the centre of the bed is therefore high, and of a bubble moving towards the wall, low. As a consequence of there being fewer bubbles close to the wall, there is a predominantly downward flow of particles close to the wall. Once established, this return flow of solids tends to maintain the tendency for bubbles to move inwards from the walls (Rowe and Everett, 1972). Both these factors are largely responsible for the characteristic bubble flow profile found in studies of bubble distribution (Werther and Molerus, 1973; Whitehead, Gartside, and Dent, 1977; Whitehead and Young, 1967; (see also Fig. 4.7 in Chapter 4).

The apparently random movement of individual particles within a fluidized bed, as shown in tracks of radioactive tracer particles, led a number of authors (May, 1959; Lewis, Gilliand and Girouard. 1961/62; Kondukov, 1964; de Groot, 1967; Miyauchi, Kaji, and Saito, 1968) to express mixing data in terms of a radial and axial dispersion (diffusion) coefficient obtained by fitting experimental results to the diffusion differential equation under specific boundary conditions.

However, due largely to the wide scatter in experimental values and the lack of correlation in terms of the bubbling phenomenon, this approach has been abandoned for some time.

Having noticed that most of the earlier work was carried out in the slugging mode (which is anyway easier to characterize than the freely bubbling mode), Thiel and others (Thiel and Potter, 1978) fitted the results of their own and earlier work in terms of a diffusive mixing model for slugging beds. Although the model relates the diffusion coefficient to actual slugging parameters and achieves a reasonable agreement, its use is limited since slugging beds are of little or no industrial importance except in pilot plant work.

5.3 VARIABLES WHICH INFLUENCE SOLIDS CIRCULATION

Whitehead and others (Whitehead, Gartside, and Dent, 1976, 1977; Whitehead and Dent, 1978) studied the circulation patterns of solids in large fluidized beds and reported them in a series of papers. The solids used were largely in Geldart's group B. Their results and those of other workers are summarized below:

(a) In a bed 1.22 m square by about 1.5 m deep, sand was fluidized at velocities up to 0.3 m/s (see Fig. 5.3). At low velocities bubble-rich

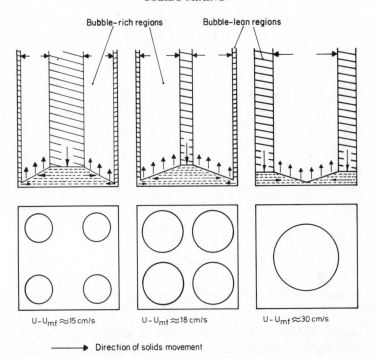

Figure 5.3   Effect of gas velocity on circulation patterns in a 1.22 × 1.22 m
bed of group B sand (Whitehead *et al.*, 1977).

areas occurred near the corners leaving bubble-free regions close to
the walls and in the centre. Solids were carried upwards with bubbles
and moved downwards elsewhere. At higher velocities the bubble-rich
regions increased in size at the expense of the bubble-free regions; at
still higher velocities the pattern changed entirely to upflow in the
centre and downflow at the walls.

(b) In shallow beds ($H/D < 0.5$), at most velocities the predominant flow
pattern is as shown in Fig. 5.4(a) with the central solids downflow
deflecting bubbles towards the walls. In deep beds ($H/D > 1$) the flow
pattern in the distributor region resembles that in a shallow bed but
changes into a central upflow–wall downflow pattern higher up (Fig.
5.4b). The relative size of the regions is a function of gas velocity, with
the 'shallow bed' zone becoming smaller as $U - U_{mf}$ increases.

(c) At equal excess gas velocities, the flow patterns and solids downflow
velocities are similar in different sizes of bed having the same aspect
ratio (H/D). This has been explained by Geldart (1980) as follows: The
ratio of bubble size at the surface divided by the bed diameter
($d_{eq,s}/D$) is approximately the same for different bed sizes at equal

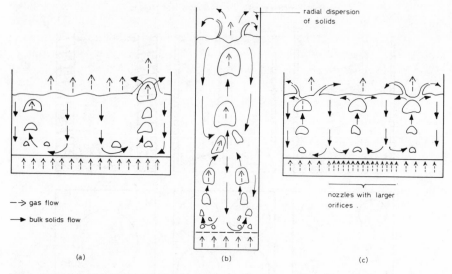

Figure 5.4 Solids circulation patterns in beds having different aspect ratios and distributors at approximately equal velocities (after Whitehead and Dent, 1982).

values of H/D and $U - U_{mf}$. This is likely to give rise to similar circulation patterns because solids are transferred laterally from bubble-rich regions to the downward-moving dense phase largely by bubbles bursting at the surface and scattering the solids sideways (Fig. 5.5).

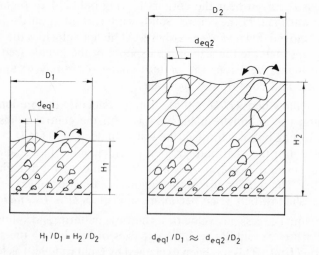

Figure 5.5 Beds containing the same powder, operated at equal values of gas velocity and of similar aspect ratio, have similar circulation patterns.

(d) The solids flow is not easily changed from its 'natural' circulation pattern, though gross deliberate 'mal-distribution' at the distributor can be effective in shallow beds (Fig. 5.4c).

(e) The presence of tubes or baffles modifies the flow patterns shown: with vertical tubes radial mixing is reduced slightly; with horizontal tubes, depending on the fraction of the cross-section occupied, vertical mixing can be inhibited to the point where the bed effectively becomes staged.

(f) Flow patterns can be time-dependent and cyclic, with changes occurring over several minutes.

(g) In group A powders bubble sizes and velocities are limited by the properties of the powder and gas (see Chapter 4, Section 4.2.3). Also, because the dense phase has a low effective viscosity, circulation velocities are similar to or higher than the isolated bubble velocities, and the absolute bubble velocities, u_A, turn out to be relatively independent of bubble size.

5.4 MODEL APPROACH TO SOLIDS CIRCULATION

5.4.1 The Parameters

It is evident from the foregoing that the mixing rate (J) ought to be defined in terms of the volumetric flowrate of bubbles through the bed, the amount of material each bubble drags up with it, the velocity at which the solids rise, and the fraction of the bed consisting of bubbles. Thus:

$$J = fn\,(\beta,\ Q_b/A,\ u_A,\ \epsilon_b) \tag{5.1}$$

where:

$$\beta = \beta_w + \beta_d \quad \text{and} \quad \beta_w = \frac{V_w}{V_b};\ \beta_d = \frac{V_d}{V_b}$$

Q_b = visible bubble flowrate

u_A = rise velocity of bubbles relative to the wall

ϵ_b = the fraction of bed consisting of bubbles

Parameter β

Rowe and Partridge (1962) showed that the amount of solids travelling in the the bubble wake V_w, expressed as a fraction of the bubble volume V_b, decreases with increasing particle size. Baeyens and Geldart (1973) studied the fraction of the solids in the drift β_d, defined as volume of the drift divided by bubble volume, and found that this also decreases with increasing particle

size. Their data are shown in Table 5.1 and expressed, somewhat speculatively, as functions of the Archimedes number in Fig. 5.6. For most group B
powders β_d is larger than β_w which averages about 0.35. For group D and

Table 5.1   Wake and drift fractions
(Baeyens and Geldart, 1973)

Powder		β_w	β_d	Y
Catalyst	47	0.43	1.00	1.00
Angular sand	252	0.26	0.42	0.50
	470	0.20	0.28	0.25
Rounded sand	106	0.32	0.70	0.82
	195	0.30	0.52	0.65

group A solids, β_w may be, respectively, 0.1 and 1. Thus, a bubble of a given
volume produces a much greater degree of particle movement in, for
example, a bed of cracking catalyst than in a bed of coal ash.

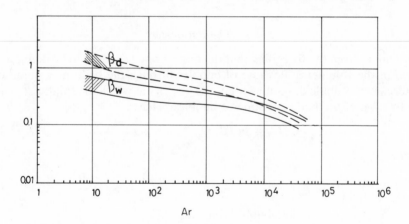

Figure 5.6   Wake and drift fractions as functions of Archimedes number.

Parameter u_A

Analysis of movie films of particles in two-dimensional beds (Baeyens and
Geldart, 1973) showed that the particles in the wake travel at a velocity
similar to that of the rising bubble while particles carried up on the drift travel
on average at about 38 per cent. of the bubble velocity, u_A, a figure later
confirmed by Ohki and Shirai (1976). Equations (4.41) and (4.9) or (4.10)
may be used to find u_A.

Parameter Q_b

The quantity of gas appearing as bubbles should be, according to the two-phase theory, equal to $(U - U_{mf})A$ (see Section 4.5.2). It is in general less than this, and Eq. (4.38) can be rewritten:

$$\frac{Q_b}{A} = Y(U - U_{mf}) \qquad (5.2)$$

where Y decreases with increasing particle size. Baeyens (1981) has correlated Y speculatively with the Archimedes number in Fig. 5.7.

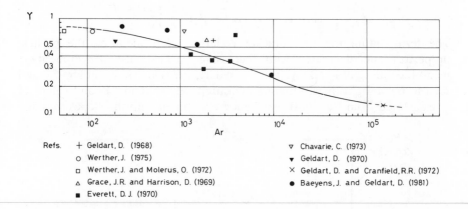

Refs.	+ Geldart, D. (1968)	▽ Chavarie, C. (1973)
	o Werther, J. (1975)	▼ Geldart, D. (1970)
	□ Werther, J. and Molerus, O. (1972)	✕ Geldart, D. and Cranfield, R.R. (1972)
	△ Grace, J.R. and Harrison, D. (1969)	● Baeyens, J. and Geldart, D. (1981)
	■ Everett, D. J. (1970)	

Figure 5.7 Deviation from two-phase theory ($Y = 1$) as a function of Archimedes number.

Parameter ϵ_b

The fraction of the bed consisting of bubbles can be obtained by measuring the bed expansion so that:

$$\epsilon_b = \frac{H - H_{mf}}{H} \qquad (5.3)$$

Alternatively, it can be predicted approximately from Eqs (4.46) and (5.2):

$$\epsilon_b = \frac{Q_b/A}{u_A} = \frac{Y(U - U_{mf})}{u_A} = \frac{Y(U - U_{mf})}{U - U_{mf} + u_b} \qquad (5.4)$$

5.4.2   Particle Velocities

Assuming that solids move upwards in the wake and drift of bubbles, and downwards elsewhere, a mass balance in any horizontal plane cutting the bed gives:

$$\left(\begin{array}{c}\text{Fraction of bed}\\\text{where solids}\\\text{move down}\end{array}\right)\times\left(\begin{array}{c}\text{average}\\\text{downwards}\\\text{particle}\\\text{velocity}\end{array}\right)=\left(\begin{array}{c}\text{fraction of}\\\text{bed where}\\\text{solids move up}\end{array}\right)\times\left(\begin{array}{c}\text{upwards}\\\text{solids}\\\text{velocity}\end{array}\right) \quad (5.5)$$

$$1 - \epsilon_b - (\beta_w + \beta_d)\,\epsilon_b\,v_p = u_A\,\beta_w\,\epsilon_b + 0.38\,u_A\,\beta_d\,\epsilon_b \quad (5.6)$$

substituting for ϵ_b from Eq. (5.4) and rearranging:

$$v_p = \left\{ \frac{\beta_w + 0.38\,\beta_d}{1 - \epsilon_b - \epsilon_b\,(\beta_w + \beta_d)}\,Y \right\}(U - U_{mf}) \quad (5.7)$$

The term within the first bracket is largely dependent on the particle properties and decreases with increasing particle size. Experimentally measured values of v_p (Baeyens and Geldart, 1973) are shown in Fig. 5.8 and seen to be straight lines; separate measurements of β and Y, and values of ϵ_b calculated from Eq. (5.4), gave good agreement with the slopes over the

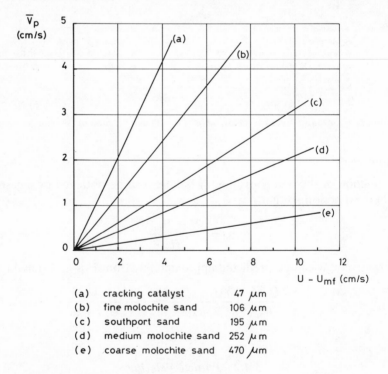

(a)	cracking catalyst	47 μm
(b)	fine molochite sand	106 μm
(c)	southport sand	195 μm
(d)	medium molochite sand	252 μm
(e)	coarse molochite sand	470 μm

Figure 5.8 Particle velocities at bed wall as a function of excess gas flow (Baeyens and Geldart, 1973).

velocity range covered. Local values of the particle velocity may be several times smaller or larger than the average (Nguyen, Whitehead, and Potter, 1978).

5.4.3 Circulation Flux

Equations (5.5) and (5.6) can be reformulated to give an expression for the solids circulation rate:

$$
\begin{pmatrix} \text{Mass} \\ \text{circulation} \\ \text{rate} \end{pmatrix} = \begin{pmatrix} \text{bulk} \\ \text{density} \\ \text{of dense} \\ \text{phase} \end{pmatrix} \times \begin{pmatrix} \text{fraction of} \\ \text{the bed where} \\ \text{solids move} \\ \text{upwards} \end{pmatrix} \times \begin{pmatrix} \text{mean} \\ \text{upwards} \\ \text{solids} \\ \text{velocity} \end{pmatrix} \times \begin{pmatrix} \text{bed} \\ \text{cross-} \\ \text{sectional} \\ \text{area} \end{pmatrix}
$$

$$(5.8)$$

and the solids flux:

$$J = \rho_p \, (1 - \epsilon_{mf}) \, (u_A \beta_w \epsilon_b + 0.38 \, u_A \beta_d \epsilon_b) \tag{5.9}$$

As before, substituting for ϵ_b:

$$J = \rho_p \, (1 - \epsilon_{mf}) \, (U - U_{mf}) \, Y \, (\beta_w + 0.38 \beta_d) \tag{5.10}$$

This can be compared with the empirical equation of Talmor and Benenati (1963) obtained in a 0.1 m diameter bed with ion exchange resin particles of 67 to 660 μm:

$$J \, (\text{kg/m}^2\text{s}) = 785 \, (U - U_{mf}) e^{-6,630 d_p} \tag{5.11}$$

The two equations agree reasonably well for group B particles but Eq. (5.11) increasingly underestimates J for larger group D materials. In Fig 5.9 experimental values from the literature are compared with predicted values (Eq. 5.10) and reasonable agreement is obtained.

At equal values of excess gas velocity, $U - U_{mf}$, the circulation flux decreases as particle size increases; however, the circulation fluxes in beds of large particles can be made large by increasing $U - U_{mf}$.

5.4.4 Bed Turnover Time

The time required to turn the bed over once is:

$$t_T = \frac{\text{mass of powder in the bed } M}{JA} \tag{5.12}$$

Substituting for M, and for J from Eq. (5.10):

$$t_T = \frac{H_{mf}}{(\beta_w + 0.38 \beta_d) \, Y \, (U - U_{mf})} \tag{5.13}$$

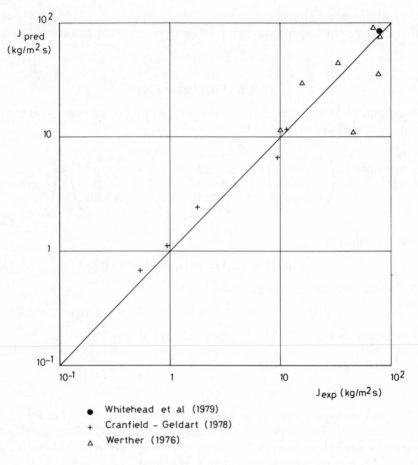

Figure 5.9: Comparison of experimental values of mixing flux with Eq. (5.10).

For a bed to which solids are continuously added and from which reacted products are removed (e.g. fluidized bed limestone calcination), the turnover time should be short compared with the solids residence time: i.e. good dispersion requires that the feed material should experience as many circulations as possible before discharge. Little research has been done on this but a ratio of residence time/turnover time of 5 to 10 seems reasonable.

5.5   GAS BACKMIXING

This is a function of the downwards velocity of the dense phase, v_p. If v_p exceeds the interstitital gas velocity, U_{mf}/ϵ_{mf}, then gas will be carried downwards with the solids. Working in a 1.22 m square bed, Nguyen,

Whitehead, and Potter (1978) injected CO_2 gas near the bed surface and measured the concentration at 49 positions below the injection point. They used two sands having U_{mf} values of 2.5 and 10.7 cm/s and found that the CO_2 concentration profiles reflected strongly the downflow patterns of the solids at various superficial fluidizing velocities (see Section 5.3 (a)). They maintain that for non-absorbing particles, backmixing commences when the ratio U/U_{mf} is about 3, and this reflects the fact that, for a given value of $U - U_{mf}$, both solids mixing and gas backmixing decreases as particle size increases. However, in view of the wide range of values for Y, β_w, and β_d as particle size changes, it would be surprising if the critical ratio at which backmixing commences remained at 3 for all solids.

On the basis of Eq. (5.7) and the criterion that gas backmixing becomes significant when $v_p > U_{mfi}$, there are grounds for believing that for group A and group D solids, the critical values of U/U_{mf} may be smaller and larger, respectively.

Gas backmixing is influenced considerably by the degree to which the gas is absorbed on the particles (Nguyen and Potter, 1974), and Bohle and van Swaaij (1978) showed that the mean residence of a strongly adsorbing gas (freon-12) injected into the bed can be a factor of ten longer than that for a weakly absorbing gas (helium). Thus the critical velocity ratio at which gas backmixing commences decreases as gas adsorption on the particle increases.

5.6 RESIDENCE TIME DISTRIBUTION

Many applications of fluidized beds involve the continuous processing of solids; e.g. the feed may be wet solids and the product dry, or the feed limestone and the product lime. In a plug flow system all particles have the same residence time t_R, while in a fully mixed system the solids have a wide range of residence times; this gives a non-uniform solid product and is inefficient for the high conversion of solids. Consider a quantity of tracer particles m kg introduced instantaneously at time zero into a bed containing M kg solids when the overall feed rate is F kg/s. If the bed is in plug flow, no tracer should appear at the solids offtake until the average residence time t_R has elapsed, where:

$$t_R = \frac{M}{F} \tag{5.14}$$

If the bed is fully mixed, tracer will appear immediately in the offtake at a concentration $C_0 = m/M$ and the concentration in the exit, C, will then fall according to:

$$\frac{C}{C_o} = e^{-t/t_R} \tag{5.15}$$

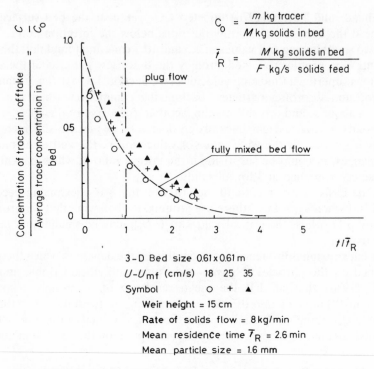

$$C_0 = \frac{m \text{ kg tracer}}{M \text{ kg solids in bed}}$$

$$\bar{t}_R = \frac{M \text{ kg solids in bed}}{F \text{ kg/s solids feed}}$$

3-D Bed size 0.61 x 0.61 m

U-U_mf (cm/s)	18	25	35
Symbol	O	+	▲

Weir height = 15 cm

Rate of solids flow = 8 kg/min

Mean residence time $\bar{t}_R$ = 2.6 min

Mean particle size = 1.6 mm

Figure 5.10 External age distribution of tracer solids in a fluid bed with constant solids feed — tracer added instantaneously at $t/t_R = 0$ (Cranfield, 1978)

Typical experimental data are shown in Fig. 5.10.

Similarly, the fraction of solids f spending less than time t in a completely mixed bed is:

$$f = 1 - e^{-t/t_R} \qquad (5.16)$$

A significant percentage of solids (18.2 per cent.) spends less than 20 per cent. of the average residence time t_R in a fully mixed bed. This problem can be solved in shallow beds (< 0.15 m) by installing vertical baffles or making the bed in the form of a long channel (see Chapter 10). Baffles are sometimes used in deeper beds but become progressively less effective in preventing short-circuiting of solids as bed depth increases. There is a need for further work on baffled deep beds.

5.7 SEGREGATION AND DEFLUIDIZATION

Researchers often use the words *segregation* and *defluidization* interchangeably and this makes for difficulties in interpretation of published data.

A bed may be 'well fluidized' in the sense that all the particles are fully supported by the gas may still be segregated in the sense that the local composition does not correspond with the overall average. Defluidization can be regarded as a special case of segregation in that some or all of the bed, usually that near the distributor, may be immobile. Segregation is likely to occur when there is a substantial difference in drag/unit weight between different particles. Particles having high drag/unit weight migrate to the surface while those with low drag/unit weight migrate to the distributor. When fluidized, the upper part of the bed will attain a fairly uniform composition, while the component that tends to sink forms a concentrated bottom layer. If the particles differ only in size, the larger particles settle to the bottom. Nienow and Rowe, who have done a considerable amount of work on segregation caused by density differences (e.g. Nienow, Rowe, and Cheung, 1978) have used the words *flotsam* and *jetsam* to describe the solids which occupy, respectively, the top and bottom of the bed. In non-technical English flotsam means 'floating wreckage' and jetsam, goods thrown out to lighten a ship and later washed ashore! A quantitative measure of segregation is required and two parameter are often used:

(a) The coefficient of segregation C_s is defined as:

$$\left(\frac{x_B - x_T}{x_B + x_T} \right) \times 100$$

where x_B and x_T are the concentration of the material of interest in the bottom and top halves of the bed, respectively. Clearly C_s can have values between -100 and $+100$ per cent., 0 per cent. being perfect mixing.

(b) The mixing index M_1 is defined as $x/\bar{x}$, where x is the concentration of the material under scrutiny at some level in the bed and $\bar{x}$ is its average concentration.

5.7.1 Methods of Investigation

A qualitative indication of defluidization can be obtained from the bed pressure drop–gas velocity curve. Consider a bed which consists of materials likely to segregate and which is either premixed outside the bed or vigorously bubbled and then collapsed rapidly. If the gas velocity is gradually increased, a pressure drop curve OAB (Fig. 5.11) is obtained; the wavy part of the curve AB occurs as the mixture becomes fluidized and then progressively segregates and defluidizes locally. All the particles are fully supported at U_{cf}, and are therefore fluidized with no fixed bed region at the distributor. Further increase in velocity produces bubbles throughout the bed, but they may not be sufficiently large at the distributor to promote good mixing, i.e. the bed may be segregated.

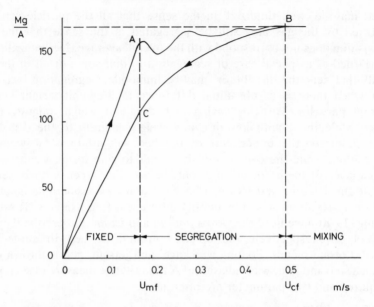

Figure 5.11   Pressure drop curve for wide size distribution powder initially well mixed (Boland, 1971).

As the gas velocity is decreased, defluidization occurs systematically so that the pressure drop is being measured across a fluidized bed zone superimposed on an increasingly deep fixed bed; this gives the curve BCO.

A quantitative method, the 'freezing' technique, has already been described in Section 5.2.1 and gives rise to curves of the type shown in Fig. 5.12.

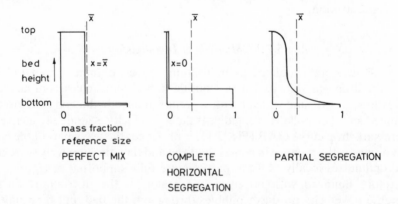

Figure 5.12   Typical data obtained from bed 'freezing' experiments.

5.7.2 Segregation By Density Difference

Segregation is much more likely to occur when the powder contains particles of different densities than when the size range is very broad; Decamps, Dumont, and Goossens (1971), Bena et al. (1968), and several other groups of workers have investigated segregation by density difference. In the papers of Nienow, Rowe, and coworkers (Rowe, Nienow, and Agbim, 1972; Nienow, Rowe, and Cheung, 1978) various combinations of heavy and light particles were fluidized. Rather complex empirical correlations were developed and have proved quite useful:

$$\frac{U_{TO}}{U_{mfS}} = \left(\frac{U_{mfB}}{U_{mfS}}\right)^{1.2} + 0.9 \left(\frac{\rho_H}{\rho_L} - 1\right)^{1.1} \left(\frac{d_H}{d_L}\right)^{0.7} - 2.2 \sqrt{\bar{x}} \, (1 - e^{-H/D})^{1.4}$$

$$(5.17)$$

where U_{mfB} and U_{mfS} are the bigger and smaller minimum fluidization velocities of the two sets of particles; ρ_H and ρ_L are the particle densitities of the denser and less dense particle; d_H and d_L are the sizes of denser and less dense particle; $\bar{x}$ is the mass fraction of the denser particles in the whole bed; H and D are the height and diameter of the bed; U_{TO} is the critical velocity at which mixing takes over from segregation. In general, $U_{TO} > U_{cf}$.

Having calculated U_{TO} it is then inserted into Eqs (5.18) and (5.19) to calculate the mixing index M_I:

$$M_I = \frac{x}{\bar{x}} = (1 + e^{-z})^{-1} \tag{5.18}$$

where:

$$z = \frac{U - U_{TO}}{U - U_{mfS}} e^{U/U_{TO}} \tag{5.19}$$

Thus z and therefore M_I increase as the gas velocity U increases.

Although developed for relatively small particles at low velocities the equations have been applied with success to large particle systems and velocities up to 5 m/s (Geldart et al., 1981).

5.7.3 Segregation By Size Difference

Providing the operating gas velocity is larger than U_{cf}, the velocity of complete fluidization, there will not be a fixed bed (defluidized) region at the distributor. However, that is not to say that the bed will be well mixed; it may be segregated with more coarse at the bottom and more fines at the top. Boland (1971), working with particles up to about 550 μm, developed a bubbling bed model to predict the value of $U - U_{mf}$ required to avoid segregation by size. Geldart *et al.* (1981) studied much larger particles fluidized at velocities up to 5 m/s and concluded that segregation by size difference appears to be an intrinsic feature of beds of large particles. It becomes larger as:

(a) The size of the fines decreases.
(b) The mean size of the bed solids increases.
(c) $U - U_{mf}$ approaches zero.

However, whereas segregation by density difference decreases progressively with increasing gas velocity, above values of $U - U_{mf}$ larger than about 2 m/s segregation by size difference becomes relatively constant. The presence of horizontal tubes appeared to make relatively little difference in the experiments cited, in which fines were reinjected near the distributor. One plausible hypothesis is that fines are carried upwards by three mechanisms and downwards by one. The mechanisms are:

1. Entrainment within the channels or bubbles at velocities which exceed the terminal velocities of the fines.
2. Bulk movement of fines in the wake/drift material dragged upwards by bubbles.
3. Stripping of fines through the voids between particles in the dense phase.
4. The downward movement is achieved by the bulk movement of fines trapped within the dense phase in the bubble-free regions.

These mechanisms explain in a qualitative way why segregation by size is relatively insensitive to velocity. Since upward mechanisms 1 and 2 and the return downward motion of mechanism 4 increase with gas velocity, they tend to be self-cancelling.

If the size of the coarse is increased, for a given superficial gas velocity and fines fraction, the rate of mechanism 3 will increase, partly because the size of the voids is bigger and also because the interstitial gas velocity, which is at least U_{mf}/ϵ_{mf}, increases. The downward movement of mechanism 4 will decrease because $U - U_{mf}$ which governs mixing has decreased. Qualitative relationships between the variable were proposed (Geldart *et al.*, 1981):

$$\frac{x_{fTOP}}{\overline{x}_f} = \left(\frac{d_p}{d_f}\right)^a \left(\frac{U_{mf}}{U - U_{mf}}\right)^b \tag{5.20}$$

Clearly much more work is required in order to develop quantitative relationships.

5.7.4 *Defluidization Through Particle Stickiness/Agglomeration*

Defluidization sometimes occurs when interparticle forces become excessive. This may be brought about by the presence of moisture or binding agents or by the softening of the particle with increasing temperature. The result is that particles may agglomerate to such a size that the gas velocity is insufficient to support them. This type of defluidization can occur very rapidly (Gluckman, Yerushalmi, and Squires, 1976). The bed is well-fluidized one moment and within 5 seconds is dead. As the bed defluidizes and the particles become loosely stuck together, the fluidizing gas blows a hole through the bed in order to escape and the pressure drop decreases rapidly.

The tendency to agglomerate depends on (a) the stickiness of the particles, which is a function of temperature, (b) the available surface area — the smaller the particles the greater the surface area — and (c) particle momentum, which is a function of particle size and gas velocity.

In general:

$$\text{Agglomerating tendency} = \frac{(a) \times (b)}{(c)} \tag{5.21}$$

At the present time it is not possible to predict from first principles the quantitative relationship between the variables, and experiments are essential to determine the conditions required for stable operation.

Defluidization can be influenced by the design of the gas distributor and it is essential to ensure that there are no dead areas at the plate. This might be taken to imply that a porous plate is best, and it is true that all the solids will be aerated; however, the small bubbles produced do not have sufficient energy to provide the vigorous movement required to remix large particles which may have started to segregate there.

On the other hand, if drilled plates having large holes are used, the large distance between them (for a given pitch/diameter ratio) allows solids to settle out. Baeyens (1973) made a study of particle movement caused by bubbles and suggested that for particle movement over the entire plate:

$$p < \lambda \, d_{eq,o} \tag{5.22}$$

where λ is a function of the particle mobility and $d_{eq,0}$ is the initial bubble size produced at the distributor. Geldart (1977) used this approach to derive Eq. (5.23), which gives the minimum gas velocity required to avoid dead areas between holes:

$$U - U_{mf} > \frac{(gp)^{0.5}}{K \, \lambda^{2.5}} \tag{5.23}$$

where $K = 1.73$ for holes on square pitch and 2.17 for triangular pitch. For conservative design, $\lambda \approx 1$ for group B solids and 1.5 for group A. Wen, Krishnan, and Kalyanraman, (1980) used thermistors to detect particle movement at distributor plates and propose that

$$U - U_{mf} > \frac{30}{p} (p - d_{or})^{0.716} \, d_p^{0.205} \qquad (5.24)$$

where the units are centimetres per second and centimetres, respectively, with d_p in micrometres. Dead zones are easier to eliminate using tuyeres with multiple horizontal holes or conical top bubble caps (sometimes called 'Chinese hats').

5.8 PROMOTION OF MIXING

Mechanical aids or suitable bed internals are sometimes used to promote mixing and/or to reduce segregation in materials which are naturally cohesive (group C) or pass through a sticky stage (coal gasification). Alternatively, they may be used in applications when reaction rate requirements or a desire to avoid elutriation favour gas velocities which are so low that mixing/heat transfer would otherwise be unacceptably low.

The use of a stirred bed (Brekken, Lancaster, and Wheelock, 1970) achieves satisfactory mixing of strongly cohesive powders due to constant breakdown of any channels formed near the distributor.

Strong particle circulation can be deliberately set up within a bed by including a draft tube. If excess gas is injected through the tube, the additional bubble stream will create a pumping action. A refined method is described by Decamps, Dumont, and Goossens (1971), who suggest using an internal airlift to generate more intense vertical mixing and prevent segregatioon at lower overall gas flowrates, Baxerres, Haewungscharren, and Gibert (1977) have proposed an alternative to the draft tube, called the whirling bed.

5.9 CONCLUDING REMARKS

It is the bubbling phenomenon which causes particle circulation patterns to be established. This circulation is essential where bed homogeneity is required and of particular interest when solid feed streams must be rapidly and totally dispersed in order to avoid localized accumulations.

Model and empirical equations for mixing, as developed above, enable the designer to estimate those operating conditions which yield the required particle circulation. However, there may be other conflicting requirements such as the degree of conversion and heat transfer. The remaining difficulty in scale-up of mixing data from small diameter columns to industrial equipment is a result of the non-equivalent circulation patterns in both sizes. Fluidized behaviour at equivalent bubble to bed diameter ratios offers a promising approach to scale-up.

Further work should be directed towards the phenomenon of segregation by size and density difference, particularly on a larger scale and at higher velocities. The whole question of mixing patterns at velocities above 1 m/s also deserves greater attention.

5.10 EXAMPLES

Example 5.1

A fluid bed catalytic reactor of 3.5 m diameter is used for a fast, exothermic reaction releasing 125 kcal/mol of product. Assuming a production of 2,000 kmol/h of product, and 90 per cent. conversion immediately above the distributor:

(a) What solid flux is required to carry the heat produced to the heat exchanger surfaces located within the bulk of the bed (average temperature difference between the bed and surface is 100°C)?

(b) What gas flowrate will produce such a flux?

Properties of catalyst: d_p : 80 μm $\qquad$ $\rho_p = 1,500$ kg/m^3

$\qquad\qquad\qquad\quad$ C_{ps} : 0.2 kcal/kg°C

$\qquad\qquad\qquad\quad$ U_{mf}: 0.3 cm/s $\qquad$ ϵ_{mf} : 0.5

Properties of gas: $\quad$ ρ_g : 1 kg/m^3 $\qquad$ μ : 4 × 10^{-5} kg/ms

Solution

(a) Production of reaction heat within the distributor zone:

$\qquad$ $0.9 \times 2,000 \times 125,000 = 225 \times 10^6$ kcal/h

Solids flux required:

$\qquad$ $\dfrac{225 \times 10^6}{0.2 \times 100} = 11,25 \times 10^6$ kg/h or 325 kg/m^{2}s

(b) The velocity required is calculated using Figs 5.6 and 5.7 in conjunction with Eq. (5.10) or Eq. (5.11).

With $\beta_w = 0.5$, $\beta_d = 1$, $Y = 1$ from Eq. (5.10):

$\qquad\qquad\qquad$ $U \simeq 0.5$ m/s

From Eq. (5.11)

$\qquad\qquad\qquad$ $U \simeq 0.7$ m/s

Example 5.2

Limestone of average particle size 2 mm and density 2,700 kg/m^3 is fed to a fluidized calciner with 1.3 m diameter (static bed height 0.8m) at a rate of

3,500 kg/h. The $CaCO_3$ concentration in the offtake must be less than 5 per cent.

(a) What velocity is required to avoid segregation of the limestone feed within the bed which will be 95 per cent. lime ($\rho_p \approx 1,600$ kg/m³)?

(b) Calculate the turnover time of the limestone at the selected velocity.

Solution

(a) Assume that the lime produced and the limestone feed have the same particle size: $d_{CaO} = 2$ mm, $\rho_{CaO} = 1,600$ kg/m³. If the offtake contains a maximum of 5 per cent. of $CaCO_3$, the fraction of limestone in the well-mixed bed is less than 0.05 ($= \bar{x}$).

For combustion gases at 850°C : $\rho_g = 0.28$ kg/m³, $\mu = 4 \times 10^{-5}$ kg/ms. U_{mf} can be calculated from the Wen-Yu equation (which slightly over-estimates at temperature), Eq. (2.27). Therefore:

$$U_{mf,CaO} = 0.8 \text{ m/s} \quad \text{and} \quad U_{mf, CaCO_3} = 1.3 \text{ m/s}$$

Calculate U_{TO} from Eq. (5.17) and use Eqs (5.19) and (5.18) to determine the mixing index M_I for various gas velocities. These are shown on Fig. 5.13. Because conversion takes place continuously, thus gradually reducing the density of the limestone, there will be relatively little pure limestone in the bed. However, for conservative design, assume the residual $CaCo_3$ is present as limestone particles. From Fig. 5.13, for good

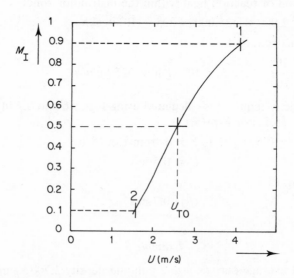

Figure 5.13 Mixing index as a function of gas velocity.

mixing ($M_I = 0.9$) the gas velocity should be 4.2 m/s. Operation at 1.6 m/s would give pronounced segregation ($M_I = 0.1$).

(b) For the limestone, $Ar = 37,082$. From Figs 5.6 and 5.7 and Eq. (5.13), operation at $U = 4.2$ m/s gives a turnover time:

$$t_T = \frac{0.8}{(0.1 + 0.38 \times 0.1) \times 0.15 \times (4.2 - 0.8)} = 11.4s$$

Example 5.3

A rectangular drier (1.5×4 m) for polymer powder ($d_p = 600$ μm, $\rho_p = 1,300$ kg/m) is fitted with a multi-hole tuyere distributor (234 tuyeres at square 0.15 m pitch, each tuyere having four orifices of 6 mm diameter). Its blower delivers a maximum gas flow of 15,000 N m^3/h at 120°C.

(a) Which is the largest polymer size that can be treated at 120°C in the equipment without dead zones on the plate?

(b) Is the superficial velocity obtained from the blower sufficient to guarantee good overall bed mixing for this size of polymer?

Solution

(a) The maximum polymer size can be calculated from Eq. (5.23) taking into account:

   (i) maximum superficial velocity available from the blower (120°C) = 1 m/s;

   (ii) $\mu = 2 \times 10^{-5}$ kg/ms and $\rho_g = 0.9$ kg/m^3;

   (iii) from the laminar part of the Wen-Yu equation:
   $$U_{mf} = 3,865 \, d_p^2 \text{ (c.g.s. units)};$$

   (iv) with $p = 0.15$ m, $U_{max} = 1$ m/s, and $\lambda = 1$, the value of $d_{p,max} = 1.14$ mm ($U_{mf} \simeq 0.5$ m/s).

(b) There is no absolute criterion to check overall mixing of this 1.14 mm powder at 1 m/s. Tentatively, however, one can imagine that optimum mixing conditions prevail when achieving the maximum heat transfer rates in the bed. The velocity required for this condition can be estimated from either:

$$0.92 \, Ar^{0.32} \leqslant (U - U_{mf})_{max} \leqslant 1.25 \, Ar^{0.36} \qquad (5.25)$$

(Baeyens and Geldart, 1981)

or

$$Re_{opt} \geqslant \frac{Ar}{18 + 5.22 \, \sqrt{(Ar)}} \qquad (5.26)$$

(Todes, 1965; see Chapter 9, Fig. 9.4)

Using these equations with $Ar = 42{,}512$:

From Eq. (5.25) : $0.28 \leqslant (U - U_{mf})_{opt} \leqslant 0.58$ m/s

From Eq. (5.26) : $U_{opt} \geqslant 0.75$ m/s

At $U = 1$ m/s and $U_{mf} = 0.5$ m/s, both equations indicate that mixing conditions are adequate.

5.11 NOMENCLATURE

A	cross-sectional area of fluidized bed	m^2
C	concentration of tracer particles	—
C_s	coefficient of segregation	—
$d_{eq,s}$	equivalent bubble diameter at bed surface	m
$d_{eq,o}$	equivalent bubble diameter initiated at distributor	m
d_{or}	diameter of distributor orifice	m
d_p	average particle diameter	m
d_H, d_L	diameter of, respectively, denser and less dense particles in binary mixture	m
d_f	average diameter of segregating fines	m
D	bed diameter	m
f	fraction of solids spending less than time t in the bed	—
g	gravitational constant	9.8 m/s^2
H	bed height	m
J	mixing flux	kg/m^{2}s
M_I	mixing index	—
p	pitch of holes in distributor	m
Q_b	visible bubble flowrate	m^3/s
t	time	s
t_T	time required to turn over the bed once	s
t_R	average particle residence time	s
u_A	rise velocity of bubbles relative to the wall	m/s
u_b	rise velocity of a single isolated bubble	m/s
U	superficial gas velocity	m/s
U_{cf}	velocity at which all particles are fully supported (Eq. 2.31)	m/s
U_{mfB}, U_{mfS}	bigger and smaller minimum fluidization velocities of two sets of particles	m/s
U_{mfi}	interstitial gas velocity at minimum fluidization	m/s

U_{TO}	takeover velocity	m/s
v_p	average downwards particle velocity	m/s
V	volume	m^3
$\bar{x}$	average mass fraction of the denser particles in a binary mixture	—
x_B, x_T	mass fraction of material of interest in, respectively, the bottom and top halves of the bed	—
$\bar{x}_f$, $x_{f,Top}$	mass fraction of fines, respectively, as bed average or in the top halves of the bed	—
Y	ratio of volumetric flowrate of bubbles, Q_b, and excess gas flowrate, $(U - U_{mf})A$	
β_w, β_D	fraction of solids carried up by a bubble within, respectively, its wake and drift	
ϵ_b	fraction of the bed consisting of bubbles	—
ρ_H, ρ_L	densities of denser and less dense particles in binary mixture	kg/m^3
ρ_p	absolute density of fluidized particles	kg/m^3

5.12 REFERENCES

Baeyens, J. (1973). Ph.D. Dissertation, University of Bradford.

Baeyens, J. (1981). 'Gas fluidization', Short Course, I. Chem. E./University of Bradford.

Baeyens, J., and Geldart, D. (1973). *Fluidization et ses Applications*, Toulouse, p. 182.

Baeyens, J., and Geldart, D. (1981). *J. Powder and Bulk Solids Tech.*, **4**, (4), 1.

Baxerres, J.L., Haewsungscharren, A., and Gibert, H. (1977). *Lebensm. Wiss. Technol.*, **10**, 191.

Bena, J., Havalda, I., Bafinec, M., and Ilowsky, J. (1968). *Coll. Czech. Chem. Comm.*, **33**, 2620.

Brekken, R.A., Lancaster, E.B., and Wheelock, T.D. (1970). *Chem. Eng. Prog. Symp. Ser.*, **66**, (101), 91.

Bohle, W., and van Swaaj, W.P.M. (1978). In *Fluidization* (Eds J.F. Davidson and D.L. Keairns), Cambridge University Press, p. 167.

Boland, D. (1971). Ph.D. Dissertation, University of Bradford.

Chavarie, C., (1973). Ph.D. Dissertation, McGill Univ.

Cranfield, R.R. (1978). *A.I.Ch.E. Symp. Ser.*, **74** (176), 54.

Decamps, F., Dumont, G., and Goossens, W. (1971). *Powder Technol.*, **5**, 299.

de Groot, J.H. (1967). *Proc. Interm. Symp. Fluidization*, Netherlands University Press, Amsterdam, p. 348.

Everett, D.J., (1970). Ph.D. Dissertation, University of London.

Geldart, D. (1977). 'Gas fluidization', Short Course, I.Chem.E./University of Bradford.

122   GAS FLUIDIZATION TECHNOLOGY

Geldart, D. (1980. 'Advances in gas fluidization', Short Course at Center for Professional Advancement, New Jersey.
Geldart, D., Baeyens, J., Pope, D.J., and Vandeweyer, P. (1981). *Powder Technol.*, **30**, 195.
Geldart, D., and Kelsey, J.R. (1968). In *'Fluidization'*, I.Chem.E. (London) Symp. Ser., p.114.
Gluckman, M.J., Yerushalmi, J., and Squires, A.M. (1976). In *Fluidization Technology* (Ed. D.L. Keairns) Vol. II, Hemisphere, p. 395.
Grace, J.R., and Harrison, D. (1969). *Chem. Eng. Sci.*, **24**, 497.
Kondukov, N.B., (1964). *Intern, Chem. Eng.*, **4**, 43.
Lewis, W.K., Gilliland, E.R., and Girouard, H. (1961/62). *Chem. Eng. Progr. Symp. Ser.*, **58** (38), 87.
May, W.G. (1954) *Chem. Eng. Progr.*, **55**, 49.
Miyauchi, T., Kaji, H., and Saito, K. (1963). *J. Chem. Eng. (Japan)*, **1**, 72.
Nienow, A.W., Rowe, P.N., and Cheung, L.Y.I. (1978). *Powder Technol.*, **20**, 89.
Nguyen, H.V., Whitehead, A.B., and Potter, O.E. (1978). In *Fluidization* (Eds J.F. (Ed. H.M. Hulbert), vol. 133, Am. Chem. Soc., p. 290.
Nguyen, H.V., Whitehead, A.B., and Potter, O.E. (1978). In *Fluidization* (Eds J.F. Davidson and D.L. Keairns), Cambridge University Press, p.140.
Okhi, K., and Shirai, T. (1976). *Fluidization Technology*. (Ed. D.L. Keairns), Hemisphere, p.95.
Rowe, P.N., and Everett, D.J., (1972). *Trans. Instn. Chem. Engrs.*, **50**, 42.
Rowe, P.N., Nienow, A.W., and Agbim, A.J. (1972). *Trans. Instn. Chem. Engrs*, **50**, 310.
Rowe, P.N., and Partridge, B.A. (1962). *Proc. Symp. Interaction between Fluids and Particles*, Instn. Chem. Engrs, p.135.
Rowe, P.N., Partridge, B.A., Cheney A.G., Henwood, G.A., and Lyall, E. *Trans. Inst Chem. Engrs.*, **50**, 310.
Talmor, E., and Benenati, R.F. (1963). *A.I.Ch.E.*, **9**, 536.
Thiel, W.J., and Potter, O.E. (1978). *A.I.Ch.E.*, **24** (4), 561.
Todes, O.M. (1965). *Appl. of fluid beds in the chemical industry*, Part 2. Izd Znanie Leningrad, p.4.
Wen, C.Y., Krishnan, R., and Kalyanaraman, R. (1980). In *Fluidization* (Eds J.R. Grace and J.M. Matsen) Plenum Press, p. 405.
Werther, J. (1975). Proceedings of CHISA Conf., Prague.
Werther, J., and Molerus, O., (1972). Proceedings of CHISA Conf, Prague.
Werther, J. (1976). In *Fluidization Technology* Vol. I (Ed. D.L. Keairns), McGraw-Hill, p. 215.
Werther, J., and Molerus, O. (1973). *Int. J. Multiphase Flow*, **1**, 103.
Whitehead, A.B., and Auff, A. (1976). *Powder Technol.*, **15**, 77.
Whitehead, A.B., and Dent, D.C. (1978). *Fluidization* (Eds J.F. Davidson and D.L. Keairns), Cambridge University Press, p.44.
Whitehead, A.B., and Dent, D.C. (1982). *A.I.Ch.E.*, **28**, 169.
Whitehead, A.B., Gartside, G., and Dent, D.C. (1976). *Powder Technol.*, **14**, 61.

CHAPTER 6

Particle Entrainment and Carryover

D. GELDART

6.1 INTRODUCTION

In most applications of fluidization the carryover of fine particles in the exit gases is regarded as a nuisance to be endured in exchange for the advantages offered by other features of the technique. In a few processes selective removal of small light solids by elutriation is advantageous, while in the recirculating (fast) fluid bed, carryover and reinjection of the whole bed material is essential.

Three processes which typify these different ways of viewing carryover are catalytic cracking, incineration of sludges from waste-water treatment plants, and calcination of aluminium trihydrate. In the first, attrition of valuable catalyst in cyclones and loss of the fines can increase the cost and controllability of the operation; in the second the solids are a coarse sand which remains in the bed, and the fine light ash from the burnt sludge is carried out continuously as an overhead product, thus providing a convenient method of removal; in the third process, high throughput and the control of residence time and bed temperature depend on achieving high carryover rates. In this chapter we shall consider the factors affecting carryover and discuss ways of predicting the amount and size distribution of solids carried over.

6.2 DEFINITIONS

The terms *carryover, entrainment,* and *elutriation* are often used interchangeably to describe the ejection of particles from the surface of a bubbling fluidized bed and their removal from the unit in the gas stream. *Elutriation* also describes the technique used to determine the size of subsieve particles or to prepare narrow cuts from a material having a wide size distribution. This involves passing air or water at a known velocity upwards through a powder sample until no more is carried out. The material collected is weighed and its size calculated using Stokes' law.

The terminology and processes occurring in *entrainment* are best illustrated through a series of experiments. Imagine a tall vertical column into which solid particles can be introduced (Fig. 6.1). First drop in a single particle. Providing it remains more than about 10 particle diameters away from the wall, its velocity will increase rapidly until it reaches a steady falling rate which is called its *terminal velocity*. Ways of calculating this are given later.

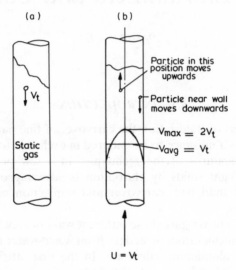

Figure 6.1   Motion of particles falling in a tube
(a) without and (b) with upward flow of gas.

Now pass air up the tube at a superficial velocity equal to the particle's terminal velocity and drop the particle in again. The way it behaves now depends on several factors. If the gas flow is laminar with respect to the column there will be the well-known parabolic velocity distribution across the tube, the maximum velocity at the centre being twice the mean superficial velocity, and zero gas velocity at the wall. Depending on its radial position the particle may move upwards or downwards (Fig. 6.1b). If the gas flow is turbulent not only does the velocity vary across the tube but there are also random velocity fluctuations with respect to time, and the particle motion is more unpredictable. Thus, allowing a particle to fall through a static gas and moving the gas upwards past the particle at the same relative velocity are by no means identical.

If now a few grams of powder are dropped into the tube with no gas flow the solids may fall as one or more pseudo-particles having effective sizes larger than those of individual particles, but lower densities. Particles tend to be shielded by those beneath them and therefore draw nearer to each other. Very small particles having intrinsically cohesive properties, or particles which are sticky or damp, also tend to cluster together, and have a higher terminal velocity than would be expected from their individual sizes.

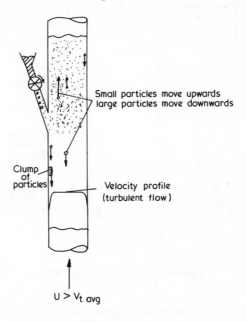

Figure 6.2   Dilute phase particle transport.

Now pass gas up the tube at a velocity considerably higher than the terminal velocity of the mean particle size, and feed in a continuous stream of particles (Fig. 6.2). They will be picked up (*entrained*) by the gas and transported upwards in dilute phase pneumatic transport. If the size distribution is so wide that some particles have terminal velocities higher than the gas velocity then a fractionation (*elutriation*) of solids occurs with the finer particles carried upwards and the coarser falling to the bottom of the tube. If the gas velocity is much higher than the terminal velocity of even the largest particle and the solids feed rate is gradually increased, the concentration of solids in the gas increases, together with the pressure drop along the tube, until a solids/gas loading is reached at which the solids start to fill up the lower section of the tube and dilute phase flow is no longer possible in the upper section. The gas/solid suspension then collapses into dense phase or slugging flow depending on the powder and the tube diameter. The loading at which this occurs is known as the *saturation carrying capacity* of the gas and the critical gas velocity is the *choking velocity*.

Now reduce the gas velocity, allow a fluidized bed to form and cut off the solids feed. Bubbles bursting at the surface of the bed will project particles into the *freeboard*, i.e. the space between the bed surface and the gas offtake (Fig. 6.3). Depending on their terminal velocities and the gas velocity, solids

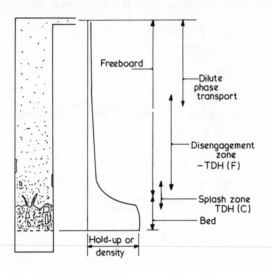

Figure 6.3   Zones in a fluidized bed.

are carried up the tube to various heights. The larger particles fall back and the smallest are carried out of the offtake tube; consequently, the *solids loading* (expressed as kilogram solids per second divided by cubic metres of gas per second of kilograms of solids per second divided by kilograms of gas per second; i.e. as kg/m^3 or kg/kg) will decline with height. The region within which the solids loading falls is called the transport disengagement height)TDH). Some confusion is caused by the application of the term TDH to two related but different circumstances:

(a) Particles having a terminal velocity larger than the gas velocity are flung upwards and then fall back. These particles are *coarses* and above the height they reach are found only *fines* — particles with terminal velocities smaller than the gas velocity.

(b) The concentration (or hold-up) of fines decreases with height and, it is said, eventually reaches a constant value above a certain height.

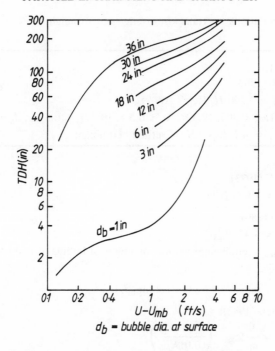

Figure 6.4 Transport disengagement height as a function of bubble size and gas velocity (Zenz, 1983).

The height which the *coarses* reach is sometimes called the *splash height* but we shall call it TDH(C). The height above which the *fines* hold-up changes little or not at all will be called TDH(F). Published correlations for the TDH are given in Table 6.1.

It is clear that the definition of a particle as *coarse* or *fine* will depend upon its terminal velocity and the superficial gas velocity, so we shall consider next how best to calculate terminal velocities.

6.3 PARTICLE TERMINAL VELOCITY

The terminal (or free-fall) velocity of a particle, v_t, figures in most of the correlations for carryover and is also of considerable importance in sedimentation, gas and liquid cyclone design, and many other areas of chemical engineering. It has therefore been a subject of study by numerous workers, and as there are several excellent comprehensive texts available (e.g. Clift, Grace, and Weber, 1978; Allen, 1981), the treatment here will be kept fairly brief.

Table 6.1   Correlations for transport disengagement height

TDH (C)

Soroko *et al.* (1969):
$$\text{TDH (C)} = 1{,}200\ H_s\ \text{Re}_p^{1.55}\ \text{Ar}^{-1.1}$$
for $15 < \text{Re}_p < 300$; $19.5 < \text{Ar} < 6.5 \times 10^5$; $H_s < 0.5$ m ; $d_p = 0.7 - 2.5$ mm
where $\text{Re}_p = \rho_g U d_p / \mu$; $H_s =$ settled bed height

---

TDH (F)

Fournol *et al.* (1973):
$$\text{TDH(F)} = 1{,}000\ U^2/g$$

Horio *et al.* (1980):
$$\text{TDH(F)} = 4.47\ d_{eq,s}^{1/2}$$
where $d_{eq,s} =$ equivalent volume diameter of bubble at bed surface

Amitin (1968):
$$\text{TDH(F)} = 0.85 U^{1.2}\ (7.33 - 1.2\ \log_{10} U)$$

Wen and Chen (1982):
$$\text{TDH(F)}_i = \frac{1}{a_i}\ \log_e\ \left(\frac{E_o - K_{i\infty}^*}{0.01 K_{i\infty}^*}\right)$$
where $\dfrac{E_o}{A d_{eq,s}} = 3.07 \times 10^{-9}\ \dfrac{\rho_g^{3.5}\ g^{0.5}}{\mu^{2.5}}\ (U - U_{mf})^{2.5\dagger}$
and measured values $3.5 < a_i > 6.5$ m^{-1}: use average a_i, ≈ 4 m^{-1}.

[†]This equation is based on beds smaller than 0.6m diameter and it is unwise to use it for large beds.

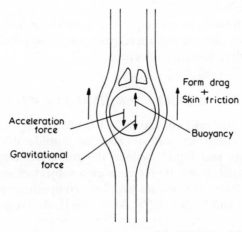

Figure 6.5   Forces acting on a particle falling in a fluid.

6.3.1 Spherical Particles

Consider a spherical particle falling freely in an infinite fluid. The forces acting on it are shown in Fig. 6.5:

Gravitational force − buoyancy force − drag force = accelerating force

$$(6.1)$$

The drag force can be expressed in terms of a drag coefficient C_D (a sort of friction factor analogous to f in the packed bed equation), the projected area perpendicular to the flow A_p, and the inertia of the fluid:

$$F = C_D A_p \cdot \tfrac{1}{2}\, \rho_f\, v^2 \qquad (6.2)$$

As in flow through pipes and packed beds, the 'friction factor' C_D turns out to be a function of the Reynolds number which in this case is defined as:

$$\mathrm{Re} = \frac{\rho_f\, v\, d_v}{\mu} \qquad (6.3)$$

For a sphere $A_p = \pi d_v^2/4$ and substituting in Eq. (6.2) gives:

$$F = C_D \frac{\pi\, d_v^2}{8}\, \rho_f\, v^2 \qquad (6.4)$$

Equation (6.1) can now be written algebraically:

$$\frac{\pi}{6}\,(\rho_p - \rho_f)g - F = \pi d_v^3\, \rho_p\, \frac{dv}{dt} \qquad (6.5)$$

When a particle reaches its free-fall or terminal velocity the accelerating force is zero, $v = v_t$, and Eq. (6.5) becomes:

$$F = \frac{\pi}{6}\, d_v^3\, (\rho_p - \rho_f)g \qquad (6.6)$$

In the general case, combination of Eqs (6.4) and (6.6) gives:

$$C_D = \frac{4}{3} \frac{(\rho_p - \rho_f)d_v\, g}{\rho_f v_t^2} \qquad (6.7)$$

By using experimental data from spheres falling in a variety of fluids (Table 6.2), log C_D is plotted against log Re to give the so-salled 'standard drag curve', which has three broad regions (Fig. 6.6):

(a) Laminar region (Re < 0.2). Stokes solved the analytical equations for the drag on a smooth rigid spherical particle moving on a homogenous viscous fluid at velocity v:

$$F = 3\, \pi\, \mu\, v\, d_v \qquad (6.8)$$

Table 6.2 Standard drag coefficients for spheres (calculated from equations in Clift, Grace, and Weber, 1978)

Re	C_D	$C_D Re^2$	C_D/Re
0.01	2,400	2.4×10^{-1}	2.4×10^5
0.02	1,204	4.8×10^{-1}	6.02×10^4
0.05	484	1.21	9.68×10^3
0.1	244	2.44	2.44×10^3
0.2	124	4.96	6.2×10^2
0.5	51.5	12.87	1.03×10^2
1	27.1	27.1	27.1
2	14.76	59	7.38
5	7.03	1.76×10^2	1.41
10	4.26	4.26×10^2	4.26×10^{-1}
20	2.71	1.08×10^3	1.35×10^{-1}
50	1.57	3.93×10^3	3.14×10^{-2}
100	1.09	1.09×10^4	1.09×10^{-2}
200	0.77	3.08×10^4	3.85×10^{-3}
500	0.555	1.39×10^5	1.11×10^{-3}
1,000	0.471	4.71×10^5	4.71×10^{-4}
2,000	0.421	1.68×10^6	2.1×10^{-4}
5,000	0.387	9.67×10^6	7.74×10^{-5}
10,000	0.405	4.05×10^7	4.05×10^{-5}
20,000	0.442	1.77×10^8	2.21×10^{-5}
50,000	0.474	1.18×10^9	9.48×10^{-6}
100,000	0.5	5×10^9	5×10^{-6}
200,000	0.497	1.99×10^{10}	2.48×10^{-6}
500,000	0.376	9.4×10^{10}	7.52×10^{-7}
10^6	0.11		

Susbtituting this expression for F in Eq. (6.6) and writing $v = v_{t,ST}$ gives:

$$v_{t,ST} = \frac{(\rho_p - \rho_f)g \, d_v^2}{18\mu} \tag{6.9}$$

Spherical/rounded sand particles ($\rho_p = 2{,}600$ kg/m^3) smaller than 33 μm falling in air are in the Stokes region; if the expression is used for larger and/or denser particles then the terminal velocity will be overestimated. Combining Eqs (6.4) and (6.8) gives:

$$C_D = \frac{24}{Re_t} \tag{6.10}$$

which is analogous to the expression for the friction factor for laminar flow in pipes ($f = 16/Re$). Particles with terminal Reynolds number > 0.2 reach their terminal velocity very rapidly (see Table 6.3).

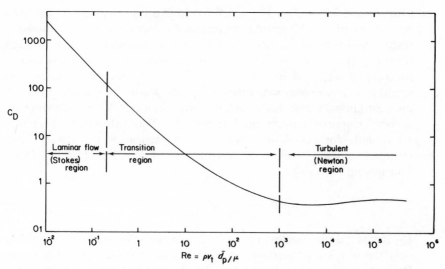

Figure 6.6 Standard drag curve for spheres versus particle Reynolds number.

(b) Turbulent flow ($Re_t > 1,000$). Spherical sand particles larger than 1,500 μm in air are in the turbulent regime (sometimes called the Newton region) and $C_D = 0.43$; substitution in Eq. (6.7) and rearrangement leads to:

$$v_{t,N} = \sqrt{\left\{ \frac{4}{3} \frac{(\rho_p - \rho_f)g\, d_v}{0.43\rho_f} \right\}} \qquad (6.11)$$

Note that the fluid viscosity does not appear in the equation at all and that v_t is much less dependent on particle size than it is in the laminar regime (Eq. 6.9). On the other hand, the time and distance taken to attain $v_{t.N}$ become considerable (Table 6.3). The procedure for calculating the acceleration time will be found in Heywood (1962).

Table 6.3 Time and distance for spheres ($\rho_p = 2,600$ kg/m³) to reach 99 per cent, of their terminal velocities falling in air at ambient conditions

d_v	v_t m/s	Time taken to reach 0.99 v_t	Distance fallen before reaching 0.99v_t	
		s	Metres	Particle diameters
30 μm	0.07	0.033	1.85×10^{-3}	61
3 mm	14	3.5	35	11,670
3 cm	44	11.9	453	15,100

(c) Transitional flow ($0.2 < Re_t < 1{,}000$). In many applications of fluidized beds the particles of interest have terminal velocities within the transition regime and a rapid method of calculation is necessary. It can be seen from Eq. (6.7) and Fig. 6.5 that v_t is a function of C_D which is itself a function of Re_t, and Re_t contains v_t. There are several empirical and semitheoretical expressions relating C_D and Re in the transition region, the best known being that of Schillar and Nauman (1933). it depends on successive approximations and is therefore tedious. Moreover, it is not at all suitable for calculating the size of a particle which has a given terminal velocity. A more convenient procedure is based on Eq. (6.7). Multiplying both sides by $\rho_f^2 v_t^2 d_v^2/\mu^2$ gives:

$$C_D\, Re_t^2 = \frac{4}{3}\, \rho_f(\rho_p - \rho_f)g\, d_v^3 \tag{6.12}$$

$C_D\, Re_t^2$ is a dimensionless group ($= 4/3$ Archimedes number) containing only the physical properties of the particle/fluid system. The C_D versus Re_t data can be recalculated as $C_D\, Re_t^2$ so that Re_t can readily be found for any value of $C_D Re_t^2$, either from a graph of from a polynomial (Fouda and Capes, 1976).

   To find the particle size corresponding to a given terminal velocity, d_v must be eliminated from Eq. (6.7) by dividing by Re_t to give:

$$\frac{C_D}{Re_t} = \frac{4}{3}\, \frac{\rho_p - \rho_f}{\rho_f^2}\, \frac{\mu g}{v_t^3} \tag{6.13}$$

C_D/Re_t is calculated and the plot of C_D/Re_t versus Re_t used to find Re_t and Re_t hence d_v.

Equations (6.12) and (6.13) and the associated graphs, Fig. 6.7, are valid over the whole range of Reynolds numbers.

6.3.2 Non-Spherical Particles

There is as yet no satisfactory way of predicting the terminal velocity of an irregular shaped particle. We have already encountered (Chapter 2) the problem of defining the size of non-spherical particles and although sphericity ψ is a useful parameter, two particles with similar values of ψ can have quite different shapes (Table 2.1); for example, a regular tetrahedron and a disc whose diameter is four times its height have ψ values of 0.67 and 0.69, respectively. A symmetrical particle, such as a sphere, cube, or tetrahedron, falls vertically in a stable manner at a constant velocity in any orientation, whereas a disc will move to the side and wobble as it falls. This will be obvious to anyone who has watched a leaf fall from a tree of tried to cover a small coin on the bottom of a bucket of water by dropping a larger coin on to it.

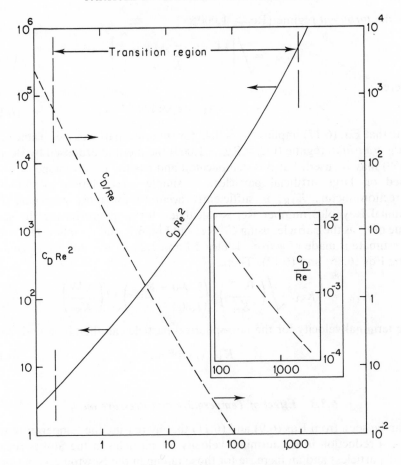

Figure 6.7 $C_D Re^2$ and C_D/Re as functions of Reynolds number.

Nevertheless, in spite of its limitations most corrections to the spherical particle equations make use of ψ. Pettyjohn and Christiansen (1948) introduced two correction factors for regular non-spherical shapes with sphericities between 0.67 and 0.906.

For the *laminar* regime ($Re_t < 0.2$):

$$v_{t.ST} = K_{ST} \; \frac{(\rho_p - \rho_f) \, g \, d_v^2}{18\mu} \tag{6.14}$$

where:

$$K_{ST} = 0.843 \log \frac{\psi}{0.065} \tag{6.15}$$

For the *turbulent* regime ($Re_t > 1,000$):

$$v_{t,N} = \sqrt{\left\{ \frac{4}{3} \frac{(\rho_p - \rho_f) \, g \, d_v}{K_N \, \rho_f} \right\}} \tag{6.16}$$

where:

$$K_N = 5.31 - 4.88 \, \psi \tag{6.17}$$

Note that Eq. (6.17) implies $C_D = 0.43$ for spherical particles (see Table 6.2). In the *transition* regime ($0.2 > Re_t > 1,000$) the method proposed by Becker (1959) may be used, but it is complicated and has the disadvantage of being based on large artificial particles. A simple interpolation to obtain a correction factor, K_{TR}, is sufficiently accurate for most purposes. The terminal Reynolds number Re_t is calculated for a sphere having the same value of d_v as the particle, using $C_D \, Re_t^2$ and Fig. 6.7, and v_t (sphere) is found. As estimate is made of ψ from Tables 2.1 and 2.2 and K_{ST} and K_N calculated using Eqs (6.15) and (6.17). Then:

$$K_{TR} \approx \left\{ K_{ST} - \sqrt{\left(\frac{0.43}{K_N} \right)} \right\} \cdot \left(\frac{1,000 - Re_t}{1,000 - 0.2} \right) + \sqrt{\left(\frac{0.43}{K_N} \right)} \tag{6.18}$$

The terminal velocity for the non-spherical particle can then be found from:

$$v_t = K_{TR} \, v_t \text{ (sphere)} \tag{6.19}$$

6.3.3 Effect of Temperature and Pressure on v_t

It can be seen from Eqs (6.9) and (6.11) that increasing the temperature will cause a reduction in the terminal velocity for particles in the Stokes region (small particles) and an increase for those falling in the Newton region (large particles); increasing the pressure has little effect in the Stokes region but reduces v_t for large particles.

6.4 MECHANISMS OF ENTRAINMENT AND DISENGAGEMENT, AND TDH

There is no general agreement concerning the mechanisms by which solids are ejected into the freeboard, though bubbles clearly have an important role to play. While on the one hand postulating that bursting bubbles act like intermittant jets, Zenz and Weil (1958) also point out that extensive tests using catalyst failed to establish a relationship between bubble size and TDH. In a later publication, Zenz (1983) provides a graph showing the variation of TDH as a function of superficial gas velocity and bubble size at the bed surface (Fig. 6.4). George and Grace (1978) did experiments which enabled them to calculate particle ejection velocities; 275 μm ballotini were found to be ejected at velocities two to three times the bubble velocity. Leva and Wen (1971) considered that stratification of fines within a bed was an essential requirement for entrainment; Geldart *et al.* (1981) showed that this may be of importance in beds of coarse solids containing small amounts of fines, but beds of fine particles which are extremely well mixed also have considerable entrainment rates. George and Grace (1978) obtained data which they interpreted as demonstrating that it is the bubble wake which is responsible for ejecting particles into the freeboard, but observations of the surface of two-dimensional beds indicate that in general it is the nose of bubble which ejects the particles; however, ejection from the wake can occur when a bubble bursts through the base of another which is itself about to break at the surface.

The picture is complicated even more by three other recent findings; Morooka, Kawazuishi, and Kato (1980) and Horio *et al.* (1980 report that the solids concentration in the freeboard is non-uniform, with a downward-moving denser stream of solids at the wall and a dilute upward-moving region in a central core; Yerushalmi *et al.* (1978) deduced that particles form clusters, or agglomerates, making it possible for fines to remain in the system even at very high velocities (see Chapter 7); and Geldart *et al.* (1979) and Geldart and Pope (1983) showed that fine particles interact with coarse in the freeboard, increasing the height to which *coarses* are lifted and causing the carryover of particles much larger than would be expected as judged by their terminal velocities.

It is therefore not surprising that entrainment and carryover cannot be calculated from first principles, nor that empirical correlations give predictions which differ by factors of over 100.

6.5 THEORETICAL BACKGROUND

Many workers have presented their results in terms of an elutriation rate constant, K_{ih}^*, where:

$$\left\{\begin{array}{l}\text{Instantaneous rate of}\\\text{removal of solids of}\\\text{size } d_i\end{array}\right\} = K_{ih}^* \left\{\begin{array}{l}\text{bed}\\\text{area}\end{array}\right\} \left\{\begin{array}{l}\text{fraction of bed}\\\text{consisting of size}\\d_{pi} \text{ at time } t\end{array}\right\} \tag{6.20}$$

$$\frac{d}{dt}(x_{Bi} M_B) = K_{ih}^* A x_{Bi} = E_{ih} A \tag{6.21}$$

where E_{ih} is the net carryover flux for a component of size d_{pi} when the gas offtake is at a height h above the bed surface.

For *continuous* operation, x_{Bi}, the concentration of size d_{pi} in the bed, and M_B, the mass of solids in the bed, are constant, and:

$$R_i = K^*_{ih} A x_{Bi} = E_{ih} A \qquad (6.22)$$

$$R_T = \Sigma R_i = \Sigma K^*_{ih} A x_{Bi} \qquad (6.23)$$

or $\qquad\qquad E_h = \Sigma E_{ih} = \Sigma K^*_{ih} x_{Bi} \qquad (6.23a)$

The solids loading of size fraction d_{pi} in the off-gases:

$$\rho_i = \frac{R_i}{UA} = \frac{E_{ih}}{U} \qquad (6.24)$$

and the total loading presented to the cyclone is:

$$\rho' = \Sigma \rho_i \qquad (6.25)$$

The concentration of size d_{pi} in the exit gas stream may be calculated from Eqs (6.22) and (6.23):

$$y_i = \frac{R_i}{R_T} = \frac{E_{ih}}{E_h} \qquad (6.26)$$

For *batch* operation, x_{Bi} changes with time and Eq. (6.21) must be integrated. Providing M_B does not change by more than about 15 per cent., $m_{i,t}$, the mass of size fraction d_{pi} carried over in time t increases asymptotically as:

$$m_{i,t} = x_{Bio} M_B \left\{ 1 - \exp \left(-K^*_{ih} \frac{At}{M_B} \right) \right\} \qquad (6.27)$$

where x_{Bio} is the initial concentration of size fraction d_{pi} in the bed.

Similarly, the concentration of size fraction d_{pi} in the bed declines exponentially according to:

$$x_{Bi} = x_{Bio} \exp \left(-K^*_{ih} \frac{At}{M_B} \right) \qquad (6.28)$$

As with any batch separation process, such as distillation, it becomes progressively more difficult to remove a given fraction as its concentration decreases.

6.6 EFFECT OF SYSTEM AND OPERATING VARIABLES ON K^*_{ih}

The rate of carryover cannot be predicted unless K^*_{ih} is known, and as yet K^*_{ih} cannot be calculated from first principles, though several hydrodynamic models have been proposed.

An approach originally proposed by Lewis, Gilliland, and Lang (1962) and developed by Kunii and Levenspiel (1969) postulated three phases in the freeboard; in the light of more recent work, we may now picture four (Fig. 6.8):

Phase 1. A gas stream carrying dispersed solids upwards.

Phase 2. Agglomerates ejected upwards by the action of bubbles.

Phase 3. Agglomerates falling downwards.

Phase 4. An emulsion, more concentrated than phase 1 moving downwards at the wall.

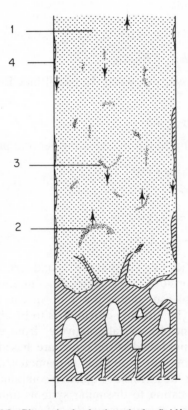

Figure 6.8 Phases in the freeboard of a fluidized bed:

There is mass transfer between these phases at rates which will depend on gas velocity, bubble action, particle size and size distribution, and bed diameter. Although algebraic equations can be written for this model, as yet the approach has not yielded any useful or realistic predictions.

For the present, the simplified approach of Large, Martini, and Bergougnou (1976) seems more promising. They picture the overall carryover flux

E_{ih}, for a given component of size d_{pi} as consisting of two partial fluxes:

(a) A continuous flux flowing upwards from bed to outlet (this corresponds to phase 1), $E_{i\infty}$.

(b) A flux of agglomerates ejected by bursting bubbles, which decreases exponentially as a function of freeboard height (this corresponds to the net algebraic sum of phases 2, 3 and 4).

Expressed algebraically:

$$E_{ih} = E_{i\infty} + E_{io} \, e^{-a_i h} \qquad (6.29)$$

where E_{io} is the component ejection flux $= E_o x_{bi}$ and

$$E_{i\infty} = K^*_{i\infty} x_{Bi}; \qquad E_{ih} = K^*_{ih} x_{Bi} \qquad (6.30)$$

The total solids carryover flux when the gas offtake is at any height h above the bed surface:

$$E_h = E_\infty + E_o \exp(-ah) \qquad (6.31)$$

Wen and Chen (1982) develop this idea further and propose:

$$E_h = E_\infty + (E_o - E_\infty) \exp(-ah) \qquad (6.32)$$

On a component basis:

$$E_{ih} = E_{i\infty} + (E_{io} - E_{i\infty}) \exp(-a_i h) \qquad (6.33)$$

$$\text{or } K^*_{ih} = K^*_{i\infty} + (E_o - K^*_{i\infty}) \exp(-a_i h) \qquad (6.33a)$$

If the column is sufficiently tall that the second term on the right-hand side of Eq. (6.33) can be considered negligible, a measurement at the exit can give $E_{i\infty}$, and two other measurements lower down can then be used to find exponent a_i and E_{io}. The ultimate objective is to be able to predict all these factors from first principles or, alternatively, from correlations. Unfortunately, many of the published correlations are based on data taken from columns which were too short, too small in diameter, or both; consequently, they correlate K^*_{ih}, not the intrinsic or ejection component rate constants, and this makes their application to dissimilar systems unreliable. However, we can make some general comments on the effect which various parameters have on the carryover of particles.

6.6.1 Size of Particles

Particles whose terminal velocities are less than the superficial gas velocity — the fines

Particles with $v_t/U < 1$ are the most likely to be carried over, and most researchers agree that as the size of the *fines* decreases, the component carryover rate constant, K_{ih}^*, increases, though closer examination of some data (George and Grace, 1978; Geldart, 1981; and Wong, 1983) suggests that K_{ih}^* may start to decrease again for fractions less than about 45 μm, possibly due to the tendency of cohesive fine particles to agglomerate into clusters.

Particles having $v_t > U$ – the coarses

Intuition suggests that particles with terminal velocities larger than the superficial gas velocity should not appear in the exit gas, or that if they do, the freeboard must be less than the TDH. However, *coarses* have been found in the cyclones of a commercial zinc roaster having a 9 m high freeboard (Avedesian, 1974) and of a fluid bed combustor with a 3.5 m freeboard (Merrick and Highley, 1972). Geldart and Pope (1983) investigated this phenomenon in a 0.3 m diameter bed 5 m tall and found that the size and quantity of *coarses* carried over depends strongly on the carryover rate of the *fines* and can be considerable at fluxes above 2 kg/m²s. The cause is believed to be collisions and momentum interchange between the fines and coarses; in effect, the TDH of the coarses, TDH(C), is a function of the *fines* carryover flux.

6.6.2   Concentration of Fines in the Bed

Equation (6.21) assumes that component carryover rates are directly proportional to their average concentration in the bed. Setting aside for the moment the question of segregation within the bed, Figs 6.9 and 6.10 show this assumption to be justified for both small (< 2 per cent) and large (> 20 per cent.) concentrations.

6.6.3   Type of Solids in the Bed

In powders which have a small mean size, say less than 300 μm, and a normal size distribution of particles of equal density, there is little evidence of segregation within the bed provided it is vigorously bubbling, and the solids presented to the freeboard by the bubbles have a composition similar to that of the bed as a whole. If, however, the mean size of the particles is large, say > 800 μm, and there is either a very wide or a bimodal distribution, then there will be vertical segregation, with more fines at the top of the bed even at high velocities (Geldart *et al.*, 1981). Similarly, if the bed is composed of solids having different densities, in general the top of the bed will be richer in the lighter solids. Under these circumstances carryover is likely to be higher than predicted by correlations based on a well-mixed bed of fine solids.

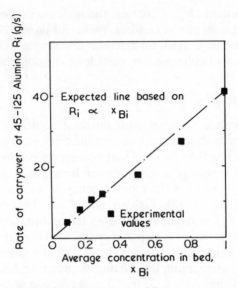

Figure 6.9 Carryover rate of fines versus average concentration of fines in the bed (from Geldart *et al.*, 1979).

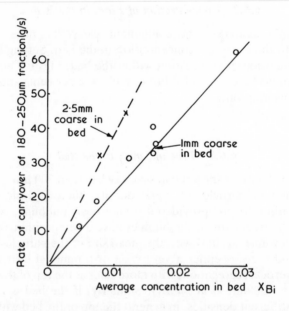

Figure 6.10 Influence of size of coarses in the bed on the carryover rate of fines (from Geldart *et al.*, 1981).

6.6.4 Gas Velocity

All correlations include the superficial gas velocity, usually in association with v_t, as $U - v_t$, $(U - v_t)/v_t$, $(U - v_t)/U$, or v_t/U. Certainly the component solids carryover flux is very sensitive to gas velocity, but although the dependency is of the form $K_{ih}^* \, \alpha \, U^n$ no single value can be assigned to the exponent which may range from 2 to 4 depending on the systems.

6.6.5 Operating Pressure

As pointed out in Section 6.3.3, terminal velocities are unaffected by gas density in the Stokes region, slightly in the transition region, and in the turbulent regime according to $\sqrt{(1/\rho_g)}$. This means that as the gas pressure is increased at constant gas velocity, the size of the particles carried over will increase slightly. More importantly, however, the rate of carryover has been found experimentally to be directly proportional to absolute pressure (May and Russel, quoted in Zenz and Weil, 1958).

6.6.6 Bed Diameter

Few data are available on the effect of scale-up; from a systematic study in beds from 0.019 to 0.146 m diameter Lewis, Gilliland, and Lang (1962) concluded that there was little change above 0.1 diameter but that the entrainment rate increased with decreasing diameter for diameters smaller than 0.04 m. These conclusions are based largely on data from beds of solids only 0.1 m deep and may not reflect a general trend, through Ryan (1970) found a similar effect of bed diameter below 0.15 m diameter. Wohlfarth (1979), fluidized catalyst in beds 0.46, 1.0 and 4 m in diameter. His results require careful interpretation as they are presented in dimensionless form, and conditions in each of the three beds were not identical. His data show that at equal velocities and distances above the bed surface, entrainment fluxes were approximately equal in both the 0.46 and 1 m diameter beds.

6.6.7 Internals in the Bed and Freeboard

There are relatively few reports on the effect which baffles, in and above the fluidized bed, have on entrainment. Geldart (1981) and Colakyan et al. (1979) found that the presence of tubes beneath and just above the bed surface had no discernible effects in beds of coarse solids containing fines. Martini, Bergougnou, and Baker (1976), working with fine sand, found that louvers situated just above the bed surface helped to reduce carryover of the larger fractions. If situated higher in the freeboard they could actually increase carryover. Similarly, filling the freeboard with tubes would increase the superficial gas velocity and increase carryover. Cylindrical mesh packings in the bed reduce carryover (due to smaller bubbles) but in the freeboard they increased carryover. (Tweddle, Capes and Osberg, 1970).

The most effective way to reduce carryover is to increase the diameter of the freeboard thus reducing the gas velocity.

6.7 CORRELATIONS FOR $K_{i\infty}^*$

In view of the number of variables and the complexity and interrelation of the mechanisms occurring in and above the bed, it is hardly surprising that the many published correlations predict widely different rates of carryover when applied to systems other than those on which they are based. Also there is no guarantee that the offtake was above TDH(F). A full discussion of the correlations is beyond the scope of this section, but Fig. 6.11 shows how much correlations differ and a list is given in Table 6.4. Another problem which complicates the prediction of carryover rates is that fines may be generated in the bed due to chemical, hydrodynamic, or mechanical breakdown of the solids; this makes it difficult to predict the concentration of any particular size fraction in the bed, and experimental work may be required if carryover is a crucial parameter. The results should then be compared with published correlations, and the one which agrees most closely used for subsequent scale-up calculations.

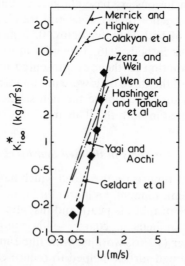

Figure 6.11 Comparison of experimental and predicted carryover rates (from George and Grace, 1978).

Table 6.4 Correlations for elutriation rate constant $K_{i\infty}^*$ (in S1 units)

Yagi and Aochi (1955):

$$\frac{K_{i\infty}^* \, g \, d_{pi}^2}{\mu(U - v_t)^2} = 0.0015 \, Re_t^{0.6} + 0.01 \, Re_t^{1.2}$$

Zenz and Weil (1958):

$$\frac{K_{i\infty}^*}{\mu_g U} = \begin{cases} 1.26 \times 10^7 \left(\dfrac{U^2}{g \, d_{pi} \, \rho_p^2} \right)^{1.88} & \text{when } \dfrac{U^2}{g \, d_{pi} \, \rho_p^2} < 3 \times 10^{-4} \\[3mm] 4.31 \times 10^4 \left(\dfrac{U^2}{g \, d_{pi} \, \rho_p^2} \right)^{1.18} & \text{when } \dfrac{U^2}{g \, d_{pi} \, \rho_p^2} > 3 \times 10^{-4} \end{cases}$$

Wen and Hashinger (1960):

$$\frac{K_{i\infty}^*}{\rho_g(U - v_t)} = 1.7 \times 10^{-5} \left\{ \frac{(U - v_t)^2}{g \, d_{pi}} \right\}^{0.5} Re_t^{0.725} \left(\frac{\rho_p - \rho_g}{\rho_g} \right)^{1.15} \left(\frac{U - v_t}{v_t} \right)^{0.1}$$

Tanaka *et al.* (1972):

$$\frac{K_{i\infty}^*}{\rho_g(U - v_t)} = 4.6 \times 10^{-2} \left\{ \frac{(U - v_t)^2}{g \, d_{pi}} \right\}^{0.5} Re_t^{0.3} \left(\frac{\rho_p - \rho_g}{\rho_g} \right)^{0.15}$$

Merrick and Highley (1974)

$$\frac{K_{i\infty}^*}{\rho_g \, U} = A + 130 \exp \left\{ - 10.4 \left(\frac{v_t}{U} \right)^{0.5} \left(\frac{U_{mf}}{U - U_{mf}} \right)^{0.25} \right\}$$

Geldart *et al.* (1979) revised (Eq. 6.34):

$$\frac{K_{i\infty}^*}{\rho_g U} = 23.7 \exp \left(- 5.4 \frac{v_t}{U} \right)$$

Colakyan *et al.* (1979):

$$K_{i\infty}^* \, (kg/m^2 s) = 33 \left(1 - \frac{v_t}{U} \right)^2$$

Wen and Chen (1982):

$$K_{i\infty}^* = \rho_p(1 - \epsilon_i)(U - v_t)$$

where

$$\epsilon_i = \left\{ 1 + \frac{\lambda \, (U - v_t)^2}{2gD} \right\}^{-1/4.7}$$

and λ is obtained from:

$$\frac{\lambda \rho_p}{d_{pi}^2} \left[\frac{\mu}{\rho_g} \right]^{2.5} = \begin{cases} 5.17 \left\{ \dfrac{\rho_g(U - v_t)d_{pi}}{\mu} \right\}^{-1.5} D^2 & \text{for } \dfrac{\rho_g(U - v_t)d_{pi}}{\mu} \leqslant \dfrac{2.38}{D} \\[3mm] 12.3 \left\{ \dfrac{\rho_g(U - v_t)d_{pi}}{\mu} \right\}^{-2.5} D & \text{for } \dfrac{\rho_g(U - v_t)d_{pi}}{\mu} \geqslant \dfrac{2.38}{D} \end{cases}$$

In the absence of an experimental unit of at least 0.1 m diameter, carryover must be predicted using correlations.

For the elutriation of particles less than about 100 μm from well-mixed beds fluidized at velocities up to about 1.2 m/s, the Zenz-Weil (1958) correlation probably gives the most consistently realistic values of K_i^*, particularly if the whole bed is potentially entrainable (i.e. even the largest particles have a terminal velocity less than the superficial gas velocity). Such materials as cracking catalyst, coal char, and other low density solids fall into this category.

For higher velocities and larger particles, averaging $K_{i\infty}^*$ predicted by the Zenz-Weil (1958) correlation and the revised Geldart *et al.* (1979) correlation will give values which are not unrealistic:

$$\frac{K_{i\infty}^*}{\rho_g U} = 23.7 \ e^{-5.4 v_t/U} \tag{6.34}$$

For beds of coarse solids containing recycled fines fluidized at up to 5 m/s (e.g. fines injected into combustors, granulators, or dryers) the carryover rate constant $K_{i\infty}^*$ of the fines ($v_t/U < 1$) can be predicted from Geldart's (1981) correlations.

The correlation is also of the form:

$$\frac{K_{i\infty}^*}{\rho_g U} = A \ e^{-B v_t/U} \tag{6.35}$$

where for a bed consisting largely of 1 mm solids:

$$A = 31.4; \qquad B = 4.27$$

and for 2.5 mm coarses in the bed:

$$A = 49.1; \qquad B = 4$$

Linear interpolation of A and B can be made for solids between 1 and 2.5 mm.

High accuracy should not be expected from any of these approaches and in the present state of knowledge agreement between experiment and prediction is unlikely to be better than $\pm$ 100 per cent.

6.8 CALCULATION OF CARRYOVER RATE

Assuming K_{ih}^* has been predicted from correlations for each size range (e.g. from Eq. 6.33a using E_o from the last equation in Table 6.1) or determined experimentally, calculation of the carryover rate requires a knowledge of the concentration of each size range in the bed. The general case is shown in Fig. 6.12 and many combinations are possible; for example:

(a) No feed; cyclone 100 per cent. efficient; total recycle of cyclone product

(b) Continuous feed; cyclone efficiency varies depending on d_{pi}; partial recycle of fines

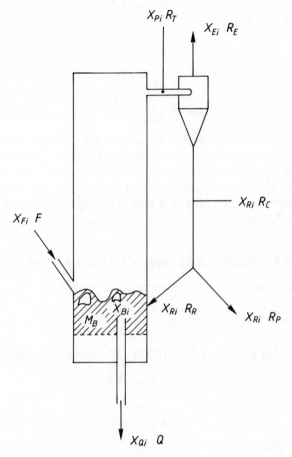

Figure 6.12 Component balance around a fluidized bed.

Whatever the arrangement, overall and component mass balances must be done. As an illustration, assume $R_E = R_R = 0$ and F and Q are non-zero. A mass balance on the size fraction d_{pi} gives:

$$x_{Fi}F = x_{pi}R_T + x_{qi}Q \tag{6.36}$$

An overall mass balance gives:

$$F = R_T + Q \tag{6.37}$$

Now:

$$x_{pi}R_T = K_{ih}^* A x_{Bi} = E_{ih}A \tag{6.38}$$

$$R_T = \Sigma K_{ih}^* A x_{Bi} \tag{6.39}$$

and in a well mixed bed:

$$x_{qi} = x_{Bi}$$

Substituting in Eq. (6.36) and rearranging:

$$x_{Bi} = \frac{x_{Fi}F}{K_{ih}^* A + F - R_T} \tag{6.40}$$

This equation cannot be solved directly because, as Eq. (6.39) shows, R_T depends on the value of x_{Bi} for each size fraction. In practice a rapidly converging trial and error loop can be set up, with $R_T = 0$ for the first trial. If the freeboard height is larger than about 2 m, and the column diameter exceeds about 0.2 m, then for size fractions having $v_t/U < 0.5$, $K_{ih}^* \approx K_{i\infty}^*$, this simplifies and shortens the calculation significantly.

6.9   PARTICLE RESIDENCE TIMES IN THE BED

The residence time of the size fraction d_{pi} in the bed is:

$$t_{Ri} = \frac{x_{Bi}M_B}{x_{Fi}F} \tag{6.41}$$

Substituting from Eqs (6.36) and (6.37) and remembering that

$$M_B = \rho_{Bmf} A H_{mf} \tag{6.42}$$

$$t_{Ri} = \frac{\rho_{Bmf} A H_{mf}}{K_{ih}^* A + F - R_T} \tag{6.43}$$

For very fine particles:

$$K_{ih}^* A \gg F - R_T \quad \text{and } K_{ih}^* \approx K_{i'}^*$$

so:

$$t_{Ri} \approx \frac{\rho_{Bmf} H_{mf}}{K_{i\infty}^*} \tag{6.44}$$

That is, the residence time is independent of the concentration and directly proportional to the bed depth.

For coarse solids:

$$K_{ih}^* \ll F - R_T \quad \text{and} \quad t_{Ri} \approx \frac{\rho_{Bmf} A H_{mf}}{F} \tag{6.45}$$

6.10   CONCLUSION

There is still a long way to go before carryover rates can be predicted with real confidence. Cold model experiments in columns at least 0.1 m in diameter can be of assistance but there is no substitute for making measurements on the plant. Unfortunately, due to the different perceptions and objectives of production and development engineers, these are never easy to make and the data are frequently incomplete or unsatisfactory.

In addition to the influence of the quantifiable variables discussed in Section 6.6, other parameters often assume importance in a bed operating at high temperature and/or pressure: e.g. production of fines by mechanical, thermal, and chemical attack on the particles at the grid, in the bed, and in the cyclones; particle stickiness caused by deposition and softening of impurities on the surface; blocking of cyclone diplegs; leakage around the distributor which gives mal-distribution of gas. These other effects may occur in many fluid beds but only become apparent when there is a serious increase in the rate of solids carryover.

6.10   EXAMPLE ON ELUTRIATION

A powder is fed to a fluidized bed 0.5 m in diameter at the rate of 540 kg/h. The column, which is 3 m tall from distributor to gas offtake, contains 200 kg of solids having a settled bulk density of 1,450 kg/m^3 and a particle density of 2,630 kg/m^3. The powder is to be fluidized at 1 m/s at 300°C by a gas of density 0.61 kg/m^3 and viscosity 2.95×10^{-5} N s/m^2. The size distribution of the feed solids is given in Table 6.5. Estimate the solids loading on the cyclone if (a) none, (b) all of the elutriated solids are returned to the bed.

Table 6.5   Composition of feed.

BS Mesh	Sieve size d_p, μm	Weight, %
− 44 + 60	300	3.3
− 60 + 85	215	22.8
− 85 + 100	165	16.4
−100 + 120	137	20.3
−120 + 150	115	16.8
−150 + 200	90	13.3
−200 + 350	60	5.2
−350	(30)	1.9

Assumptions

Particles are sufficiently rounded that $d_v \approx d_p$. The bed is well mixed and the solids removed from the bottom of the bed have the same composition as the bed.

Step 1. Preliminary calculations:

(a) From Eq. (2.8), d_p = 125 μm.

(b) From Eq. (2.27), U_{mf} = 0.01 m/s.

(c) From data given, static bed depth = 0.7 m and this is approximately equal to H_{mf}.

(d) We need to know the height of the freeboard when the bed is in operation. Using the same approach illustrated in Example 4.4, the bubble size at the bed surface $d_{eq,s}$ is calculated to be 0.5 m, which means that the bed would be near slugging under ambient conditions. The average expanded bed depth is calculated (Eq. 4.7) to be 1.5m, giving a freeboard height of about 1.5 m.

(e) Critical particle size corresponding to U = 1 m/s is calculated using Eq. (6.13) and Fig. 6.7 to be 164 μm, so almost 70 per cent of the solids are potentially elutriable.

Step 2. Estimate TDH. Since most of the particles are elutriable we are interested in TDH(F). Values calculated using equations in Table 6.1 are listed below. Correlations of Horio and Wen and Chen require a knowledge of bubble size at the bed surface $d_{eq,s}$ which in this case has been calculated as 0.5 m.

	TDH(F), m
Fournol *et al.*	102
Horio *et al.*	3.2
Amitin	6.3
Zenz and Weil	4.3

Clearly the available freeboard height of 1.5 m is below TDH(F) and (a) particles larger than the critical size (164 μm) will be carried over, (b) the carryover rate will be larger than the value based on $K^*_{i\infty}$.

Step 3. Since the powder will be well mixed and d_p is less than 300 μm, calculate $K^*_{i\infty}$ for each size fraction from Zenz-Weil (1958) and Geldart *et al.* (1979) revised, and take the average (Table 6.6).

Step 4. Calculate equilibrium concentration of each size fraction in the bed using Eq. (6.40). In order to simplify the calculation it will be assumed that $K^*_{ih} = K^*_{i\infty}$. This will enable us to calculate values of x_B; $E_{i\infty}$ can then be found (Eq. 6.30). In the first trial assume R_T = 0:

$$x_{Bi(1)} = \frac{x_{Fi} F}{K_{i\infty}^* A + F}$$

e.g. for a 30 μm fraction, $x_{Fi} = 0.0188$ and $K_{i\infty}^* = 7.34$ kg/m^2 s;
$F = 540$ kg/h $= 0.15$ kg/s; $A = 0.196$ m^2. Thus:

$$x_{Bi(1)} = \frac{0.15 \times 0.0188}{7.34 \times 0.196 + 0.15} = 0.00177$$

This is done for each size fraction (see Table 6.7, column 3) and so the first estimate of R_T is found. This is then inserted into Eq. (6.4;) and the second estimate of x_T is found. Three trials are sufficient; it can be seen from the equilibrium values of x_{Bi} that compared with the feed, the bed is deficient in the finest sizes and richer in the coarser.

Table 6.6 Calculation of elutriation rate constants $K_{i\infty}^*$

d_{pi}, μm	30	60	90	115	137	165	195
v_t, m/s	0.044	0.19	0.35	0.55	0.74	1	1.24
$K_{i\infty}^*$, kg/m^{2}s (Zenz and Weil, 1958)	3.28	1.26	0.59	0.37	0.27	0.19	0.13
$K_{i\infty}^*$, kg/m^{2}s (Geldart et al., 1979)	11.4	5.18	2.18	0.74	0.27	0.065	0.018
$K_{i\infty}^*$ (avg), kg/m^{2}s	7.34	3.22	1.38	0.56	0.27	0.13	0.07

Table 6.7

d_p, μm	x_{Fi}	x_{Bi} (First trial)	$E_{i\infty} A$, kg/s $\times 10^{-3}$	x_{Bi} (Second trial)	$E_{i\infty} A$, kg/s $\times 10^{-3}$	x_{Bi} (Third trial)	$E_{i\infty} A$, kg/s $\times 10^{-3}$
30	0.0188	0.00177	2.55	0.00183	2.63	0.00183	2.63
60	0.0524	0.01	6.31	0.01	6.31	0.01	6.31
90	0.1328	0.0473	12.8	0.053	14.3	0.054	14.6
115	0.1684	0.0972	10.6	0.117	12.8	0.121	13.3
137	0.203	0.15	7.94	0.191	10.0	0.2	10.6
165	0.164	0.14	3.57	0.187	4.76	0.197	5.0

$R_T = E_\infty A = 43.8 \times 10^{-3}$ $R_T = 50.8 \times 10^{-3}$ $R_T = 52.4 \times 10^{-3}$

Step 5.

(a) *From Table 6.7, $E_\infty \approx 52.4 \times 10^{-3}/0.196 = 0.267$ kg/m²s.*

(b) From the last equation in Table 6.1, the total ejection flux at the bed surface, E_0, is calculated to be 35.4 kg/m²s, using $d_{eq,s} = 0.5$ m (step 1d.)

(d) From Eq. (6.32) at $h = 1.5$ m and taking $a = 4$ m⁻¹:

$$E_h = 0.267 + (35.4 - 0.267)\, e^{-4 \times 1.5}$$
$$= 0.354 \text{ kg/m}^2\text{s}$$

Therefore total carried over $= 0.354 \times 0.196 = 0.0694$ kg/s. Of 0.15 kg/s fed in, 0.0694 kg/s (46 per cent.) is elutriated and the solids loading to the external cyclone is:

$$\frac{0.0694}{1 \times 0.196} = 0.35 \text{ kg/m}^3$$

Step 6. If all the solids from the cyclone are returned, the bed will have the same composition as the feed and the loading on the cyclone is calculated to be 0.746 kg/m³.

6.12 NOMENCLATURE

A	cross-sectional area of bed at surface	m²
a, a_i	overall and component decay constants	m⁻¹
Ar	Archimedes number	—
$d_{eq,s}$	equivalent volume diameter of a bubble at bed surface	m
d_p	mean sieve size of a powder	m or μm
d_{pi}	arithmetic mean of adjacent sieve apertures	m
d_v	volume diameter	m
D	diameter of bed	m
E_h	entrainment flux of solids at gas offtake $(= \Sigma\, E_{ih})$	kg/m²s
E_{ih}	entrainment flux of size fraction d_{pi} at gas offtake $(= x_{pi}\, E_h)$	kg/m²s
E_{io}	entrainment flux of size fraction d_{pi} at bed surface $(= x_{Bi}\, E_o)$	kg/m²s
$E_{i\infty}$	entrainment flux of size fraction d_{pi} when offtake is above TDH(F) $(= x_{pi}\, E_\infty)$	kg/m²s

E_o	entrainment flux of solids at bed surface $(= \Sigma\, E_{io})$	kg/m²s
E_∞	entrainment flux of solids when offtake is above TDH(F) $(= \Sigma\, E_{i\infty})$	kg/m²s
F	rate at which fresh solids are added to bed	kg/s
g	acceleration of gravity	9.81 m/s²
h	distance above bed surface	m
H_s	height of settled bed	m
$K_{i\infty}^*$	elutriation rate constant for size fraction d_{pi} when $h >$ TDH(F)	kg/m²s
K_{ih}^*	elutriation rate constant for size fraction d_{pi} when offtake is at height h	kg/m²s
$m_{i,t}$	mass of solids size d_i carried out of bed in time t	kg
M_B	mass of solids in bed	kg
Δp	pressure drop across a section of the freeboard	kg/m²
Q	rate at which solids are removed from bed	kg/s
R_i	rate of removal of size fraction d_{pi} by gas	kg/s
R_T	rate of removal of solids by gas	kg/s
Re_t	terminal velocity Reynolds number $\rho_g\, v_t d_v/\mu$	—
Re_p	superficial velocity Reynolds number $\rho_g\, U d_p/\mu$	—
t, t_{Ri}	time, residence time of size d_{pi}	s
TDH (C)	distance above bed surface required for disengagement of all particles having $v_t > U$	m
TDH(F)	distance above bed surface beyond which entrainment rate becomes relatively unchanging	m
TDH(F)$_i$	distance above bed surface beyond which rate for size d_{pi} becomes relatively unchanging	m
U	superficial gas velocity	m/s
v_t	freefall (terminal) velocity of an isolated particle	m/s
x_{Bi}	equilibrium concentration of size d_{pi} in bed	—
x_{Bio}	initial concentration of size d_{pi} in batch system	—
x_{Bit}	concentration of size d_{pi} in batch bed after time t	—
x_{Ei}	concentration of size d_{pi} in solids leaving the cyclone with the gas	—
x_{Fi}	concentration of size d_{pi} in feed entering the bed	—
x_{pi}	concentration of size d_{pi} in solids entering the cyclone	—

x_{Ri}	concentration of size d_{pi} in fines returned to the bed from the cyclone	—
y_i	mass fraction of size fraction d_{pi} in carryover	
ϵ	voidage	—
ϵ_{mf}	voidage in bed at incipient fluidization	—
μ	gas viscosity	kg/ms
ρ_f	density of fluid	kg/m^3
ρ_g	density of gas	kg/m^3
ρ_i	solids loading of size fraction d_{pi} in offtake gas	kg solids/m^3
ρ_p	density of particle, including pores	kg/m^3

6.13   REFERENCES

Allen, T. (1981). *Particle Size Measurement*, 3rd ed., Chapman Hall.
Amitin, A.V., Martyushin, I.G., and Gurevich, D.A. (1968). *Khim. Tk. Top. Mas.*, **3** 20.
Avedesian, M.M. (1974). Private commumication.
Becker, H. (1959). *Can. J. Chem. eng.*, **37**, 85.
Clift, R., Grace, J.R., and Weber, M.E. (1978). *Bubbles, Drops and Particles*, Academic Press.
Colakyan, M., Jovanovic, G., Catipovic, N., and Fitzgerald, T. (1979). *AIChE* 72nd Ann. Meeting, Los Angeles.
Fouda, A.E., and Capes, C. (1976). *Powder Technol.*, **13**, 291.
Geldart, D. (1981). 'Behaviour of fines in a fluidized bed of coarse solids', Electric Power Research Institute, Palo Alto, Final Report EPRI CS2094.
Geldart, D., Baeyens, J., Pope D.J., and Van de Wijer, P. (1981). *Powder Technol.*, **30**, 195.
Geldart, D., and Pope, D. (1983). *Powder Technol.*, **34**, 95.
George, S., and Grace, J.R. (1978). *A.I.Ch.E. Symp. Ser.*, **74** (176), 67.
Heywood, H. (1962). *Proc. of Symp. on Interaction between Fluids and Particles*, Inst. Chem. Engrs (Lond.), p.1.
Horio, M., Taki, A., Hsieh, Y.S., and Muchi, I. (1980). In *Fluidization* (Eds J.R. Grace and J.M. Matsen), Plenum Press, p. 509.
Kunii, D., and Levenspiel, O. (1969). *Fluidization Engineering*, J. Wiley, New York.
Large, J.F., Martini, Y., and Bergougnou, M.A. (1976). Int. Powder and Bulk Handling Conf., Chicago.
Leva, M., and Wen, C.Y. (1971). In *Fluidization* (Eds J.F. Davidson and D. Harrison), Academic Press, London.
Lewis, W.K., Gilliland, E.R., and Lang, P.M. (1962). *Chem. Eng. Progr. Symp. Series, 58* (38), 65.
Martini, Y., Bergougnou, M.A., and Baker, C.G.J. (1976). In *Fluidization Technology* (Ed. D.L. Keairns), vol . 2, Hemisphere, Washington.
Merrick, D., and Highley, J. (1972). *A.I.Ch.E. Symp. Ser.*, **70**, (137), 366.
Morooka, S., Kawazuishi, K., and Kato, Y. (1980). *Powder Technol.*, **26**, 75.

Pettyjohn, E.A., and Christiansen, E.B., (1948). *Chem. Eng. Progr.,* **44**, 157.
Ryan, W.J. (1970). *Chemeca 70,* Chemical Eng. Conf. Australia, Butterworh and Co. (Australia) Ltd.
Schillar, L., and Nauman, A. (1933). Z. *Ver. Dtsch. Ing.* **77**, 318.
Tanaka, I., and Shinohara, H. (1978). *Int. Chem. Eng.,* **18**, No. 2, 276.
Tweddle, T.A., Capes, C.E., and Osberg, G.L. (1970). *Ind. Eng. Chem. Proc. Des. Dev.,* **9**, 85.
Wen. C.Y., and Chen, L.H., (1982). *A.I.Ch.E.J.,* **28**, 117.
Wen, C.Y., and Hashinger, R.F., (1960). *A.I.Ch.E.J.,* **60**, 220.
Wohlfarth, W. (1979). Private communication.
Wong, A.C.Y. (1983). Ph. D. Dissertation, University of Bradford.
Yagi, S. and Aochi, T., (1955). Paper presented at Soc. Chem. Engrs. (Japan) Spring Meeting.
Yerushalmi, J., Cankurt, N., Geldart, D., and Liss, B. (1978). *A.I.Ch.E. Symp. Ser.,* **74** (176), 1.
Zenz, F.A. (1983). *Chem. Engng.,* Nov. 28.
Zenz, F.A., and Weil, N.A. (1958). *A.I.Ch.E.J.,* **4**, 472.

Gas Fluidization Technology
Edited by D. Geldart
Copyright © 1986 John Wiley & Sons Ltd.

CHAPTER 7

High Velocity Fluidized Beds

J. YERUSHALMI

7.1 INTRODUCTION

Consider a bubbling fluidized bed (Fig. 7.1a). As the velocity of the gas is slowly raised, the heterogeneous, two-phase character of the bed first peaks, then gradually gives way to a condition of increasing uniformity culminating in the *turbulent* state (Fig. 7.1b) in which large discrete bubbles or voids are on the whole absent. In the turbulent fluidized bed, there is an upper bed surface, though it is considerably more diffuse than in a bubbling fluidized bed because of the greater freeboard activity attending operation at higher gas velocities,

The turbulent regime extends to the so-called *transport velocity*. As the

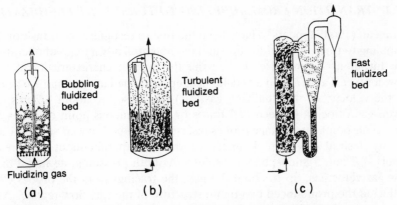

Figure 7.1 Schematics of (a) bubbling, (b) turbulent, and (c) fast fluidized beds.

155

transport velocity is approached, there is a sharp increase in the rate of particle carryover, and in the absence of solid recycle, the bed would empty rapidly. Beyond the transport velocity, particles fed to the bottom of the column or vessel traverse it in fully entrained transport flow, and the concentration or density of the resulting suspension depends not only on the velocity of the gas but also on the flowrate of the solids. If the solids are fed to the column at a sufficiently high rate — e.g. by circulating solids carried over from the column to its bottom via external cyclones and a standpipe — then it is possible to maintain in the column a relatively large solids concentration typical of the *fast fluidized bed* condition (Fig. 7.10).

The turbulent and the fast fluidized beds which are the subject of this chapter are referred to as *high velocity fluidized beds*. As their names suggest, these regimes lie beyond the bubbling range, yet they preserve all the underlying elements associated with the fluidized state (e.g. uniformity of temperature throughout the bed's volume).

Fundamental research into the underlying phenomena of high velocity fluid beds and systematic compilation of design data lag well behind commercial applications which are developing rapidly. This is due in part to some real technical difficulties which stand in the way of research in this area. Though experiments in small equipment may be revealing and useful, all fluidization research requires, at some point, sufficiently large scale, and this is equally true for turbulent and fast beds as for bubbling systems. High gas velocities and large scale fast beds also require a tall structure which may rise beyond 10 m. Nevertheless, the last few years have seen an increase in the number of investigators attracted to this area.

In this chapter the body of existing data on high velocity fluidizing beds is presented and discussed. Attention is drawn to unknowns and uncertainties and, accordingly, to research needs.

7.2   TRANSITION FROM BUBBLING TO TURBULENT FLUIDIZATION

Lanneau (1960) appears to have been the first to recognise the transition from bubbling to turbulent fluidization and the potential advantages of operating in the turbulent regime. He studied the fluidization characteristics of a fine powder (see Table 7.1) in a bed 4.6 m deep contained in a column 7.62 cm i.d. as the velocity of the fluidizing gas (air) was raised to about 1.7 m/s. At a given gas velocity, he recorded traces of instantaneous point densities from two small-point capacitance probes inserted in the bed spaced vertically apart at any desired elevation. Lanneau conducted experiments at two pressure levels, 1.7 and 5 atmospheres absolute. At both pressures, he noted that at low gas velocities, up to about 0.3 m/s, the tracings from the probes clearly reflected the pronounced two-phase structure of the slug flow regime. As the velocity was raised above 0.3 m/s, the heterogeneous character of the bed

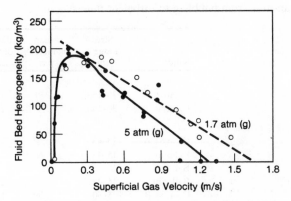

Figure 7.2 Lanneau's heterogeneity — the mean deviation of the instantaneous point density from the time-average bed density (from Lanneau, 1960).

gradually changed to a condition of increasing homogeneity. Lanneau suggested that the change was accompanied by the breakdown of slugs into smaller bubbles containing a higher proportion of solid particles. At gas velocities beyond about 1 m/s, the tracings indicated, in his words, that a condition of 'almost uniform or "particulate" fluidization' was approached.

To represent the degree of non-uniformity of a fluid bed, Lanneau defined the parameter 'heterogeneity' as the mean deviation of the instantaneous point density from the time-averaged bed density (Fig. 7.2). Noting that in his experiments the heterogeneity peaked at around 0.3 m/s, he argued that this point was the worst from the standpoint of the efficiency of contact between gas and solid. The more uniform regime that lay beyond a velocity of 0.3 m/s promised a greater degree of interaction between gas and solid, and a higher efficiency of contact.

Lanneau's paper seems to have gone generally unnoticed. More than a decade passed before Kehoe and Davidson (1971), working in transparent equipment, described the transition from bubbling (slugging, in fact) to turbulent fluidization, and it remained for Massimilla (1975) to provide a laboratory demonstration of the higher contacting efficiency of the turbulent fluidized bed (Fig. 7.3).

What Kehoe and Davidson saw was a breakdown of the slugging regime into a 'state of continuous coalescence — virtually a channelling state with tongues of fluid darting in zigzag fashion through the bed'. They used the word 'turbulent' to describe this state. Working with several powders and utilizing capacitance probes, they reported that the transition to turbulence occurred at:

$$U > 3v_{t,min} \tag{7.1}$$

158

Table 7.1 Transition from bubbling to turbulent fluidization;

Solids	Size range, μm	Mean diameter[a], μm	Particle density, g/cm^3	Particle sphericity
Catalyst	30–150	70	2.0	NA
Catalyst	30–150	70	2.0	NA
Catalyst A	15–43	22[c]	1.1	NA
Ballotini	15–43	22[c]	2.2	NA
Ballotini	15–43	22[c]	2.2	NA
Ballotini	15–43	22[c]	2.2	NA
Catalyst B	15–90	26[c]	1.1	NA
Catalyst B	15–90	26[c]	1.1	NA
Catalyst B	15–90	26[c]	1.1	NA
Catalyst C	40–90	55[c]	1.1	NA
Catalyst C	40–90	55[c]	1.1	NA
Catalyst C	40–90	55[c]	1.1	NA
Catalyst	NA	50	1.0	NA
Dicalite	0–160	33	1.67	0.4
FCC[e]	0–130	49	1.07	1.0
HFZ–20	0–130	49	1.45	1.0
HFZ–20	0–130	49	1.45	1.0
Alumina	40–200	103	2.46	1.0
Alumina	40–200	103	2.46	1.0
Sand	80–670	268	2.65	0.8
Sand	80–670	268	2.65	0.8
Glass	105–210	157	2.42	1.0
Glass	550–750	650	2.48	1.0
Glass	2,500–2,700	2,600	2.92	1.0
FCC	0–180	60	0.93	NA
FCC	0–180	60	0.93	NA
FCC	0–180	60	0.93	NA
Catalyst	NA	60	0.94	NA
Alumina	NA	95	1.55	NA
Ludox	NA	60	1.4	NA

[a] Unless otherwise indicated, the volume surface mean diameter, d_p is used.

[b] C = capacitance probes; X = x-ray photography; PDF = pressure drop fluctuations; PT = pressure transducers; V = visual.

[c] The mean diameter was computed from 1/(0.5/smallest cut + 0.5/largest cut).

Summary of test conditions and results

Experimental methods[b]	Apparatus	Pressure level (atm)	v_t, cm/s	U_c, cm/s	U_k cm/s	U_k/v_t
Lanneau (1960)						
C	7.5 cm i.d.	1.7	27	30	Uncertain	—
C	7.5 cm i.d.	5.0	21	30	Uncertain	—
Kehoe and Davidson						
X	10 cm i.d.	1.0	1.6	—	11.0	6.8
X	10 cm i.d.	1.0	3.2	—	35.0	11.0
C	5 cm i.d.	1.0	3.2	—	40.0	12.5
C	6.2 × 0.6 cm	1.0	3.2	—	35.0	11.0
C,X	10 cm i.d.	1.0	2.3	—	18.0	7.8
C	6.2 × 0.6 cm	1.0	2.3	—	17.0	7.4
C	10 cm i.d.[d]	1.0	2.3	—	32.0[d]	14.0
X	10 cm i.d.[d]	1.0	10.0	—	44.0	4.4
C	5 cm i.d.	1.0	10.0	—	50.0	5.0
C	10 cm i.d.[d]	1.0	10.0	—	≫ 50.0[d]	≫ 5.0
Massimilla (1973)						
C	15.6 cm i.d.	1.0	9.5	—	30–40.0	3.1–4.2
The City College studies						
PDF	15.2 cm i.d.	1.0	2.3	53	107.0	47.1
PDF	15.2 cm i.d.	1.0	7.8	61	61.0	7.8
PDF	15.2 cm i.d.	1.0	10.6	91	137.0	12.9
PT	5 × 51 cm	1.0	10.6	61	107.0	10.9
PDF	15.2 cm i.d.	1.0	58.2	122	274.0	4.7
PT	5 × 51 cm	1.0	58.2	137	255.0	4.4
pdf	15.2 cm i.d.	1.0	116.0	274	550.0	4.7
PT	5 × 51 cm	1.0	116.0	150–215	Uncertain[f]	—
PT	5 × 51 cm	1.0	95.0	150–215	Uncertain	—
Canada, McLaughlin, and Staub						
PT,C	0.3 × 0.3 m and 0.61 × 0.61 m	1–10	Varies with pressure	NA		0.5–0.65
PT,C	0.3 × 0.3 m and 0.61 × 0.61 m	1–10	Varies with pressure	NA		0.3–0.35
Thiel and Potter (1977)						
V	5.1 cm i.d.	1.0	10.0	—	41	4.1
V	10.2 cm i.d.	1.0	10.0	—	22	2.2
V	21.8 cm i.d.	1.0	10	—	22.5	0.25
Carotenuto, Crescitelli, and Donsi (1974), Crescitelli et al. (1978)						
PT,C	15.2 cm i.d.	1.0	10.0	—	20.0	2.0
PT,C	15.2 cm i.d.	1.0	36.0	—	100.0	2.8
PT,C	15.2 cm i.d.	1.0	15.0	—	33.0	2.2

[d] The bed depth was 225 cm; in all other experiments conducted by Kehoe and Davidson, bed height ranged below 270 cm.

[e] FCC = fluid cracking catalyst.

[f] Owing to wall effects.

where $v_{t,min}$ is the terminal velocity of particles belonging to the smallest cut.

Since then, the transition from bubbling to turbulent fluidization has been the subject of a number of investigations (Carotenuto, Crescitelli, and Donsi, 1974; Thiel and Potter, 1977; Crescitelli *at al.*, 1978; at General Electric, Canada, McLaughlin, and Staub, 1976, and Staub and Canada 1978; and at the City College of New York, Yerushalmi, Turner, and Squires, 1976, Cankurt and Yerushalmi, 1978, Turner, 1978, and Yerushalmi and Cankurt 1978, 1979). The test conditions, experimental methods, and the results of these investigations are summarised in Table 7.1.

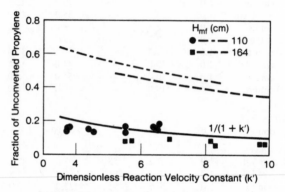

Figure 7.3 Ammonoxidation of propylene to acrylonitrile in a turbulent fluid bed (from Massimilla, 1973). $k' = k H_{mf}/U$ where k is the reaction velocity constant (s^{-1}). The upper dashed curves are predictions based on a slug flow model (Hovmand and Davidson, 1971).

Much of the work focused on fine powders belonging to Geldart's group A. The City College studies also encompassed solids belonging to group B, but the data often showed some scatter owing to strong wall effects in the form of piston-like movement of solid and some channelling in the two-dimensional bed. Still more severe wall effects precluded altogether experiments with solids coarser than sand with a mean size of 268 μm (see Table 7.1).

Larger equipment at General Electric afforded tests with coarse solids belonging to Geldart's group D, uniformly sized glass beads 650 and 2,600 μm in diameter. One of the experimental units, with a square cross-section measuring 0.3 by 0.3 m, was designed for tests at pressures up to 10 atms. With the work oriented towards study of fluid bed boilers, extensive tests were also conducted with tube banks immersed in the beds.

Even a casual inspection of Table 7.1 reveals a lack of coherence and several apparent inconsistancies; all the work to date does point to several conclusions.

7.2.1 Transition Velocities

The transition from bubbling to turbulent fluidization is gradual and spans a range of gas velocities which depend on the properties of gas and solids and also on equipment scale. A consistent delineation of the stages of the transition has not yet been established. Nor is it certain whether the transition can be clearly defined. The City College investigators, measuring pressure

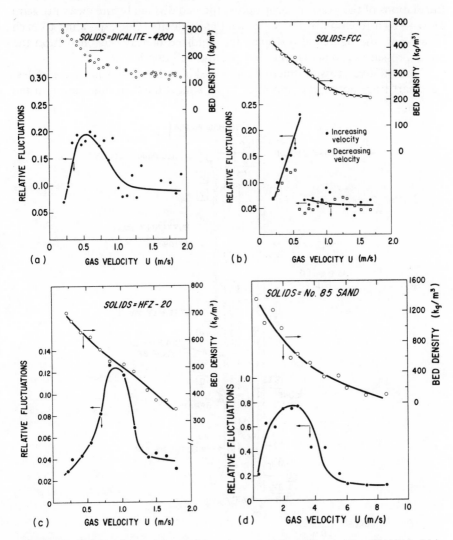

Figure 7.4 Pressure fluctuations relative to the mean pressure drop across a fluid bed of (a) Dicalite, (b) fluid cracking catalyst (FCC), (c) HFZ-20, and (d) sand. The upper curve gives the corresponding mean pressure drop across the bed.

fluctuations, observed that the transition occurs between two velocities: U_c, the velocity at which the pressure fluctuations peak, and U_k, the velocity at which the fluctuations, having decayed from the peak value, begin to level off. U_k marks the end of the transition and the onset of the turbulent regime.

Results for four solids are illustrated in Fig. 7.4. The data were obtained in a 15.2 m diameter column incorporated in the apparatus shown schematically in Fig. 7.5. Figure 7.4 gives the relative pressure fluctuations — i.e. the fluctuations of the pressure drop across the bed divided by the mean pressure drop — as a function of the superficial gas velocity. Corresponding to each data point, the figures also display the fluidized density reckoned from the mean pressure drop across the bed and the bed depth.

These relative pressure fluctuations are akin to Lanneau's heterogeneities, but apart from noting the approach to an almost uniform fluidization in the

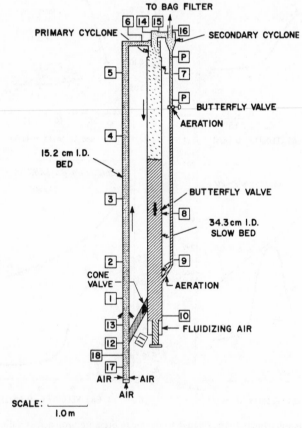

Figure 7.5 Schematic of the City College 15.2 cm fluidization system.

range of 1 to 1.7 m/s, Lanneau does not recognize any clean onset of a different regime.

Like the City College investigators, Canada, McLaughlin, and Staub (1976) characterize the transition in terms of the relative pressure fluctuations and similarly mark the onset of the turbulent regime, U_k, as the velocity at which the fluctuations begin to level off. With the exception of Thiel and Potter the remaining investigators cited in Table 7.1 report only a single transition velocity while stressing the progressive character of the transition. Kehoe and Davidson note that the transition to turbulent fludization was chosen rather arbitrarily as 'the velocity above which it was impossible to detect slugs' on traces from capacitance probes. In a similar fashion, Massimilla, Carotenuto et al., and Crescitelli et al., report the velocity at which the traces from capacitance probes no longer exhibit any strong and clear periodic element as the onset of the turbulent regime.

The decay of the relative pressure fluctuations from their peak, on the one hand, and the gradual loss of a strong periodic component in the traces from capacitance probes, on the other, reflect essentially the same physical phenomena, the same progressive changes in the structure of the bed. For this reason, I have listed in Table 7.1 under U_k all the transition velocities reported by Kehoe and Davidson, Massimilla, Carotenuto et al., and Crescitelli et al. I have similarly listed the results of Thiel and Potter although, apart from stating on the basis of visual observations that the transition velocity marked the breakdown of slugs into smaller bubbles, these authors are vague on precisely how the transition was determined.

It is apparent marking off U_k from either the relative pressure fluctuations or the traces from capacitance probes entails a large degree of arbitrary judgement. Canada, McLaughlin, and Staub, plotting the superficial gas velocity U/ϵ versus U (Fig. 7.6), noted a clear change of slope near the transition velocity. A different approach was taken by Kehoe and Davidson and Crescitelli et al. Both utilized capacitance probes to deduce the bubble or lean phase velocity in narrow beds of fine powders undergoing transition from slugging to turbulent fluidization. The results are shown in Figs 7.7 and 7.8. In both cases, the experimental results are compared to the familiar expression:

$$u_A = U - U_{mf} + 0.35(gD)^{1/2} \tag{7.2}$$

arising from the two-phase theory. Crescitelli et al. also compare the results with:

$$u_A = U - U_{mf} + 0.35(2gD)^{1/2} \tag{7.3}$$

In both cases, the actual lean phase velocities exceed the rise velocity predicted by the two-phase theory (a point to which I shall return presently), and there is a sharp change in slope at U_k. However, there is one essential difference between the results of Kehoe and Davidson, and those of Cresitelli

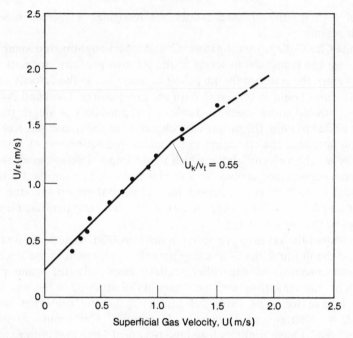

Figure 7.6 Correlation of the expansion of a bed of 650 μm glass particles at a pressure of 10 atm (from Canada, McLaughlin, and Staub 1976).

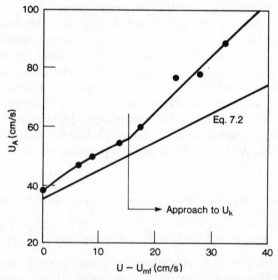

Figure 7.7 Lean phase velocities in a fluid bed undergoing transition to turbulent fluidization (from Kehoe and Davidson, 1971). Experiments were conducted in a 10 cm column with catalyst B (see Table 7.1).

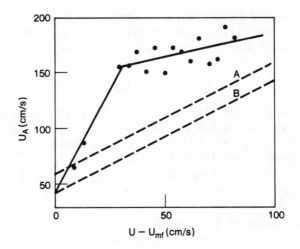

Figure 7.8   Lean phase velocities in a fluid bed undergoing transition to turbulent fluidization (from Crescitelli *et al.*, 1978). Experiments were conducted with Ludox catalyst (see Table 7.1).

et al. The former show an increase in the lean phase velocity past U_k; the latter show a decrease. An inconsistency? Or are the two transition velocities somehow dissimilar?

7.2.2   The ratio U_k/v_t

The last column in Table 7.1 shows the ratios of the transition velocity U_k to the terminal velocity v_t of a particle of size d_p. The ratios are strongly dependent on the particle's size and density. The ratios approach and even exceed 10 for the finer solids, and decrease sharply with particle size and density. For the coarse glass beads tested by Canada, McLaughlin, and Staub, U_k/v_t falls below unity and lies in the range of 0.5 to 0.65 for the 650 μm glass, and 0.3 to 0.35 for coarser 2,600 μm solid. The significance of these results is discussed later.

7.2.3   Effect of particle size and size distribution

Although the results summarized in Table 7.1 do not lend themselves to correlations, they clearly indicate that both U_c and U_k increase with particle size and density. Little information, however, is available on the effect of particle size distribution. Kehoe and Davidson added their finest catalyst A to catalyst C to obtain catalyst B, and noted that as a result the transition velocity was lowered significantly. This is not surprising since the mean size of catalyst B is considerably smaller than that of catalyst C. Information on the effect of size distribution for solids having comparable mean size is lacking.

7.2.4   Effect of bed height

Canada, McLaughlin, and Staub report no apparent effect of bed height over a range of 25 to 70 cm. Kehoe and Davidson, on the other hand, report that (for catalyst B and C) a bed 2.55 m in height gave rise to transition velocities higher than those recorded in beds of the same solids ranging in heights to only 1.7 m. The shallower beds, they argue, support a greater degree of slug coalescence and the approach to the 'breakdown' velocity is accordingly faster.

Lanneau conducted his experiments in a bed 4.6 m deep, and he noted that the structure of the bed could vary significantly with bed height. When operating at 5 atm and at a gas velocity of 0.15 m/s the pressure traces showed that the bubbles forming immediately above the distributor were still small. In the middle of the bed, owing to coalescence, the bubbles had grown large and the two-phase structure of the bubbling (slugging) regime dominated, but in the upper section of the bed the bubbles appear to have become smaller again. The implication that the structure of a fluid bed can vary significantly with height, especially in tall beds, ought to be taken into account in reactor modelling.

7.2.5   Effect of pressure

Canada, McLaughlin and Staub report that, at the same gas velocity, the fluidization grows smoother, as reckoned from the relative pressure fluctuations, as the pressure is raised. For a given solid, the transition velocity U_k decreases with increasing pressure, but over the range of 1 to 10 atm the ratio of U_k to the particle's terminal velocity v_1 remained essentially constant, 0.5 to 0.65 for the 650 μm glass and 0.3 to 0.35 for coarser 2,600 μm solid. Lanneau also notes that at a given gas velocity, the fluidization is more uniform at the higher pressure level. This is reflected in the lower heterogeneity measured past the peak at 5 as compared to 1.7 atm (Fig. 7.2).

Other workers have also noted the 'smoothing' effect of pressure, particularly in fluidized beds of group A powders.

7.2.6   Effect of bed size

The results of Kehoe and Davidson, and especially of Thiel and Potter, indicate that for a given solid the transition velocity U_k decreases with increasing bed size. The City College results corroborate this if comparison is drawn between the data from the 15.2 cm diameter column and that from the two-dimensional bed. In the latter, the transition generally came into play at lower gas velocities. The two-dimensional bed is comparable in cross-sectional area to the 15.2 cm column, but as it affords growth bubbles larger that 15.2 cm in that respect it could be regarded as effectively larger.

That bed size may affect the transition in question should not be surprising. Essentially all the experiments summarized in Table 7.1 were conducted in equipment of relatively small size which gave rise to slugging behaviour over the range of gas velocities preceding the transition and overlapping its earlier stages. Over that range, bed size directly affects the shape, size, and rise velocity of the slugs, and the fluidized density at a given gas velocity. Equation (7.2), which is a statement of the two-phase theory for the slugging regime, is a reminder of that. Since the transition to turbulent fluidization involves changes in the structure of the bed, it may be expected to show some dependence on bed size. In beds of sufficient size, where freely bubbling behaviour prevails, the transition to turbulent fluidization ought to be no longer influenced by bed size.

The dependence of bed size of the results described here provides a clue to one of the underlying elements to which the mechanism of the transition must ultimately be linked — the absolute rise velocity of the bubble (or slug) u_A. All the investigators who have commented upon the transition's mechanism agree that it entails a breakdown, gradual or otherwise, of the slugs into smaller bubbles or voids. Clearly, the size the slug attains, its rise velocity, and its interaction with the emulsion just before the transition must be the focus of those seeking to understand the forces that drive the transition and the attendant changes in the bed's structure.

Invoking the two-phase theory, one may gather from Eq. (7.2) that over the slugging range two different powders in the same bed will give rise to slugs of comparable size and speed at a given superficial gas velocity. The process of slug disintegration would accordingly commence earlier for the lighter and/or smaller powder. This generally accords with the study of Clift, Grace and Weber, (1974) on the splitting of bubbles.

On the other hand, given two beds of different size holding the same solid, the transition will commence at a lower gas velocity in the larger because it supports a larger slug rising at higher velocity. The results of Thiel and Potter (1977) corroborate this point.

It should be noted that although the results of Kehoe and Davidson do show a dependence on bed size, the dependence is slight.

7.2.7 Bed expansion

Bed expansion affords more insight into the mechanism of the transition to turbulent fluidization. If the two-phase theory holds and slug flow persists, the bed expansion should obey:

$$\frac{H}{H_{mf}} = 1 + \frac{U - U_{mf}}{0.35(gD)^{1/2}} \tag{7.4}$$

Massimilla (1973) reports data which strongly deviate from Eq. (7.4) beginning with a superficial gas velocity around 15 cm/s (Fig. 7.9).

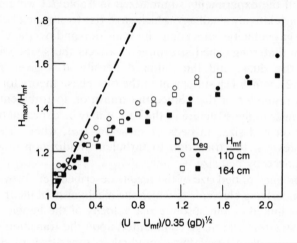

Figure 7.9 Bed expansion data (from Massimilla, 1973).

Massimilla attributes the departure from ideal slug flow behaviour to slug velocities exceeding the rise velocity predicted by Eq. (7.1), owing to the acceleration effect brought about by coalescence of slugs throughout the column. Above the transition velocity (30 to 40 cm/s), the higher velocity of the lean phase, he argues, arises from the continuous process of slug breakdown and the coalescence and rearrangement of the resulting small bubbles or voids into new slugs. This picture of higher lean phase velocities accords well with the results of Kehoe and Davidson and Crescitelli *et al.* presented earlier (Figs 7.7 and 7.8). An equally plausible explanation is the expansion of the dense phase which affords higher interstitial flow and smaller slugs or voids.

7.2.8 Mechanism

All this does not quite add up to a complete picture of the transition to turbulent fluidization, but several elements of the mechanism do emerge more clearly. These I attempt to recapitulate briefly below.

The transition to turbulent fluidization involves the gradual breakdown of large bubbles (or slugs) into smaller bubbles and voids. The process of bubble splitting is to some extent counterbalanced by coalescence of the resulting smaller bubbles, but the net effect is a progressive change towards a structure of greater homogeneity, culminating in the turbulent state where on the whole large discrete bubbles are absent, and traces for capacitance probes no longer register a clear periodic element, To be sure, a relatively large void or slug will occasionally form, travel some distance up, and soon disperse, This and the recurring process in which the small voids split, coalesce, and accelerate do lend the phenomena some apparent periodic character. To one viewing a turbulent fluid bed this impression may be particularly strong when the bed is narrow and where accordingly the voids tend to segregate towards the centre, leaving the wall region fairly dense, and when the bed material is coarse.

Over the transition and well into the turbulent regime, bed expansion falls below the levels expected from ideal (two-phase) slug behaviour. This is due at least in part to the high lean phase velocities measured over the same range. In turn, the high lean phase velocities (which also depart from ideal slug flow behaviour) appear to arise from the swift acceleration which attends the recurring coalescence of slugs, in the early stages of the transition, and of the smaller voids towards the end and past the transition.

The observed bed expansion also arises from the expansion of the dense phase. This and the active process of splitting and coalescence of voids also contribute a higher proportion of solids in the lean phase.

The phenomena that underlie the transition from bubbling to turbulent fluidization remain rather obscure. The key event of course is the breakdown of the bubbles or slugs, but although this matter has received some attention (Harrison, Davidson, and de Kock, 1961; Clift and Grace, 1972; Clift, Grace, and Weber, 1974) no clear and acceptable explanation has been advanced on why bubbles break. Some evidence was offered earlier showing that the point or range around which bubbles begin to break may be related, for a given solid, to some critical value of the absolute rise velocity of the bubble.

7.3 THE TURBULENT REGIME

The structure of a turbulent fluid bed depends strongly on the transition velocity U_c. If U_c is small, as in beds of small particles, the sizes of the bubbles (slugs) attained before the onset of the transition is relatively small, and the scale of demixing of gas and solid in the resulting turbulent bed is small. In beds of coarse solids, on the other hand, U_c is relatively high, the size of the bubbles attained before the transition commences is correspondingly larger, and the structure of the resulting turbulent bed is more open, marked by wider channels and by a grosser scale of the demixing patterns of gas and solid.

These observations are borne out by photographs of the transition in two-dimensional beds presented by Kehoe and Davidson (1971).

Consider a fluidized bed of closely sized coarse particles, such as those used in the experiments of Canada, McLaughlin, and Staub (1976). The turbulent regime for the 2,600 μm glass extends for around (0.3 to 0.35)v_t up to a velocity at least equal to v_t. Over this range, since the gas velocity lies below the terminal velocity of the individual particle, no carryover occurs provided the freeboard is sufficiently high. In a bed of fine powder, however, the turbulent regime start at values of U_k/v_t much larger than 1, and the ratio increases as the mean size of the powder decreases (Table 7.1).

The concept that particles combine into loose clusters can be used to explain the range of the turbulent regime for a fine powder. The mean effective size and density of these clusters at any gas velocity is such that their terminal velocities remain greater than the velocity of the gas. This concept can explain why a bed of fine solid can be maintained at gas velocities that are 10 to 20 times the terminal velocity of its median particles. Entrainment will, of course, take place as particles are swept from the lean phase near the top of the bed, and by the erosion of clusters by the surrounding lean phase.

The turbulent regime extends from U_k to the transport velocity U_{TR}. The approach to the transport velocity is accompanied by a sharp increase in the rate of carryover. It is difficult, without having made suitable measurements, to associate the transport velocity with some mean cluster size. By analogy to the behaviour of beds of closely sized coarse particles, one may surmise that when the transport velocity is reached, the terminal velocities of a large proportion of the clusters are smaller than the velocity of the gas. For fine solids, the transport velocity is typically an order of magnitude greater than the terminal velocity of the individual median particle. For coarse particles, the transport velocity lies closer to the terminal velocity of the median particle.

Though they differ in structure, the turbulent and the bubbling fluid beds are similar in several respects. Over the turbulent regime, though particle carryover rates are relatively high, the fluidized density of the bed remains quite high and an upper bed level is present, though it is, of course, more diffuse than in a bubbling fluidized bed. As in the bubbling bed, when solid is fed to the turbulent bed at a rate matching the saturation carrying capacity, the concentration of solid above the bed decays through the transport disengaging height and approaches a constant value. If solid is fed to the bed at a rate above the saturation carrying capacity, the bed level rises until the new solid rate is matched by solid carried out of the bed. The gas exit now lies within the transport disengaging height. However, unless the slip velocity — i.e. the relative velocity of gas and solid — changes significantly, the fluidized density of the bed itself remains virtually the same.

Some of the general points made above may be illustrated by experimental results reported by Yerushalmi and Cankurt (1979) for fluid cracking catalyst

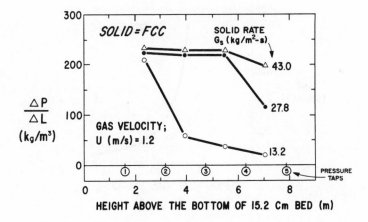

Figure 7.10   Pressure drop profiles across a fluid bed of cracking catalyst (FCC).

(FCC). The experiments were conducted in the 15.2 cm column of the apparatus shown in Fig. 7.5. Figure 7.10 displays three vertical pressure drop profiles recorded across the 15.2 diameter column. Each profile corresponds to a different solid rate. Each data point represents the pressure gradient between two adjacent pressure taps. The positions of the pressure taps are indicated along the horizontal axis. At 1.2 m/s, the saturation carrying capacity is about 13 kg/m^{2}s. The corresponding pressure drop profile reflects the presence of a dense bed (of density around 200 kg/m^3) at the bottom of the column and the decay of solid concentration along the transport disengaging height. The profile also indicates that the solid hold up begins to level off from about the column's midpoint. As the solid rate is raised beyond the saturation carrying capacity, the upper bed level rises, and the solid concentration near the top of the bed increases, until a new equilibrium condition is reached at which solid leaves the 15.2 cm column at a rate matching the new feed rate. Figure 7.10 also indicates that the bed density increases slightly with the solid rate. This arises from the fact that corresponding to each bed density and solid rate, one may regard the bed as circulating at a mean solid velocity that is given by the ratio of the solid rate G_s to the bed density ρ_B. This mean solid velocity increases with the solid rate, and accordingly the slip velocity, upon which the bed density ultimately depends, decreases with the solid rate, giving the somewhat higher bed densities.

7.4   THE TRANSPORT VELOCITY AND THE FAST BED REGIME

The transport velocity may be regarded as the boundary which divides vertical gas solid flow regimes into two groups of states. Below it lie the

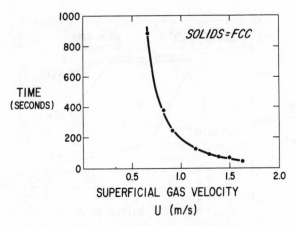

Figure 7.11   Time required for the pressure drop across a
bed of fluid cracking catalyst (FCC) maintained in a 15.2 cm
column to decay to 2.5 cm of water as a function of the
superficial gas velocity.

bubbling and the turbulent fluid beds where, save for some carryover, the bed
in general experiences no net flow and remains at the bottom of the holding
vessel. These are, to use Lanneau's description, the *captive* states. Above lie
the transport regimes which encompass a wide range of states from dilute
phase flow to the fast bed condition.

The transport velocity for the fluid cracking catalyst lies around 1.2 to 1.5
m/s (which is nearly 20 times the terminal velocity of a single particle with a
size equal to the mean diameter of the powder, 49 μm). As this velocity is
approached, there is a sharp increase in the rate of particle carryover and in
the absence of solid recycle the bed would empty rapidly. This may be
illustrated by the results, shown in Fig. 7.11, of a simple experiment in which
the flow of recycle solid to the 15.2 cm diameter column was suddenly halted,
and the time it took the bed pressure drop to fall to 2.5 cm of water gauge
(w.g.) was measured at different gas velocities. Because of the continuous
nature of the curve in Fig. 7.11, it could hardly serve to pinpoint the transport
velocity. To do that, other approaches, to be described shortly, proved more
useful.

Beyond the transport velocity, solids traverse the column in fully entrained
flow, and the density or concentration of the resulting suspension depends not
only on the gas velocity but also on the solid rate. If the solid rate is low,
dilute phase flow results (Fig. 7.12). As its name is meant to suggest, solid
concentration is low in the dilute phase flow regime, the particles stream
upwards in relatively straight paths, and slip velocities lie in the neighborhood
of the free-fall velocities of the individual particles. As the solid rate

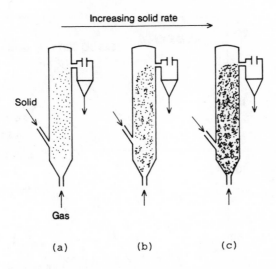

Figure 7.12 The concentration of a gas/solid suspension as a function of the solid rate.

increases, the suspension becomes progressively denser (Fig. 7.12b), and at a sufficiently high solid rate, the fast fluidized bed (Yerushalmi, Turner, and Squires, 1976) is established (Fig. 7.12c). The fast bed condition is marked by relatively high solid concentrations, aggregation of the particles in clusters and strands which break apart and reform in rapid succession, extensive backmixing of solids, and slip velocities that are in order of magnitude greater than the free-fall velocities of the individual particles. As noted earlier, a way to create the dense suspension typical of the fast bed condition is simply to circulate the solids emerging from the top of the column back to its bottom by means of cyclones and a standpipe (Fig. 7.1c).

Whereas the distinction between dilute phase flow and the fast bed condition is clear-cut, the suspension density or solid rate at which the fast bed may be said to come into play (at a given gas velocity) has not been established. As in the bubbling fluid bed, the extensive solid backmixing in the fast bed is responsible for the uniformity of temperature throughout the bed. The study of thermal communication may serve perhaps to delineate the onset of the fast bed at a given gas velocity. For the cracking catalyst, extensive backmixing of solid was noted over a range of gas velocities at a loading around 40 kg/m^3.

Figure 7.13 presents fluidization data for FCC obtained in the 15.2 cm column, mostly over the fast bed range. The figure gives the pressure gradient measured across the middle section of the column as a function of the solid rate at different gas velocities. Over this range, wall solid friction is relatively

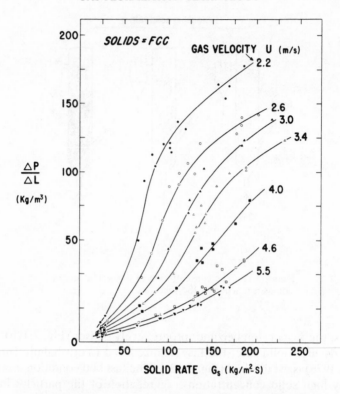

Figure 7.13 Pressure gradient versus the solid rate at different gas velocities. Data were taken with the fluid cracking catalyst (FCC) in the 15.2 cm column shown in Fig. 7.5 across a section extending from 3.14 to 6.28 m from the bottom of the column.

small (Van Swaaji, Burman, and Van Breugel, 1970; Turner, 1978), and the pressure gradient can essentially be regarded as the fluidized density $\bar{\rho}$ in the section of the bed in question. Figure 7.13 attests to the high solid concentrations that can be maintained in the fast bed at gas velocities that are an order of magnitude greater than those normally employed in a bubbling fluidized bed of the same solid.

The fluidized density is not uniform along the height of the fast fluidized bed. The density is normally higher at the bottom, owing in part to acceleration effects, and decays towards the top. In beds of light powders (such as the fluid cracking catalyst), the decay is very gradual. In beds of coarser or heavier solids the decay could be rather sharp.

Slip velocities are high in the fast bed. Those corresponding to the data shown in Fig. 7.13 and normalized by dividing by v_t for the mean size of FCC are given in Fig. 7.14. The slip velocity used in Fig. 7.14 is a mean value based on the mean solid velocity.

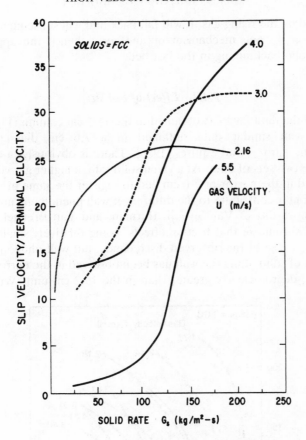

Figure 7.14 Slip velocities corresponding to the data shown in Fig. 7.13.

$$\text{Slip velocity} = \frac{U}{\epsilon} - v_s \qquad (7.5)$$

where ϵ is the bed voidage and v_s is the mean solid velocity; v_s is turn is defined by:

$$v_s = \frac{G_s}{\rho_B} = \frac{G_s}{\rho_p(1 - \epsilon)} \qquad (7.6)$$

The large slip ratios can be explained by the aggregation of the solid in the fast bed into relatively large dense clusters of particles. A large cluster would naturally have an effective free-fall velocity that is considerably greater than that of a single particle in isolation. If a cluster is sufficiently large, it cannot be sustained by the rising gas; it will fall back and will subsequently undergo disintegration by one mechanism or another — hence the apparent high degree of solid backmixing in the fast bed.

7.4.1 Effect of bed size

In Fig. 7.15 the data far FCC obtained in the 15.2 cm column (Fig. 7.13) are compared with similar data obtained in a 7.6 cm diameter column (Yerushalmi, Turner, and Squires, 1976). There is obviously, a discrepancy between the two sets of data. At a given solid rate, a higher gas velocity must be employed in the narrower 7.6 cm bed to register the same pressure drop. Two factors may contribute to the difference: wall phenomena and wall solid friction. The results of Van Swaaji, Burman, and Van Breugel (1970) and Turner (1978) indicate that friction effects, being relatively small, could not alone be the cause of the observed discrepancy, but wall phenomena could. Segregation of solid along the wall has been observed in the narrower 7.6 cm column to a degree clearly greater than in the 15.2 cm unit. With the core

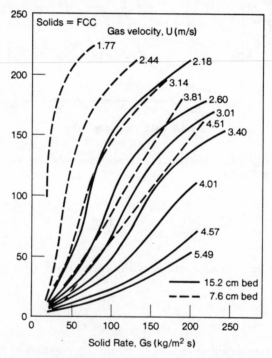

Figure 7.15 Comparison between the data for FCC obtained in the 15.2 cm column and shown in Fig. 7.13 and corresponding data for the same solid obtained in a 7.6 cm column.

somewhat more dilute, gas in the 7.6 cm bed can stream relative to the solid at higher velocities than in the 15.2 cm bed, for the same solid rate and solid hold up. That is, slip velocities are higher in the 7.6 cm bed. In the extreme case, if the entire solid segregated along the wall in a dense layer, leaving the core empty, slip velocities would be even higher. The bed diameter at which wall phenomena ceases to exert a significant influence, for a given solid, has not been determined.

In principle the transport velocity should not be dependent on bed size. However, in laboratory experiments, it might appear to be dependent to the extent that wall phenomena influence the structure of the bed in narrow columns and in tests with coarse or heavy particles.

7.4.2 Effect of pressure

There is essentially no fundamental information on the effect of pressure on the phenomena discussed here. Pressure may exert its well-known smoothing effect, and in one known instance involving highly concentrated suspensions maintained at velocities around 7 to 10 m/s, this has indeed been reported to be the case (Horsler, Lacey, and Thompson, 1969). Pressure may also diminish the degree of aggregation in the fast bed, and thus reduce the extent of solid backmixing. By stability considerations Grace and Tuot (1979) verify the concept of cluster formation in the fast bed, and their theory predicts that pressure should lead to a reduction in clusters size. Similar arguments lead to the speculation that the transport velocity will decrease with pressure.

7.4.3 The flooding point

Following the practice of students of vertical gas liquid flow, Yerushalmi and Cankurt (1979) used the data shown in Fig. 7.13 to prepare plots of the volumetric gas flux, in cubic metres per square metre, versus the volumetric solid flux at constant bed densities. The results are shown in Fig. 7.16. Note that U is in fact the volumetric gas flux and $(1 - \epsilon)v_s$ is the volumetric solid flux G_s/ρ_p; $(1 - \epsilon)$ is the volumetric solid concentration.

The curves of Fig. 7.16 are straight lines converging, with one exception, on the point defined by $U = 1.37$ and $v_s(1 - \epsilon) = 0.02$ m/s. The line corresponding to $\rho = 32$ kg/m^3 does not converge on this point probably due to predominance of wall phenomena at this relatively low suspension density. The intersection is the so-called *flooding point* (Wallis, 1969), the onset of concurrent flow of both phases, and $U = 1.37$ m/s is acordingly the transport velocity. The flooding point will always have a positive coordinate for $v_s(1 - \epsilon)$ because of particle carryover from a turbulent bed. Indeed, $v_s (1 - \epsilon) = 0.02$ m/s reflects the fluidized density (about 192 kg/m^3) and the

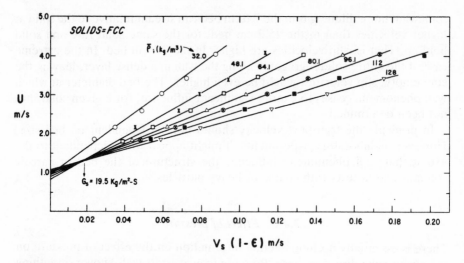

Figure 7.16 Volumetric flux of gas U versus the solid flux v_s $(1-\epsilon)$ at constant bed densities. Drawn from data shown in Fig. 7.13.

saturation carrying capacity (about 20 kg/m²s) from a turbulent bed maintained just below the transport velocity.

Determination of the flooding point thus provides one method for fixing the transport velocity.

7.5 MORE ON THE TRANSPORT VELOCITY

Further experiments to delineate the boundary between the turbulent and the transport regimes (i.e. the transport velocity) were conducted in the modified version of the 15.2 cm diameter system shown in Fig. 7.17. The U-tube solid transfer line, connecting the standpipe ('slow bed') and the 15.2 column, was replaced by a 60° (from vertical) transfer line also measuring 15.2 cm i.d. Control of the solid flow in the modified system was afforded by a conical valve installed at the entrance from the slow bed to the lateral transfer line.

In a typical experiment, air was supplied to the 15.2 cm column through the ports located at its bottom. At the start of the experiment, the solid control value was closed; then, with the velocity of the gas set at some desired value, the solid rate was slowly raised in small increments. After each increment, the character of the gas solid flow along the column was observed and the pressure drops across the various sections of the column were recorded. Attention was primarily focused on two sections: the bottom section of the 15.2 cm column below the point where solids from the slow bed enter and the section of the column between the pressure taps numbers 1 and 2 in Fig. 7.17. We shall concentrate on section 1–2.

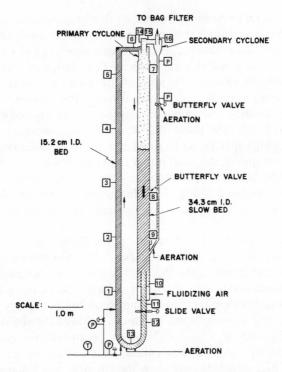

Figure 7.17 Schematic of the modified 15.2 cm fluidization system.

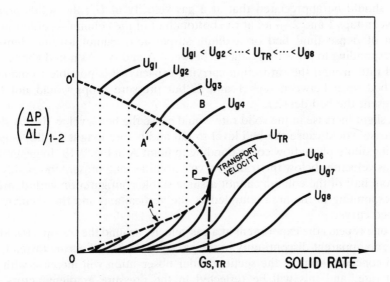

Figure 7.18 Qualitative representation of the pressure gradients measured in the modified 15.2 cm system between pressure taps 1 and 2.

It might be useful to consider the results first in qualitative terms as shown in Fig. 7.18. Pressure drop is shown as a function of the solids rate at different gas velocities both lower and higher than the transport velocity U_{TR}. Consider an experiment conducted at a gas velocity U_{g3}; that is at a typical velocity below the transport velocity. So long as the solids rate remains below the saturation carrying capacity, solids will traverse the column in dilute phase flow, and the resulting low pressure drop would correspond to some point along the curve OA. At point A, the solids rate becomes equal to the saturation carrying capacity, and a small increase in the solid rate beyond that produces the collapse of the solid into a fluid bed which forms the bottom of the column. The bed level will stabilize at some point above or below the pressure tap numbered 2 in Fig. 7.17. In the latter case, an additional small increment in the solids rate can bring the bed's upper surface past pressure tap 2.

The density of the fluid bed at this point corresponds to the pressure gradient at point A' of Fig. 7.18. The curve $A'B$ shows further increase in the pressure drop, and hence in the fluidized density, with increasing solid rate. This comes about since, at a given gas velocity, an increase in solids rate produces a corresponding increase in the mean solids velocity, and hence a decrease in the slip velocity upon which the fluidized density depends.

If the relative velocity, $U_{g3} - v_s$, between gas and solid is smaller than U_k, the bed will be bubbling or, in a column of small diameter, slugging, and pressure drop measurements will show the characteristic fluctuations attending this condition. In this case, the pressure gradients represent mean values.

It should be appreciated that at a gas velocity of U_{g3} the solids present between taps 1 and 2 — ie. at the bottom half of the column — either in the form of dense fluid bed or a dilute suspension, cannot assume densities corresponding to pressure gradients between A and A'. As noted above, at a solid rate around the saturation carrying capacity, it is possible to maintain the bed level between taps 1 and 2, but pressure drops would not then represent the bed density.

A slight increase in the solid rate would cause the bed level to rise. A slight decrease would cause the bed level to fall and before long it would 'empty', leaving dilute phase flow corresponding to point A in Fig. 7.18. In general, at all gas velocities below the transport velocity, the solid present in roughly the bottom half of the column cannot assume stable configuration with densities corresponding to the area bounded by the vertical axis and the semicircular dashed curve.

If one repeats the experiment at a gas velocity around the transport velocity U_{TR}, no apparent discontinuity in the gas solid states will occur. Instead, the solid concentrations in the section under observation will increase with the solid rate, and this will be reflected in the pressure gradient across this section. The same holds for all gas velocities above the transport velocity.

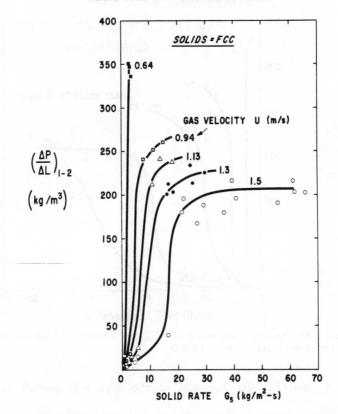

Figure 7.19 Pressure gradients measured in the modified 15.2 cm system between points 1 and 2. Solid = fluid cracking catalyst (FCC).

Figure 7.19 gives the pressure gradients measured between taps 1 and 2 using FCC. During the experiments the solid rate suffered continuous exursions which introduced a scatter in the data. Notwithstanding, the results shown in Fig. 7.19 support the earlier findings that for FCC, the transport velocity lies around 1.5 m/s and that the corresponding saturation carrying capacity G_{TR} is in the region of 20 kg/m^2 s.

Figure 7.20 shows the corresponding data for HFZ-20, and indicates that the transport velocity is about 2.1 m/s and that G_{TR} lies around 34 to 39 kg/m^{2}s. Table 7.1 shows that HFZ-20 is similar in size distribution to FCC, but its particle density is about 50 per cent greater.

Experiments with fine alumina ($d_p = 103$ μm, $\rho_p = 2.46$ g/cm) reported elsewhere (Graff and Yerushalmi, 1978) indicate that for this solid the transport velocity lies around 3.7 to 4 m/s.

The few values of transport velocity presented in this section obviously do not afford a basis for any correlations, though, like the other transition

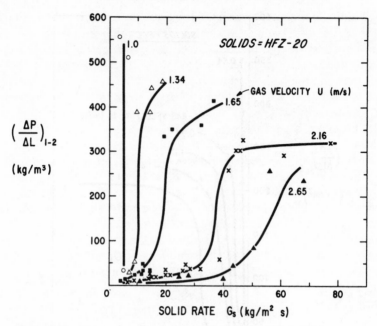

Figure 7.20 Pressure gradients measured in the modified 15.2 cm system between points 1 and 2. Solid = HFZ-20.

velocities, U_c and U_k, the transport velocity U_{TR} with particle size and density.

The phenomena of choking, described in the pneumatic conveying literature, is associated closely with the typical experiments described above (Yang, 1976). To the practising engineer, choking is an undesirable clear-cut event in which a stable smoothly flowing suspension collapses into a slugging condition with the attendent large pressure fluctuations (Zenz and Othmer, 1960).

In a typical experiment aimed at determining the choking conditions (i.e. the choking velocity and the solid concentration at a given solid rate) the solid rate is set at a given value, and the velocity of the gas is then slowly decreased until choking occurs. Reducing the gas velocity at a fixed rate amounts to moving vertically upwards in Fig. 7.18. If the solid rate G_s is lower than the saturation carrying capacity G_{TR} (corresponding to the transport velocity), then as the gas velocity is decreased choking will occur at the velocity for which G_s is the saturation carrying capacity, and the characteristics of the suspension just before its collapse are represented by a point on the dashed curve OAP vertically above G_s.

Having collapsed, the solid will form a fluid bed at the bottom of the pipe. Upon a slight further decrease in the gas velocity, the bed level would begin to rise until it reaches a point where a new equilibrium condition is established, i.e. until the rate at which solid leaving the pipe becomes equal to G_s. The character of the fluid bed, which has now acquired some net upward flow, will depend on the relative velocity $U - v_s$. If the relative velocity is smaller than U_k, the bed will be slugging. If the relative velocity lies beyond U_k, the bed will be a turbulent fluid bed. In the latter case, a further decrease in the gas velocity will bring about slugging conditions when the relative velocity becomes smaller than U_k.

Consider now the case in which the solid rate G_s is greater than G_{TR}. In this case, a continuous decrease in the gas velocity will not produce any collapse of the suspension. Instead, as may be gathered from Fig. 7.18, the suspension will grow denser and denser and will gradually assume the condition typical of the fast fluidized state, with considerable solid refluxing in effect. When the relative velocity becomes smaller than the transport velocity, the solid will be transported through the line in the form of a turbulent fluidized bed, and only when the relative velocity becomes smaller than U_k will choking come into play in the form of slugging, with its undesirable pressure fluctuations.

7.6 REGIMES OF FLUIDIZATION

In Fig. 7.21 the data obtained in the 15.2 cm column over the bubbling, turbulent, and fast bed regime, have been assembled in the form of a fluidization map depicting the slip velocity versus the solid volumetric

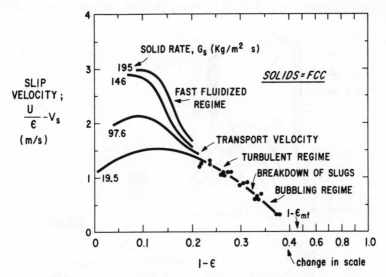

Figure 7.21 Fluidization data for a fluid cracking catalyst (FCC): the slip velocity versus the solid volumetric concentration. Data obtained in the 15.2 cm system.

concentration at the bottom of the bed. The slip velocity is used rather than the superficial gas velocity since it is the former which provides a relevant measure of the interaction of gas and solid. The figure attempts to highlight some of the points made above. Over the bubbling and the turbulent regimes, the relation between the slip velocity and the solid concentration appears unique. Not so in the fast bed where, corresponding to each solid rate, a different function connecting these variables exists. If the particles were uniformly dispersed in the fast bed, then the relation between these functions would be unique, as it approximately is, for example, in hydraulic transport of any but very fine solids (Price, Lapidus, and Elgin, 1959). However, the structure of a fast bed of fine solids is characterized by the presence of clusters and streamers of particles. The hydrodynamic behaviour of the fast bed accordingly reflects the interaction of the gas not so much with individual particles as with clusters of various sizes. A change in gas velocity or solid rate, at a given solid concentration, produces a change in the spectrum of cluster sizes — hence the multiple functions relating the slip velocity to the solid concentration.

Figure 7.22 is a qualitative fluidization map for fine solids. It is patterned after Fig. 7.21 but extends further to the dilute phase flow regime. The figure also displays schematics of equipment typical of the various regimes, each sized approximately for the same gas-treating capacity.

The difference between dilute phase flow and the fast bed condition was enumerated in Section 7.4. The distinction between the fast fluidized bed and the riser transport reactor used in many modern cat cracking plants is less

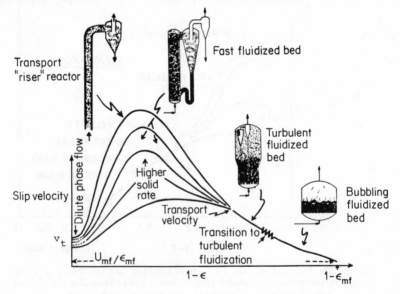

Figure 7.22 A qualitative fluidization map for fine solids.

clear. As in the fast bed, relatively high solids concentrations (typically around 160 kg/m^3), large slip velocities, solid backmixing, and particle aggregation are also manifest in the riser reactor (Matsen, 1976). The difference, which is yet to be fully explored, may lie in the disposition or structure of the solid. Saxton and Worley (1970) report several radial density profiles, in the form of contour plots, obtained by radiation attenuation measurements across a riser reactor. These reveal gross solid segregation. Some of the contour plots show concentrated zones, most likely representing solid flowing downward, along the walls, while the core remains relatively more dilute. Others display large concentrated regions extending into the core, or a few isolated pockets of high solid concentration occupying positions away from the wall and surrounded by leaner areas.

If the fast bed is to suit a wider range of physical and chemical process applications, it should entail a greater degree of homogeneity. Observations of a two-dimensional bed with powders belonging to Geldart's group A indicate that a fine scale of gas solid demixing patterns, which extends across the bed, can be maintained at gas velocities up to 4 to 5 m/s. In contrast, gas velocities in an FCC riser typically exceed 10 m/s beyond some short distance above the oil feed nozzles.

Observations of two-dimensional beds, however, may be misleading. When a round 7.6 cm fast bed of HFZ-20 was photographed by X-rays the core was revealed to be considerably more dilute than the wall region (Gajdos and Bierl, 1978). This is not altogether surprising. In columns of narrow diameter, even the bubbling and turbulent fluidized beds give rise to a structure marked by dense regions along the wall and a leaner core. This will be illustrated in Section 7.7. The work shows that at lower velocities (around 3 m/s) and at high solid loadings, the area of the leaner core diminishes and its density increases.

Figure 7.15 indicates that at the same gas and solid flowrates, segregation of solid along the walls is more pronounced in the 7.6 diameter unit than in the 15.2 cm diameter column. Whether this trend extends to columns of larger diameter remains to be seen since relevant research has yet to be undertaken. What would be the radial density profile in a fast bed of larger diameter? Will a relatively thick solid annulus surround a single leaner core? Or will the profile, apart from some significant wall phenomena, be more uniform, consisting of a cell-like structure of alternating dense and leaner regions? What is the effect of gas velocity, solid rate, the properties of solid and gas, and equipment scale on the structure of a fast bed?

Resolution of these questions must rank high among the research goals that lie ahead in this area of investigation. Resolution of these questions will place the riser in better perspective vis-à-vis the fast bed. On the basis of our laboratory work, and commercial practice to date (Reh, 1971; Yerushalmi and Cankurt, 1978), it appears that a more efficient contact of solid and gas is enhanced by the use of fine solids, gas velocities around 3 to 5 m/s, and high solid circulation rates.

7.7 GAS MIXING IN HIGH VELOCITY FLUID BEDS

Cankurt and Yerushalmi (1978) conducted experiments in the 15.2 cm diameter system, shown in Fig. 7.5, on the extent of gas backmixing in a bed of a fluid cracking catalyst fluidized at gas velocities spanning the bubbling, turbulent, and fast bed regimes. In these experiments, a tracer gas, methane, was continuously injected at a known rate through a tube entering the bed at any desired elevation. Simultaneously, a stream of sample gas was withdrawn from locations below the injection point, and its tracer concentration was analysed and recorded. Tracer could be injected and sampled at any radial position in the bed. The experiments were conducted at room conditions.

Figure 7.23 gives the radial concentration profile across the bed 5.1 cm below the injection point at different gas velocities. The tracer was injected at the centre. Those results lead to the conclusion that the degree of gas backmixing is high in the bubbling regime (0.21 m/s); it diminishes, though it

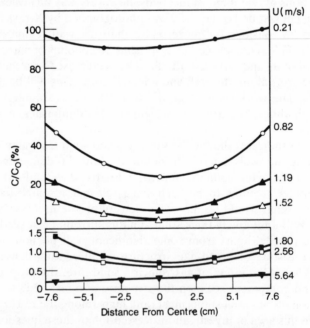

Figure 7.23 Radial concentration profiles measured 5.1 cm upstream of the injection point at different gas velocities. Tracer was injected at centre. Fluidized densities in the order of increasing velocities are 397, 266, 213, 189, 170, 64, and 19 kg/m^3.

may still be appreciable, over the turbulent range (0.82, 1.19, and 1.52 m/s) and, in the absence of a pronounced gas velocity profile, gas is essentially in plug flow in the fast fluidized bed (1.80 m/s and beyond).

This conclusion is reinforced by Figs 7.24 and 7.25, which provide complete concentration profiles (radial and axial) upstream of the injection point at gas velocities of 0.21 and 1.52 m/s. A striking aspect of these profiles is the increased concavity of the radial profiles with gas velocity. This suggests a physical picture of a mixing pattern dominated by downwards movement of dense phase solid along the wall, and a fast-moving gas flowing upwards through a leaner core.

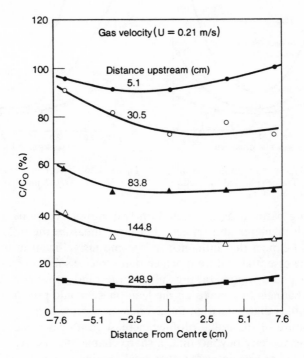

Figure 7.24 Tracer concentration profiles at 0.21 m/s. Tracer injected at centre. Fluidized density = 397 kg/m^3.

Figure 7.26 and 7.27 add further details to this picture. Figure 7.26 shows radial profiles obtained at three locations below the injection point (i,e, upstream) when the tracer was injected near the (left) wall of the bed fluidized at 1.52 m/s. Figure 7.27 gives the corresponding axial profiles measured along the left wall, the right wall, as well as along the centre of the bed.

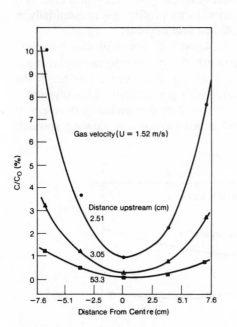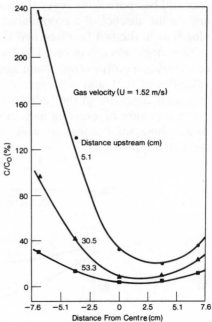

Figure 7.25 Tracer concentration profiles at 1.52 m/s. Tracer injected at centre. Fluidized density = 189 kg/m^3.

Figure 7.26 Tracer concentration profiles at 1.52 m/s. Tracer injected near the left wall.

This mixing pattern, as was noted on two earlier occasions, is typical of beds of small diameter and reflects to a large measure the influence of the wall. I would expect that influence to become insignificant in beds of large diameter; the essential feature in which dense phase solids descend against a rising leaner phase would still prevail, but the random flow of the bubbles, voids, and channels that make up the lean phase would preclude the sort of constant profiles that persist on the average in narrow beds.

The curvature of the radial profiles increases with gas velocity up to 1.52 m/s. At some velocity beyond that, a sharp change occurs. At 1.80 m/s, the curvature of the profile appears rather small, and at higher gas velocities, the profile is practically flat. This is illustrated very clearly in Fig. 7.28, in which the ratio of the concentration near the wall to that at the centre is plotted versus the superficial gas velocity. The sharp change that is bracketed by 1.52 and 1.80 m/s may be laid to a fundamental change in the fluidization regime — to a change from turbulent to fast fluidization. The transition velocity (i.e. the transport velocity) is ostensibly higher than the value determined by other methods simply because the solid itself circulated at a mean velocity around 0.3 m/s, owing, as noted earlier, to the use of tall beds and the high solid carryover rates which that entailed.

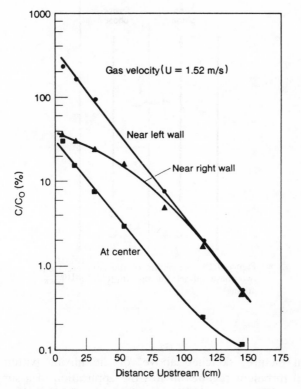

Figure 7.27 Axial concentration profiles at 1.52 m/s. Tracer injected near left wall.

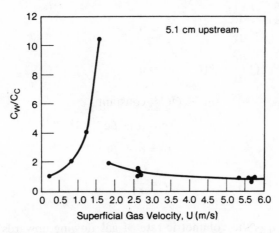

Figure 7.28 Ratio of the tracer concentration near the wall to that at the centre, measured 5.1 cm upstream of the injection point. Tracer injected at centre.

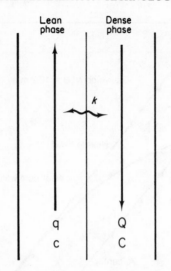

Figure 7.29 Schematic of the countercurrent two-phase model for the study of gas back-mixing.

Theory

The physical picture, sketched above, of the mixing patterns over the bubbling and turbulent regimes invites the application of a simply counter current model. The model is illustrated in Fig. 7.29, and the governing equations and their solutions are given below:

Lean phase:
$$q \frac{dc}{dx} + k(C - c) = 0 \qquad (7.7)$$

Dense phase:
$$Q \frac{dc}{dx} - k(C - c) = 0 \qquad (7.8)$$

$$qc - QC = \text{constant} \qquad (7.9)$$

$$c = \alpha + \beta e^{-k(r-1)x/q} \qquad (7.10)$$

where:
$$C = \alpha + \beta r e^{-k(r-1)x/q} \qquad (7.11)$$

$$r = \frac{q}{Q} \qquad (7.12)$$

In the above, q is the volumetric rate of gas flowing upwards and Q is the corresponding rate for the gas flowing back; α is the tracer concentration far upsteam.

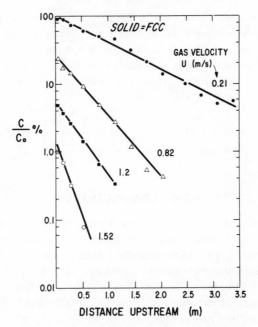

Figure 7.30 Axial concentration profile at the centre line at different gas velocities. Tracer injected at centre.

The relevance of the model is borne out in Fig. 7.30 where semilog plots of the axial concentration profiles at 0.82, 1.19, and 1.52, and also at 0.21 m/s do indeed form straight lines. From the intercepts, one may deduce the ratio of Q to q. Figure 7.31 gives the results. The figure shows the diminishing percentage of gas backflow with gas velocity over the turbulent range.

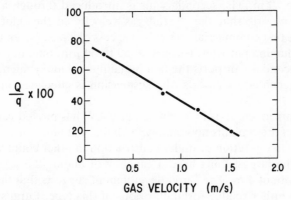

Figure 7.31 Ratio of backflow to upflow (Q/q) versus the superficial gas velocity.

Yang, Huang, and Zhao (1983) have recently studied axial and radial dispersion of helium tracer gas in a fast bed. They found virtually no gas backmixing in the central part of the riser but some near the walls. Radial dispersion was found to decrease slightly with increase in gas velocity and to increase with solid circulation rate. The relationships were expressed in an empirical corelation:

$$E_r = 43.4 \left(\frac{U}{U/\epsilon - v_s} \right) \left(\frac{1 - \epsilon}{\epsilon} \right) + 0.7 \qquad (7.13)$$

7.8 APPLICATIONS

High velocity recirculating beds are proving increasingly attractive for certain applications. Lurgi Chemie and Huttentechnik GmBh of Frankfurt, West Germany, was the first to appreciate the broad potential of the fast fluidized bed. Lurgi realized this potential in its successful process, developed jointly with Vereinigte Aluminum Werke (VAW), for calcining aluminium hydroxide to supply cell-grade alumina (Reh, 1971). The heart of the process is a high velocity fluid bed furnace (Fig. 7.32), which Lurgi often refers to as the highly expanded circulating fluid bed furnace.

Calcination is conducted at a temperature between 1,000 and 1,100°C. Heat is provided by direct injection of oil or fuel gas into the lower section of the furnace. That section is considerably denser owing to the staging of the combustion. Primary air supplied through a distributor is substoicheiometric and the combustion is completed with secondary air in the upper portions of the reactor. The temperature, however, remains remarkably uniform all along the height of the reactor which, in commercial installations, is typically around 15 to 20 m. The secondary air is introduced through a few tuyeres. Reh (1971) reports that the overall gas velocity in the pilot plant which preceded the first commercial unit was in excess of 3 m/s. Given that the mean size of the alumina particles lies around 50 to 80 μm, one may presume that below the secondary air ports the bed is maintained in turbulent fluidization. Above the secondary air ports, the suspension is circulated in the fast bed mode.

Since its introduction in 1970, the Lurgi/VAW has proved very successful, and several of these plants now supply the bulk of alumina in West Germany; they are also in operation or under construction in other countries in Europe and Asia. The largest units have a capacity of about 12.5 kg/s and have a diameter of about 4 m. The Lurgi development suggests that fast beds could be scaled up rather readily for applications of this type. Lurgi's scale-up was in two steps: a 12.5 cm i.d. unit was followed by a pilot 90 cm diameter and 8,5 m tall unit. Both the pilot and the first commercial plant (which was over 3

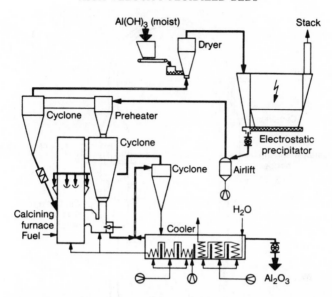

Figure 7.32 Flow diagram of the Lurgi/VAW process for production of alumina.

m in diameter) achieved design operation within a few weeks of startup.

In the only commercial Fisher-Tropsch plant, in Sasolburg, South Africa, the synthesis reaction takes place in a circulating fast fluid bed. The plant has seen successful operation since 1954, and is currently undergoing a large expansion to a capacity many times its present one. The so-called Synthol reactor (Fig. 7.33) operates at a pressure around 20 atms. The existing reactor is about 2.3 m in diameter and about 30 m tall. The gas solid suspension flows upwards through two heat exchange sections where heat liberated by the synthesis reaction is removed. The temperature near the bottom of the reactor, following the mixing of the synthesis gas and the catalyst descending from the standpipe, is about 315°C, and approximately 340°C in the settling hopper (Hoogendoorn, 1973). The gas velocity is about 6 m/s in the heat exchange section and about 2 m/s in the reactor. The solid circulation rate is about 400 kg/m²s. The fine iron catalyst used in the Fischer-Tropsch synthesis belongs to those cohesive solids which Geldart has classified as group C. These solids are simply difficult to fluidize at low velocities, either in the bubbling or even the turbulent regimes. The success of the fast bed Synthol reactor is due in a large measure to the better solid mixing and the more efficient gas solid contact brought about by the high gas velocities typical of the fast bed.

Since the late 1970s, circulating fluid bed boilers have been in commercial service burning a variety of fuels. A recent survey (Schweiger, 1985)

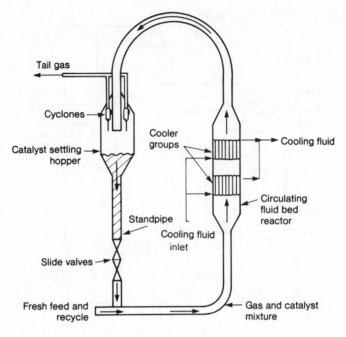

Figure 7.33 SASOL's Synthol reactor for Fischer–Tropsch synthesis.

demonstrates that a major and still growing portion of the market of fluid bed boilers has been captured by circulating fluid bed boilers. These boilers typically operate at higher combustion efficiencies and at lower excess air levels than bubbling-type fluid bed steam generators. They utilize a simpler and more forgiving fuel-feed system and, when burning high-sulphur fuels, require less limestone for SO_2 capture. Also, they are more adaptable to staged combustion affording NO_x control as well as a better load-following capability.

There are indications that some well-established successful processes operate in the turbulent regime e.g. acrylonitrile synthesis, roasting of sulphide ores (Queneau, Bracken, and Kelly, 1958); others are under development, e.g. Mobil's methanol-to-gasoline process (Penick, Lee, and Mazink, 1982). The coming years will see the catalogue of new processes expanding, but if these processes are to fully reap the potential of high velocity fluidization, they must be able to draw from a richer reservoir of fundamental research than that available at present.

There is a need for research at a scale that goes beyond the few inches that has characterized much of the work to date — for systematic study of solids spanning a wide range of properties and of the influence of temperature and pressure. Diagnostic tools that can probe the beds must be brought to bear, and chemical reactions and tracer studies are required to elucidate the interaction of gas and solid.

7.9 NOMENCLATURE

c	concentration of tracer in the lean phase
C	concentration of tracer in the dense phase
d_p	volume surface mean diameter
D	column diameter
g	acceleration of gravity
G_s	solid rate per unit cross-sectional area
$G_{s,TR}$	saturation carrying capacity corresponding to U_{TR}
H	expanded bed height
H_{mf}	bed height at minimum fluidization
q	volumentric flowrate of gas flowing upwards
Q	volumetric flowrate of gas flowing downwards
r	ratio of q/Q
U	superficial gas velocity
u_A	absolute rise velocity of bubble, slug, or lean phase gas
U_c	velocity at which pressure fluctuations across a bubbling bed begin to diminish from their peak value
U_k	velocity at which turbulent regime commences
U_{mf}	minimum fluidization velocity
U_{TR}	transport velocity
v_s	mean solid velocity
v_t	terminal velocity of a single particle
ϵ	voidage
$1 - \epsilon$	volume fraction of solid
ρ_p	particle density
ρ_B	bed density

7.10 REFERENCES

Canada, G.S., McLaughlin, M.H., and Staub, F.W. (1976). *A.I.Ch.E. Symp. Ser.*, **74** (176), 27.

Cankurt, N.T., and Yerushalmi, J. (1978). In *Fluidization* (Eds J.F. Davidson and D.L. Keairns), Cambridge University Press, p. 387.

Carotenuto, L., Crescitelli, S., and Donsi, G. (1974). *Ing. Chim. ital.*, **10**, 185.

Clift, R., and Grace, J.R. (1972). *Chem. Eng. Sci.*, **27**, 2309.

Clift, R., Grace, J.R., and Weber, M.E. (1974). *Ind. Eng. Chem. Fund.*, **13**, 46.

Crescitelli, S., Donsi, G., Russo, G., and Clift, R. (1978). Chisa Conference, Prague.

Gajdos, l.J., and Bierl, T.W. (1978). Topical report for the US Department of Energy, Contract No. EX-C-76-01-2449.

Grace, J.R., and Tuot, J. (1979). *Trans. Inst. Chem. Engrs.,* **57**, 49.

Graff, R.A., and Yerushalmi J. (1978). Seventh quarterly report, prepared for the US Department of Energy, Contract No. EX-76-S-01-2340.

Harrison, D., Davidson, J.F., and de Kock, j.W. (1961). *Trans. Instn. Chem. Engrs.,* **39**, 202.

Hoogendoorn, J.C. (1973). *Clean Fuels from Coal.,* Symposium Proceedings, Institute of Gas Technology, Chicago, Illinois, p. 353.

Horsler, A.G., Lacey, J.A., and Thompson, B.H. (1969). *Chem. Eng. Prog,* **65** (10), 59.

Kehoe, P.W.K., and Davidson, J.F. (1971). *Inst. Chem. Eng. (Lond.) Symp. Ser.,* **33**, 97.

Lanneau, K.P. (1960) *Trans. Instn Chem. Engrs,* **38**, 125.

Massimilla, L. (1973). *A.I.Ch.E. Symp. Ser.,* **69** (128), 11.

Matsen, J.M. (1976). *Fluidization Technology* (Ed. D.L. Keairns), Vol. 11, Hemisphere Publishing Corporation, p. 135.

Price, B.G.L., Lapidus, L., and Elgin, J.C. (1959). *A.I.Ch.E.J.,* 505, 93.

Penick, J.E., Lee, W., and Mazink, J. (1982) *Inst. Symp. on Chem. React. Engng,* ISCRE - 7, 1, Boston, Oct. 4.

Queneau, P., Bracken, E.H., and Kelly, D. (1958). *J. of Metals,* **10**, 527.

Reh, L. (1971). *Chem. Eng. Prog.* **67**, 58.

Saxton, A.L., and Worley, A.C. (1970). *Oil and Gas J.,* **68** (20), 84.

Schweiger, B. (1985). *Power,* Feb. 5–1.

Thiel, W.J., and Potter, O.E. (1977). *Ind. Eng. Chem. Fund.,* **16**, 242.

Turner, D.H.L. (1978), Ph. D. Dissertation, The City College, City University of New York.

Van Swaaij, W.P.M., Burman, C., and Van Breugel, J.W. (1970). *Chem. Eng. Sci.,* **25**, 1818.

Wallis, G.B. (1969). *One-Dimensional Two-Phase Flow,* McGraw-Hill, Chapter 4.

Yang, G., Huang, Z., and Zhao, L. (1983). *Fluidization* IV, Japan, May 29 – June 8, Eng. Foundation.

Yang, W.C. (1976). *Proc. of Pneumo Transport,* Vol. 3, BHRA Fluid Engineering (April).

Yerushalmi, J., and Cankurt, N.T. (1978). *CHEMTECH,* **8**, 564.

Yerushalmi, J., and Cankurt, N.T. Geldart, D., and Liss, B. (1978) *A.I.Ch.E. Symp. Series,* **74** (176), 1.

Yerushalmi, J., and Cankurt, N.T., (1979) *Powder Technol.,* **24**, 187.

Yerushalmi, J. Cankurt, N.T., Geldart, D., and Liss, B. (1978). *A.I.Ch.E. Symp. Ser.* **74** (1976). *Ind. Eng. Chem. Proc. Des. & Dev.,* **15**, 47.

Zenz, F.A., and Othmer, D.F. (1960), *Fluidization and Fluid–Particle Systems* Reinhold, New York.

Gas Fluidization Technology
Edited by D. Geldart
Copyright © 1986 John Wiley & Sons Ltd.

CHAPTER 8

Solid–Gas Separation

L. SVAROVSKY

8.1 INTRODUCTION

The elutriation of particles is an inevitable consequence of fluidized beds, and the equipment required to separate the gas and powder provides a significant contribution to the capital and operation costs of the process. In some applications the fine powder carried out is valuable and process economics cannot allow its loss; in others, emission of the particles constitutes a nuisance or a health hazard; in many applications all of these considerations apply.

Efficient gas cleaning and recovery of the solids form an essential part of the process requirements for the fluidized bed and are the subject of this chapter. Because cyclones are the most common device used in fluidized bed systems they receive the greatest attention, but some of the other techniques used as alternatives or in addition are also discussed briefly.

Solid-gas separation as a title may be interpreted to mean both *degassing* of solids (as a direct analogy with solids dewatering in solid-liquid separation) and *gas cleaning*, i.e. *dedusting* of the gas. Only the latter is to be considered here and the terms *solid-gas separation* and *gas cleaning* will be used interchangeably because the term *gas cleaning* is quite commonly used, even in cases when the solids represent the product.

Three phases may be distinguished in any gas-cleaning operation: tranport of particles onto a surface (separation), collection of separated particles from the separation surface into discharge hoppers (or particle fixation), and the disposal of the collected material from the gas-cleaning equipment. The following account of equipment deals only with the first two phases.

197

8.2 PRINCIPLES OF PARTICLE SEPARATION AND CLASSIFICATION OF EQUIPMENT

In the first phase of gas cleaning, forces are applied to the particles in order to bring them to a collecting surface; the principles of particle separation are usually classified according to the nature of the forces involved. These may be:

(a) *External forces* due to fields of acceleration which are external to the gaseous suspension, such as gravity, electrostatic, or magnetic forces, or

(b) *Internal forces* due to fields of effects which take place within the suspension itself, e.g. inertial or centrifugal forces, diffusion, coagulation, electrostatic effects of charged particles, thermophoresis, diffusiophoresis, and piezophoresis.

The process of screening, in which particles are classified in relation to their ability to pass through an aperture in the screen, does not lend itself to the above classification, but its role in gas cleaning is relatively minor.

Gas-cleaning equipment often combines two or more of the above-mentioned principles in one unit; the classification of equipment therefore does not necessarily follow the same pattern. The most common classification is into four groups, as follows:

(a) *Aeromechanical dry separators* in which gravity and/or inertial effects prevail. This group includes (Svarovsky, 1981b) cyclones, settling chambers, intertial separators, dual vortex separators, and fan collectors (or 'mechanical cyclones'). Only the first will be considered in this chapter.

(b) *Aeromechanical wet separators* (scrubbers) which make use of diffusional and inertial effects.

(c) *Electrostatic precipitators* which depend on electrostatic and gravity forces.

(d) *Filters* which use inertial and diffusional effects.

Many gas-cleaning systems combine two or more of the above groups either by using different equipment in series or by combining these in a single unit.

The above classification of equipment will be followed in the sections on equipment and a separate section will be devoted to each of the four groups.

8.3 GENERAL CHARACTERISTICS OF EQUIPMENT

There are several factors affecting the choice of gas-cleaning equipment for any particular application. These are:

Flowrate–pressure drop relationship; efficiency; economic criteria; suitability for different conditions (the nature of both the dust and the gas), solids concentration, method of disposal, reliability, etc.

8.3.1 Flowrate–pressure drop relationship

Most gas-cleaning devices have a fixed relationship between static pressure drop Δp and gas flowrate Q, depending on the configuration of the gas cleaner. Most frequently, the relationship is expressed in the same way as with other flow devices as:

$$\Delta p = \tfrac{1}{2}\, \text{Eu}\, \rho_g\, U^2 \qquad (8.1)$$

where Eu, the Euler number, is a resistance coefficient (analogous to C_D in Eq. 6.2) which may be a function of the Reynolds number and other operational variables such as the feed concentration of solids, specific water consumption with scrubbers, etc.; U is a characteristic velocity calculated from Q/A, where A is a characteristic area in the separator (e.g. the cross-section of the cylindrical body in cyclones); and ρ_g is gas density.

One exception in the application of Eq. (8.1) is in air filters, where the pressure drop is also a function of the amount of dust deposited on the filter.

8.3.2 Efficiency

Efficiency is an important criterion, since it governs the degree of cleaning. It is best expressed as gravimetric grade efficiency $G(d)$ (see Svarovsky, 1981a, for details of its evaluation). Comparison of typical grade efficiencies of aeromechanical dry (D) and wet (W) separators, electrostatic precipitators (E), and filters (F) is made in Fig. 8.1.

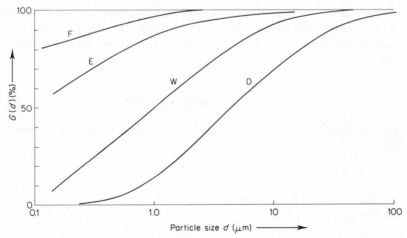

Figure 8.1 Typical grade efficiency curves of dry separators (D), wet separators (W), electrostatic precipitators (E), and filters (F).

8.3.3   Economic criteria

Economic criteria consist of the capital and running costs of the dust-arresting plant. Capital cost is normally expressed per thousand cubic metres of cleaned gas per hour; it may be further split into the cost of the construction material, cost of labour, erection, design, etc. There are other criteria in this category, such as specific volume of the plant ($m^3/1,000 \ m^3$), specific floor area taken ($m^2/1,000 \ m^3$), etc., the importance of which varies with different applications.

Running costs include cost of power, maintenance, water, etc; power needed for running the plant consists of the power for pumps, electricity for cleaning (electrostatic precipitators), and also the power for blowing the gas through the plant.

Whenever the total power requirements consist solely of the power needed for passing the given flowrate Q through the separator, the theoretical power can be calculated from the product of the required pressure drop Δp and gas flowrate: $Q \times \Delta p$ (W). To allow relative comparisons between different separators, the theoretical power may be expressed as specific energy per unit flowrate, and this is usually in watt-hours per thousand cubic metres. Thus, each newton per square metre of pressure drop represents 0.28 Wh/1,000 m^3 or in practical engineering units, 1 mm w.g. (column of water) represents 2.73 Wh/1,000 m^3. The actual power can only be derived from the theoretical power requirements if the efficiency of the fan and the electric motor are known or assumed.

8.3.4   Suitability for different conditions

There are a number of other factors, such as gas temperature and humidity, the cohesiveness and abrasiveness of the dust, reliability, limits in dust concentration, etc., which may exert an overiding influence on the final choice.

8.4   CYCLONES

The advantage of cyclones, and indeed all aeromechanical dry separators, include:

  simple design,
  low capital cost,
  suitability for higher temperatures,
  low energy consumption,
  product is dry,
  reliability,

The most important disadvantage is their relatively low efficiency for very fine particles which leads to their frequent role as a pre-cleaner.

Cyclones are now used in many different fields of technology, but their most extensive application still remains in gas cleaning, where they are employed for the separation of relatively coarse dusts.

There are basically two main designs of cyclones available: the reverse-flow cyclones and the 'uniflow' cyclone., the former being more frequently used. In both types, the inlet gas is brought tangentially into a cylindrical section and a strong vortex is thus created inside the cyclone body. Particles in the flow are subjected to centrifugal forces which move them radially outwards, towards the inside cyclone surface on which the solids separate. The two types of cyclone differ in the direction in which the clean gas leaves the main body, as is shown in the following.

A typical reverse-flow cyclone (Fig. 8.2) consists of a cylindrical section joined to a conical section and the clean gas outlet is through a pipe which extends some distance axially into the cyclone body through the top. The gas inlet may be tangential, spiral, helical, or axial. Different inlet types lead to some variations in performance, but the effect is not strong and, in most cases, the tangential inlet is preferred for its simple construction.

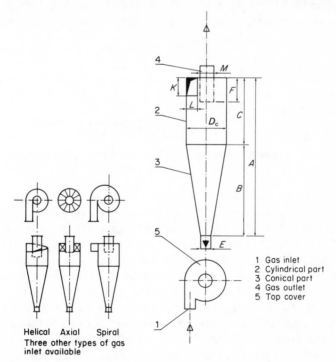

1 Gas inlet
2 Cylindrical part
3 Conical part
4 Gas outlet
5 Top cover

Helical   Axial   Spiral
Three other types of gas
inlet available

Figure 8.2 Schematic diagram of a reverse-flow cyclone with tangential inlet.

The outer vortex created by the tangential entry is helical, moving downwards — particles in the flow settle into a dust layer on the wall and this is pushed down into the apex. Hence, the removal of the dust from the collection surface is due to the gas flow and not to gravity; gravity has been found to have little effect on the separation efficiency of cyclones.

The outer vortex reverses its axial direction in the apex and creates the inner vortex going upward, which carries the gas into the outlet pipe.

As the vortex reaches very far into the apex, it is advisable not to put a rotary valve or a sliding valve there; it is better to leave the dust outlet clear and use a discharge hopper underneath the cyclone with the necessary valve on the hopper outlet. There is evidence of the vortex reaching even into the hopper itself, but the layer of dust reentrainment is much reduced by using a discharge hopper.

In the 'uniflow' cyclone, often also called the 'straight-through' cyclone (Fig. 8.3), the axial direction of the vortex is not reversed but the flow continues in the same direction and the gas leaves through an outlet pipe which is at the end opposite to the inlet. While in the reverse-flow cyclone the separated solids are in powder form, the uniflow cyclone only functions as a concentrator, i.e. the separated solids leave the unit still suspended in gas which is usually 5 to 10 per cent of the main flow. A second-stage separator has to be provided to treat the 'underflow' if the solids are required in powder form.

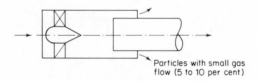
Particles with small gas flow (5 to 10 per cent)

Figure 8.3   Schematic diagram of 'uniflow' cyclone.

It can be shown theoretically (see Eq. 8.6 below) that for a given pressure drop the cut size of a cyclone is proportional to the square root of the cyclone diameter; i.e. the smaller the cyclone the higher will be its efficiency. It is therefore theoretically sound to build multi-cyclone arrangements which use several smaller units in parallel, with common inlet and exit manifolds. Great care must be taken in designing the systems to ensure even distribution of both the gas and the solids between the individual cyclones, because, if this is not achieved, the resulting blockages (and even backflow in some units) reduce the overall efficiency and the advantage of using multi-cyclones may thus be completely lost. Generally, the overall grade efficiency of any multi-cyclone is never as good as that of the individual cyclone units.

Both reverse-flow and uniflow cyclones are used in multi-cyclone arrange-

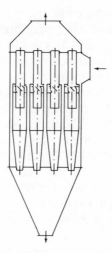

Figure 8.4 Schematic diagram of a multi-cyclone arrangement.

ments, the axial, vane inlet being most commonly employed. Figure 8.4 gives a schematic diagram of a typical multi-cyclone unit.

Multi-cyclones are generally more elaborate in construction and therefore more expensive than single cyclones; they are also more liable to abrasion and blocking of the dust discharge orifice due to the smaller diameter of the individual units. The latter disadvantage makes multicyclones unsuitable for cohesive dusts and leads to a lower limit in feed concentration of dust.

8.4.1 Flow characteristics

The static pressure drop measured between the inlet and the gas outlet of a cyclone is usually proportional to the square of gas flowrate Q: this means that the resistance coefficient defined as the Euler number, Eu, in Eq. (8.1) is practically constant for a given cyclone geometry or 'design', independent of the cyclone body diameter. The characteristic velocity U can be defined for gas cyclones in various ways but the simplest and most appropriate definition is based on the cross-section of the cylindrical body of the cyclone, so that:

$$U = \frac{4Q}{\pi D_c^2} \qquad (8.2)$$

where Q is the gas flowrate and D_c is the cyclone inside diameter.

The two other alternatives to the definition of characteristic velocity, the average inlet or outlet velocities, are not recommended because neither of them would lead to a sensible comparison of different designs; it can be argued that the cyclone body diameter is the most important dimension, determining the manufacturing costs, the space occupied, headroom, etc. As an example to demonstrate the superiority of the definition of body characteristic velocity in Eq. (8.2) consider two cyclones, identical in diameter and in all other dimensions except their gas inlet and outlet diameters. One has a large inlet and small outlet while the other has a small inlet and large gas outlet: the relative size of the two openings may be such as to result in an identical pressure drop – flowrate relationship for both cyclones. Using the body velocity defined in Eq. (8.2), the resistance coefficient Eu in Eq. (8.1) would be the same for both cyclones; this is to be expected as the cyclones are of the same size and give identical flowrates for the same pressure drops. If, however, either the inlet or the gas outlet velocities are used (and some authors still insist on using those), the resistance coefficients thus obtained would be very much different for the two cyclones, thus apparently favouring strongly (and wrongly) one of the two designs depending on which of the two alternative definitions of U is used.

The resistance coefficient Eu increases a little at high pressure drops and reduces at high concentrations, but this can usually be compensated for by plant adjustment, and clean air data are normally taken in design calculations.

8.4.2   Efficiency of separation

The second dimensionless group which characterizes the separation performance of a family of geometrically similar cyclones is the Stokes number Stk_{50} defined as:

$$\text{Stk}_{50} = \frac{d_{50}^2 \, \rho_\text{p} \, U}{18 \, \mu \, D_\text{c}} \qquad (8.3)$$

where d_{50} is the cut size (equiprobable size). The cut size and its relationship to separation efficiency are defined below.

In general, the separation efficiency of gas cyclones depends on particle size (it is then called the 'grade' efficiency) and increases from zero for ultra-fine particles to 100 per cent for very coarse particles (see curve D in Fig. 8.1). The particle size recovered at 50 per cent. efficiency is referred to as the 'cut' size d_{50} and can be understood as equivalent to the aperture size of an ideal screen that would give the same separation performance as the cyclone. The total solids recovery in a particular case then depends on the grade efficiency (or cut size) which characterizes the cyclone operated under given conditions and on the size, density, shape, and dispersion of the particles (i.e. the characteristics of the feed material).

As the grade efficiency does not rise very steeply with increasing particle size, some particles in the feed coarser than the cut size will pass through the cyclone while some particles finer than the cut size will be separated.

As can be seen from Eq. (8.3), the separation efficiency is described there only by the cut size d_{50} and no regard is given to the steepness of the grade efficiency curve. If the whole grade efficiency curve is required in design or performance calculations, it may be generated around the given cut size using plots or analytical functions of a generalized grade efficiency function available from the literature or from previously measured data. The knowledge of the exact form of the grade efficiency is usually not critical in solid-gas separation applications because only total mass recovery is of interest, and this is not much affected by the shape of the curve. Consequently, very little is known about how the shape of the grade efficiency curve is affected by operating pressure drop, cyclone size or design, and feed solids concentration.

In powder classification applications, however, including the case of de-gritting, the shape of the curve determines the amount of the 'misplaced' material, such as the amount of grit reporting to the gas outlet.

The Stokes number Stk_{50} defined in Eq. (8.3) is usually constant for a given cyclone design (i.e. a set of geometric proportions relative to the cyclone diameter D_c), when the cyclone is used to separate granular material at feed concentrations of less than about 5 g/m^3.

Particle size d is best measured as the equivalent Stokes diameter by sedimentation or elutriation methods. This equivalence is based on the assumption that, if a spherical particle and an irregularly shaped particle settle at the same velocity (in gravity or centrifugal fields), they will separate at the same efficiency. This assumption does not hold for flat or needle-shaped particles which assume different orientation in a cyclone than under gravity or centrifugal settling. Problems are also encountered when the particles undergo the separation process in an agglomerated state and the agglomerates are subsequently redispersed into single particles before particle size analysis.

8.4.3   Reverse-flow cyclone designs

Equations (8.1), (8.2), and (8.3) form the basis of gas cyclone design and scale-up. There is a whole host of different cyclone designs available today, which are usually divided into two main groups according to their geometrical proportions relative to the body diameter: the 'high efficiency' designs and the 'high rate' designs. Table 8.1 gives the geometry of the two best-known cyclone designs (refer to Fig. 8.2 for dimensions). Note that only the reverse-flow cyclone is considered here because that is the type used most widely in industrial practice.

The so-called 'high efficiency' cyclones are characterized by relatively small inlet and gas outlet orifices, and a long body, and give high recoveries. The 'high rate' designs give medium recoveries but offer low resistance to flow so

Table 8.1 Cyclone proportions for two different designs based on cyclone diameter $D_c = 1$ (see Fig. 8.2)

Cyclone type	Proportion relative to diameter D_c	A	B	C	E	F	L	K	M
Stairmand, H.E.[a]		4.0	2.5	1.5	0.375	0.5	0.2	0.5	0.5
Stairmand, H.R.[b]		4.0	2.5	1.5	0.575	0.875	0.375	0.75	0.75

[a] High efficiency cyclone
[b] High flowrate cyclone.

that a unit of given size will give much higher air capacity than a high efficiency design of the same body diameter. The high rate cyclones have large inlets and gas outlets, and are usually shorter. In order to prevent the incoming jet of air impinging on the gas outlet pipe, the inlet is spiral (wrap-round type) while the high efficiency units can (and often do) have to a simple tangential entry.

It is interesting to find that, for well-designed cyclones, there is a direct correlation between Eu and Stk_{50}: high values of the resistance coefficient usually lead to low values of Stk_{50} (therefore low cut sizes and high efficiencies), and vice versa. This is shown in Fig. 8.5 where the corres-

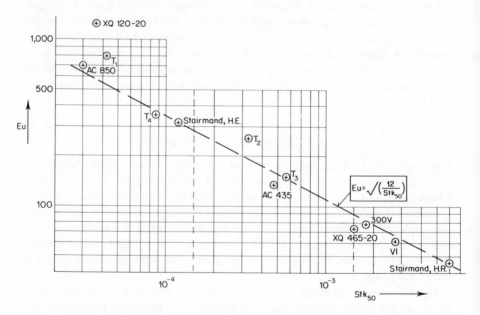

Figure 8.5 Experimental values of Eu and Stk_{50} for some commercial cyclone designs.

ponding values of Eu and Stk_{50} are plotted for several commercial and other well-known designs. The points are well scattered but a line can be drawn through them to show a general trend. The line drawn in Fig. 8.5 can be described by the following approximate equation:

$$Eu = \sqrt{\frac{12}{Stk_{50}}} \qquad (8.4)$$

This equation may be used for estimates of cut size of unknown cyclone designs (of 'reasonable' proportions) from the cyclone flow characteristics and is intended for guidance only.

Note that the scale-up of cyclones based on Eu and Stk_{50} works well for near ambient conditions but also predicts the performance reasonably well at high absolute pressures and high temperatures; this means that there is no effect of high pressures and temperatures other than that accounted for in the definitions of Eu and Stk_{50} on gas viscosity and density.

8.4.4 Recommended range of operation

One of the most important characteristics of gas cyclones is the way in which their efficiency is affected by pressure drop (or flowrate). Correctly designed and operated cyclones should give pressure drops within a recommended range, and this for most cyclone designs operated at ambient conditions is between 2 and 6 inches of water gauge (w.g.) (approximately from 500 to 1,500 Pa). Within this range, the recovery increases with applied static pressure drop. At higher absolute pressures the limits increase to higher pressure drops and the equivalence is based on the same inlet velocity.

At pressure drops below the bottom limit, the cyclone represents little more than a settling chamber, giving low efficiency due to low velocities within it which may not be capable of generating a stable vortex. Above the top limit the mass recovery no longer increases with increasing pressure drop and it may actually decline; it is therefore wasteful to operate cyclones above this limit, the value of which depends very much on the cyclone design, and particularly on the geometry at and below the dust outlet orifice (Svarovsky, 1984). This top limit can be as high as 15 in H_2O (3,740 Pa).

These recommended limits for pressure drop are demonstrated in Fig. 8.6. If a given dust is fed into a cyclone and the input dust concentration is kept constant but the flowrate Q is varied, the total efficiency E_T first increases until a maximum is reached (point B) beyond which it falls off. Theoretically, it should carry on increasing as indicated by the dashed line in Fig. 8.6 since higher tangential velocities due to higher flowrates cause stronger centrifugal fields and should improve separation. In practice, increasing turbulence at higher flowrates causes reentrainment of particles and, at very high flowrates, the efficiency falls rapidly. Although, with different dusts, different curves of

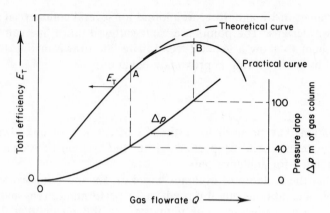

Figure 8.6   Typical operational characteristics of gas cyclones.

$E_T = f(Q)$ would be obtained, the position of the maximum is approximately constant, and from it a general range of optimum operation can be derived. Thus, a most desirable range of operation is clearly just before the optimum, between points A and B, where high separation efficiency is achieved at reasonable pressure drops Δp (or pressure heads H). It should be noted that as the pressure head H increases with Q (also plotted in Fig. 8.6), operation beyond the maximum would mean higher energy consumption and hence higher running costs.

Points A and B are usually found to correspond to H of 40 to 100 m of gas column, respectively, and this is the recommended operation range. There are, inevitably, slight variations between values recommended by different manufacturers, since different cyclone designs have somewhat different characteristics.

The effect of flowrate Q on grade efficiency $G(d)$ follows the same pattern as with the total efficiency — the best grade efficiency curve corresponds to the maximum E_T.

The increase in separation efficiency with the pressure drop Δp, within the recommended range of Δp and for a given cyclone, can be demonstrated using Eqs (8.1), (8.2), and (8.3). For a given cyclone, gas, and solids, the equations can be combined to show the dependence of d_{50} on Δp as follows:

$$d_{50} \propto (\Delta p)^{-1/4} \qquad (8.5)$$

Alternatively, for the same pressure drop and variable cyclone diameter, Eqs (8.1), (8.2), and (8.3) give another proportionality:

$$d_{50} \propto (D_c)^{1/2} \qquad (8.6)$$

which provides the argument in favour of multi-cyclone units as mentioned earlier.

The effects on the two most important operating characteristics, flowrate Q and cut size d_{50}, can be summarized as follows. The flowrate Q (capacity) is increased by:

(a) Increasing the pressure drop

(b) Increasing the feed solids concentration

(c) Design changes, leading to lower resistance to flow (Larger inlet and gas outlet orifices, for example)

Operating cut size d_{50} is reduced (i.e. mass recovery is increased) by:

(a) Increasing the pressure drop (up to a limit, as discussed above)

(b) Increasing the feed concentration (particle surface properties and gas humidity play an important role here)

(c) Design changes, usually leading to higher resistance to flow, like reduction in the inlet and outlet orifices: smaller diameter of cyclone body D_c reduces the cut size

As can be seen from the above points, flowrate and cut size are not independent and most of the operating and design variables affect them both.

The use of the knowledge of the two dimensionless groups Eu and Stk_{50} for cyclone scale-up and design is demonstrated in the following example.

Example 8.1

Determine the diameter of a gas cyclone and, if necessary, the number of the cyclones to be operated in parallel to treat 2.5 m^3/s of ambient air (viscosity is 18.25×10^{-6} N s/m^2, density 1.2 kg/m^3) laden with solids of density 2,600 kg/m^3 at a pressure drop of 1200 N/m^2 and cut size not more than 6 μm.

Take the resistance coefficient Eu (related to the superficial velocity in the cyclone U) equal to 46 and the Stokes number, corresponding to the cut size, equal to 0.006.

Solution

Cyclone diameter can be calculated from Eqs (8.1) and (8.2):

$$D_c^2 = \frac{2.5 \times 4}{\pi} \sqrt{\frac{1.2 \times 46}{1,200 \times 2}}$$

Therefore:

$$D_c = 0.695 \text{ m}$$

This is for the case of one cyclone to treat the whole of the flow. Check on cut size: from Eq. (8.3):

$$d_{50} = \frac{0.006 \times 18 \times 18.25 \times 10^{-6} \times \pi \times 0.695^3}{2,600 \times 4 \times 2.5}$$

$$d_{50} = 8.94 \ \mu m$$

This is greater than 6 μm and therefore unsatisfactory. The calculation can be repeated by using $Q/2$, $Q/3$, etc., until the cut size becomes less than 6 μm. This is reached for $Q/5$, so the solution is that there should be 5 cyclones operating in parallel, each 0.31 m in diameter.

It should be stressed here, however, that the scale-up procedure shown above is only approximate because there are effects other than inertial separation which determine the collection efficiency, such as dust reentrainment and electrostatic effects, which do not necessarily scale up in the same way.

The feed concentration of solids is known to affect the separation efficiency. Concentrations higher than about 1 g/m^3 usually lead to higher efficiency because of the increased rates of interparticle collisions and the resulting agglomeration. Most cyclones serving fluidized beds process gas having concentrations much higher than 1 g/m^3 in some cases. It is therefore fortunate that, if a cyclone is designed on the basis of the scale-up described here, the estimate of the expected collection efficiency at higher concentrations is usually conservative.

8.4.5   Abrasion and attrition in cyclones; clogging

Abrasion in gas cyclones is an important aspect of cyclone performance and is affected by the way cyclones are installed and operated as much as by the material construction and design. Materials of construction are usually steels of different grades, sometimes lined with rubber, refractory lining, or other material. Abrasion is worst in two critical zones within the cyclone body: in the cylindrical part just beyond the inlet opening and in the conical part near the dust discharge. Abrasion is a serious problem in particular in the power generation industry and in the production of building materials, although separation of apparently soft materials like wood chips is also accompanied by some abrasion.

There has been little published on the effect of operating variables of life expectancy of cyclones, the most comprehensive study being a Czech investigation published recently (Storch, 1979), but there is a great need for more information. This is particularly imperative in applications where external gas cyclones are operated at high absolute pressures and thus have to be insured as pressure vessels. The risk of explosion because of thinning of the cyclone wall due to erosive wear is eliminated in many fluid catalytic cracking units by housing the cyclones in a common pressure vessel which is itself not being eroded by the dust.

Erosive wear by a given dust is affected by five most important variables: design velocity U, direction of particle motion (radius of curvature of its trajectory R), number of impinging particles, particle size, and time. For the same feed material and same cyclone geometry, the abrasion of a cyclone is proportional to an abrasion index AI:

$$AI = \frac{U^n c}{R} \qquad (8.7)$$

where U is a characteristic velocity, c is mass concentration of solids, and R is the radius of the cyclone body. The exponent n is usually between 4 and 5 for gas cyclones, which indicates a very strong effect of the flow velocity; more than one cyclone in series, with the first one run at low pressure drop, is a better alternative than only one stage at a high pressure drop (= high velocity). The first stage is used as a pre-collector collecting oversize particles to reduce the concentration and particle size for the second stage, which can then be run at a higher pressure drop. Reduced concentration and increased cyclone diameter also reduce abrasion according to Eq. (8.7). The requirement of low velocity is stronger, however, and to achieve a given recovery, smaller diameter cyclones operated at low pressure drops are actually better than large diameter units which, for the same recovery, would have to be operated at higher pressure drops.

The above abrasion index can be used in life expectancy predictions from tests with the same material on a smaller cyclone, but the actual value of the exponent would have to be ascertained. Cyclone manufacturers probably have the best chance to obtain enough data for abrasion studies but they usually do not publish such information.

Attrition of solids is known to take place on collection in gas cyclones but little is known about how it is related to particle properties which particle sizes are most affected, and how attrition can be related to abrasion of the cyclone walls. Clearly, large particles are more likely to be affected by attrition so that finer fractions are generated by knocking off corners or complete breakage of the larger particles.

Attrition is most detectable in recirculating systems such as fluidized beds where cyclones are used to return the carryover material back to the bed. The complete inventory of the bed may pass through the cyclones many times per hour and the effect of attrition is thus magnified and therefore easily measured.

Even in single-pass systems, however, attrition is sometimes detected and it demonstrates itself by severe discrepancies in mass balance of different particle size fractions around the cyclone. It is not uncommon to find that what is supposed to be the coarse product, i.e. the collected solids, is in fact just or nearly as fine as the fine product in the gas overflow. In isolated cases, when the feed solids are very fragile or have a shape which is easily broken (thin-shelled hollow spheres, scale particles, or such like), the coarse product may even be finer than the solids remaining in the gas flow and reporting to overflow. Any meaningful evaluations of grade efficiency curves or any other assessments of the separation performance of the cyclone are then impossible.

There is clearly a need to investigate the mechanism of attrition with the view to relating it to the fracture properties of the solids, and to develop a realistic attrition 'index', similar to that used for abrasion in cyclones (Eq. 8.7). An empirical correlation for abrasion, specific to fluid catalytic cracking applications, has been proposed by Zenz (1975) and there are several similar tests used by catalyst manufacturers and users. These are usually based on attrition in a high velocity jet which enters a small bed of the powder.

Clogging of cyclones is the greatest single source of failures in their operation. This is usually caused by plugging or overloading of the dust outlet orifice and as soon as this happens, the cyclone cone quickly fills up with dust, pressure drop increases, and recovery reduces or ceases altogether. Clogging may be caused by a variety of reasons such as mechanical defects (bumps on the cyclone cone, non-circularity of the dust outlet, weld or gasket protrusion, etc.), changed chemistry, or physical properties of the dust: condensation of water vapour for example, is a sure cause.

Smaller diameter cyclones are more likely to clog than the large ones; there is a top limit in feed solids concentration (above which clogging is likely to occur) for each cyclone and, for given solids, it is roughly proportional to the cyclone diameter (Svarovsky, 1981b).

8.4.6   Discharge hoppers and diplegs

The design of the dust discharge end is very important for correct functioning of a gas cyclone. Any inward leakages of air at that end lead to sharp deterioration of separation efficiency (due to dust reentrainment) while if the cyclone is under pressure, outward leakages may marginally improve recovery but lead to pollution and loss of product. It is therefore best to keep the dust underflow as airtight as possible.

The strong vortex inside a cyclone reaches into the space underneath the dust outlet and it is most important that no powder surface is allowed to build within at least one cyclone diameter below the underflow orifice; the greater the distance the better. In order to reduce the headroom necessary to fulfil this requirement, a conical vortex breaker is sometimes used just under the dust discharge orifice which stops the vortex intruding into the discharge hopper below. This design practice is particularly common in Germany. Some companies use a 'stepped' cone to counter the effects of reentrainment and abrasion, and a recent investigation (Svarovsky, 1984) has proved the value of this design feature.

In fluid catalytic cracking beds with internal cyclones, diplegs are used to return the collected carryover particles into the fluidized bed. These are vertical pipes of 'chutes' connected directly to the dust discharge orifice of the cyclone (a stepped cone is usually used in the lower part of the cyclone) or to the bottom of its discharge hopper (if such a hopper is fitted), extending down into the fluidized bed below. Returning particles flow down the dipleg in the lean phase but in the dense phase flow into the lower part before entering the bed. The level of the dense phase in the dipleg is always higher than the bed surface and provides a necessary resistance to reduce the upflow of the air which would reduce cyclone efficiency. Trickle valves are usually fitted to the bottom end of diplegs in order to provide the seal at start-up, when no powder is yet returning into the bed; such valves do not allow any return of the solids into the bed until a sufficient head of solids has accumulated in the dipleg (see Chapter 12).

As to the designs of diplegs, Zenz (1975) recommends a clear dipleg height of at least 7 feet (2.1 m) and a minimum inside diameter of 100 mm. The solids downflow capacity of a dipleg pipe can be calculated from an existing correlation (Zenz, 1975), and this may be used to size the pipe from the properties of the solids and the amount of solids returned by the cyclone.

8.4.7 Cyclones in series

The solids recovery of a single cyclone does not continue to rise with applied pressure drop above a maximum. The recovery can be further increased by connecting cyclones in series, and this is often done in practice. No more than two cyclones should be used, however, unless steps are taken so that the cut size of the subsequent stages is made progressively lower by 'tightening up' on the design (primary stage of medium or low efficiency design and further stages of progressively more efficiency design or smaller diameter). The first stage often has a high dust loading and the operating cut size is thus reduced by agglomeration; the second stage, if identical in design, may operate at a greater cut size on account of it receiving much more dilute feed. This does not make good utilization of the second stage as its recovery is small; it is like using a fine screen followed by a coarse screen, and design changes in the second stage should therefore ensure that its cut size is as small or smaller than that of the first stage.

Apart from the resulting gain in recovery, two-stage systems are also advantageous for separation of fragile, agglomerated or abrasive dusts in that the first stage is then designed to operate at low inlet velocity. A large diameter primary cyclone may be used to collect the grit which would plug or erode the high efficiency cyclone in the second stage. The two-stage systems also offer additional reliability in that if the primary cyclone plugs, the secondary still collects.

The series connection of cyclones is not always necessarily in the direction of the gas overflow; sometimes it is advantageous to draw off 5 to 15 per cent of the gas flow through the dust outlet orifice and separate the concentrated suspension in a small secondary cyclone. Such an arrangement does not improve the overall recovery of the plant, however, other than through the beneficial effect it has on the recovery in the first stage.

Several pressurized fluidized bed combustors under test worldwide use a conventional series connections on overflow, however, with up to four stages (NCB/CURL 1981; Ernst *et al.*, 1982; Fluidized Combustion Contractors Ltd, 1982). Cyclones are also widely used in series connections with other gas-cleaning devices such as filters, electrostatic precipitators, or scrubbers, usually as precollectors, to reduce the load of the high efficiency units that follow.

8.5 AEROMECHANICAL WET SEPARATORS (SCRUBBERS)

The separation of particles from gases is aided in scrubbers by the addition of a liquid, usually water. The presence of water plays an important role in all of the three phases of particle collection outlined earlier. In the first phase, particles separate on collecting surfaces, which are either in the form of water droplets or the wet surfaces of the scrubber structure; in either case, particle reentrainment is virtually eliminated and this means that scrubbers are invariably more efficient than the dry separators such as cyclones.

In the second phase, the collected particles are washed off the collecting surfaces in a film of water and this again is done without any danger of particle reentrainment into the flow; any reentrained water droplets are separated in a spray eliminator which is incorporated in most scrubbers.

Finally, in the third phase, the solids are removed from the scrubber in the form of a slurry, which is more easily handled than dry powders.

All of the above advantages in using water to enhance particle separation lead to one major disadvantage in that the collected solids may then have to be separated from the water. Thus what was initially a solid–gas separation problem becomes a solid–liquid one.

The most important collection mechanisms employed in scrubbers are the inertial principle, interception, and diffusion; secondary mechanisms include diffusiophoresis (i,e, mass transfer in the form of condensation leads to a force on particles causing their deposition), thermophoresis (deposition due to a temperature gradient), and particle growth due to the condensation of water on the particles.

The grade efficiency function for those types of scrubbers where the inertial principle is the prevailing separation mechanism was shown by Calvert *et al.* (1972) to be of the following general form:

$$G(d) = 1 - \exp(-Ad^B) \qquad (8.8)$$

where A and B are constants. Packed towers, sieve plate columns, and Venturi scrubbers (for $2 <$ Stk < 8) follow the above equation with $B = 2$, while for centrifugal scrubbers B is about 0.67. The constant A depends on the scrubber type, specific water consumption and other variables, and on the nature of the dust (particularly its affinity to water).

In the other types of scrubbers, where diffusion and other additional principles contribute appreciably to the separation, more complicated mathematical models have to be sought (Calvert *et al.*, 1972)

8.6 ELECTROSTATIC PRECIPITATORS

The electrostatic precipitator is an apparatus which cleans gases by using electrostatic forces to remove the solid or liquid particles suspended in a gas stream. The dust-laden gas is passed through an intense electrostatic field set up between electrodes of opposite polarity. The field charges the particles and generates the separation forces leading to particle collection. The separation forces are applied directly to the particles, without any need to accelerate the gas, and this results in a very low power requirement. Furthermore, the separation efficiency is less dependent on particle size than with the mechanical, wet or dry, separators; hence electrostatic precipitators achieve high efficiencies even with extremely fine particles. However, they are found on few fluidized bed installations and will be discussed no further here. A thorough discussion can be found in Svarovsky (1981b).

8.7 FILTERS

Gas filtration may be defined as the separation of particles from gases by passing a gaseous suspension through a porous, permeable medium which retains the particles. Depending on where in the medium the particle will separate, two extreme cases can be recognized:

(a) *Dust cake filtration*, where the particles are deposited in the form of a cake on the upstream side of a relatively thin filter medium (in single-layer filters).

(b) *Depth filtration*, in which particle deposition takes place inside the medium (in packed filters).

In many practical applications, such as fabric filtration, both cake and depth filtrations take place, with the latter prevailing in the initial stages of a cyclic filtration operation and the former prevailing towards the end. In other cases, such as in sand filters or high inertia fibrous filters, depth filtration alone takes place throughout the whole cycle. A great majority of the available texts on gas filtration completely neglect the importance of cake filtration in industrial fabric filtration and build the necessary theory purely on the basis of inertial separation and diffusion taking place within the filter medium.

The first possible classification of gas filters is according to the dust concentration or 'loading' they are designed for. The 'primary' gas filters (mostly bag filters or gravel bed filters) are designed for high dust loading, usually from 0.2 to about 200 g/m³ because no gas filters can be used above this limit. The 'secondary' filters are generally employed in lighter dust loadings of no more than 100 or 200 mg/m³, such as the air-conditioning, and high efficiency filters.

High dust loadings and high gas volumes can be treated in bag filters. The filter medium is made from a relatively thin cloth in the shape of a cylinder, tube, pocket, or bag, operated in a vertical position to facilitate dust removal. The typical length/diameter ratio of the bags is 20:1, with the diameter from 120 to 150 mm. In exceptional circumstances, the bag length may be up to 10 m.

The filtration mechanism is predominantly that of dust cake filtration, the cake being formed either on the outside or on the inside of the bag depending on which way the gas is arranged to flow. In order to be able to treat larger volumes of air in the smallest possible space, multi-bag arrangements are used in which many bags are operated in parallel.

The design velocity for a given duty is usually selected from experience with the same or similar material. The most common range for separation of fine dusts in bag filters is from 1 to 1.5 cm/s, except for pulse-jet systems which allow higher velocities up to 3 cm/s. Higher velocities can be used with coarser dusts. It is not uncommon to find that some installations simply cannot be operated above a certain limiting value of filtration velocity, when extensive blocking occurs.

Filtration in bag houses is a cyclic operation and within each cycle the resistance of the filter changes to the gradual deposition of dust on the filter medium. A curve which is characteristic for each operating system is the relationship between pressure drop and the specific mass of the dust collected on the medium (this is proportional to the volume of gas filtered, at constant dust loading).

The cleaning cycle in bag filters is triggered either by a timing circuit or more often by a pressure switch.

8.8 NOMENCLATURE

A	cyclone body length	m
AI	abrasion index	(see Eq. 8.7)
B	length of the cone	m
c	dust concentration	kg/m³
C	length of cylindrical part	m
d_{50}	cut size (equiprobable size)	m

D_c	cyclone diameter	m
E	dust outlet diameter	m
Eu	Euler number	—
F	length of gas outlet pipe	m
K	inlet height	m
L	inlet width	m
M	gas outlet diameter	m
n	exponent	
R	radius of cyclone body	m
Q	gas flowrate	m^3/s
Stk_{50}	Stokes number for d_{50}	—
U	superficial gas velocity based on D_c	m/s
Δp	static pressure drop	N/m^2
μ	gas viscosity	kg/ms
ρ_g	gas density	kg/m^3
ρ_p	particle density	kg/m^3

REFERENCES

Calvert, S., Goldschmid, J., Leith, D., and Mehta, D. (1972). *Scrubber Handbook*, Natl Techn. Info. Service, Springfield, Va., PB 213–016.

Ernst, M., Hoke, R.C., Siminski, V.J., McCain J.D., Parker, R., and Dremmel, D.C. (1982). 'Evaluation of a cyclone dust collector for high temperature, high pressure particle control', *Ind. Eng. Chem. Proc. Dex. & Dev.*, **21**, 158–161.

Fluidized Combustion Contractors Ltd (1982). 'Observation and analysis work associated with a 1000–hour test program in a pressurized fluidized-bed conbustion facility', EPRI CS–2582, Project 979–3, Topical Report, Sept.

NCB/CURL (1981). 'Fluidized bed combustion, 1000 hour test program', Vols 1, 2, 3, US/DOE/ET–10423–1101, Sept.

Storch, O., *et al.* (1979). *Industrial Separators for Gas Cleaning,* Elsevier, Ansterdam.

Svarovsky, L. (Ed.) (1981a). *Solid–Liquid Separation,* 2nd ed. Butterworths, London.

Svarovsky, L. (1981b). *Solid–Gas Separation,* Elsevier, Amsterdam.

Svarovsky, L. (1984). 'Some notes on the use of gas cyclones for classification of solids', in *Proc. First European Symposium on Particle Classification in Gases and Liquids*, Nuremberg, May 9–11, Dechema.

Zenz, F.A. (1975). 'Cyclone separators', Chapter 11 in *Manual on Disposal of Refinery Wastes, Volume on Atmospheric Emissions,* API Publication 931, May.

CHAPTER 9

Fluid Bed Heat Transfer

J.S.M. BOTTERILL

9.1 INTRODUCTION

Among the advantageous properties of gas fluidized systems is the extremely large area of solid surface exposed to the gas. Thus, a cubic metre volume of 100 μm diameter particles has a surface area greater than 30,000 m^2, which is of a similar magnitude to that of the surface area of the Great Pyramid of Cheops. Such colossal solid surface area greatly facilitates solid-to-gas heat and mass transfer operations even though it may not all be accessible to fresh gas. Additionally, because of the solids mixing generated within the bulk of a bubbling gas fluidized bed, temperature gradients are reduced to negligible proportions and the bed possesses a very high effective thermal conductivity in the vertical direction. High rates of heat transfer are obtainable between the fluidized solids and an immersed transfer surface. This is also primarily a consequence of the mixing within the bulk of the bed and of the high thermal inertia of the solids. The latter property has been used in many industrial operations, e.g. in the Cat Cracker to transport heat between the regenerator and reactor, although rarely, as yet, has it been explicitly exploited for its own sake. Indeed, the advantageous thermal properties of gas fluidized solids as a high temperature heat storage and transport medium of very low vapour pressure has still to find application beyond some metallurgical heat treatment process (Baskakov, 1968; Virr, 1976).

Under the right operating conditions, a gas fluidized bed constitutes a very satisfactory thermal system. Nevertheless, there have been disappointments — a consequence of the complexity of bubbling fluidized beds (see particularly Chapters 4 and 5). Theoretical models descriptive of the heat transfer mechanism are of limited application because they require knowledge of parameters which are not generally available. Finally, the reader should be warned against using empirical correlations without first checking

219

back to see that they have been determined from tests of relevance to the particular problem.

9.2 PARTICLE-TO-GAS HEAT TRANSFER COEFFICIENTS

Because of the very large surface possessed by a mass of small particles, particle-to-gas heat transfer is only rarely a limiting factor in fluidized bed heat transfer despite the fact that, based on the total particle surface area, coefficients are typically only of the order of 6 to 23 W/m^2 K. (The exceptions are such instances as drying, where the need to provide the latent heat of vapourization can give the bed material a very high effective heat capacity, or with particles which are reacting and where the reaction is very temperature sensitive). There is much scatter in the range of the published coefficients and, indeed, there are many uncertainties involved in their experimental determination. There is the problem of measuring the relevant gas and solid temperatures (Botterill, 1975). Singh and Ferron (1978) have reported differences between gas and particle temperatures using a 'light-pipe probe' and thermistor to measure the radiant energy from the particles. Thus, for a bed of 230 μm catalyst being cooled from 200°C by fluidizing air at ambient temperature, differences as large as 70°C were detected between gas and particle temperatures. They suggest that the temperature indicated by an immersed thermocouple corresponds to a weighted mean which is 80 per cent. of the particle temperature and 20 per cent. of that of the gas. Apart from problems in making the actual measurements, there is also the problem of interpreting them. Assumptions have to be made as to the nature of the gas flow through the bed. It was pointed out by Kunii and Levenspiel (1969) that the most realistic model for this is a plug flow model. Measurements interpreted on this basis by different investigators are generally consistent with each other whereas those interpreted on the basis of a model involving complete gas mixing are widely scattered and give excessively low particle Nusselt numbers. (The measured temperature gradient at the entry to the bed is too steep to permit direct distinction between the two models; thus Richardson and Ayers (1959) noted that the gradient was undetectable after the first 2 mm or so from the point of entry into beds composed of mean particle diameters of 200 μm and less; see also Example 9.1 below.) Experimental results, covering particle Reynolds numbers up to 50 and analysed on a plug flow model (Kunii and Levenspiel. 1969), are correlated by the equation:

$$Nu_{gp} = 0.03 Re_p^{1.3} \qquad (9.1)$$

The low limiting Nusselt number obtained can be accounted for by the fact that not all of the total particle surface area is available for heat transfer with the gas. Zabrodsky (1963) proposed a model involving 'microbreaks', pointing out that the gas did not necessarily flow uniformly through the continuous phase of the bed. Littman and Sliva (1970) showed that, particularly in packed beds, there is a strong dependence on Reynolds number because the region near the point of contact between particles is not fully accessible to the flowing fluid. Reference is made (Botterill, 1975) to two studies which report that higher coefficients are obtained when the system is operated at elevated pressure.

For a well-mixed bed of particles of low Biot number (i.e. negligible internal thermal resistance) operating at a steady temperature, a simple heat balance relates the change in temperature of the fluidizing gas to its penetration into the bed. Thus, on the basis that the gas exchanging heat with the bed is in piston flow through the continuous phase, then:

$$C_g \, U \, \rho_g \, \Delta T_g = h_{gp} \, S_B \, (T_g - T_p) \, dl \qquad (9.2)$$

for an element of bed dl m deep and of unit cross-sectional area where ΔT_g is the change in temperature of the gas flowing through the element of bed and S_B is the surface area of solids per unit volume of bed. This can be integrated directly to give:

$$\ln \left(\frac{T_g - T_p}{T_{g,in} - T_p} \right) = - \left(\frac{h_{gp} \, S_B}{U \, \rho_g \, C_g} \right) l \qquad (9.3)$$

where:

$$S_B = \frac{6(1 - \rho)}{d_p} \quad m^2$$

and

$$h_{gp} = \frac{0.03 \, d_p^{0.3} \, \rho_g^{1.3} \, U^{1.3} \, k_g}{\mu^{1.3}} \qquad \text{W/m K from Eq. (9.1)}$$

The distance l_n, in which the gas-to-particle temperature difference falls by a factor:

$$n = \frac{T_{g,in} - T_p}{T_g - T_p}$$

is given by:

$$l_n = \frac{5.5 \ln n \, \mu^{1.3} \, d_p^{0.7} \, C_g}{\rho_g^{0.3} \, U^{0.3} \, k_g \, (1 - \epsilon)} \qquad (9.4)$$

Example 9.1

A bed of 450 μm particles is operating at 150°C. the temperature and superficial velocity of the incoming gas are 550°C and 0.4 m/s, respectively. Approximately how far will the incoming gas have penetrated into the bed before it is cooled to 350°C?

Solution

This estimate can be made using Eq. (9.4) and the gas physical properties for the average temperature over the specified range, i.e. between 550 and 350°C. At this temperature of 450°C the appropriate physical properties are:

$$\rho_g = 0.68 \text{ kg/m}^3$$
$$k_g = 0.04 \text{ W/m K}$$
$$C_g = 1025 \text{ J/kg K}$$
$$\mu = 2.8 \times 10^{-5} \text{ N s/m}^2$$

The gas velocity at 450°C for an inlet velocity of 0.4 m/s at 550°C will be 0.35 m/s:

$$n = \frac{550 - 150}{350 - 150} = 2$$

$$l_2 \sim \frac{5.55 \ln 2 (2.8 \times 10^{-5})^{1.3} (450 \times 10^{-6})^{0.7} \, 1{,}025}{0.68^{0.3} \times 0.35^{0.3} \times 0.04 (1 - 0.4)} \sim 1.4 \text{ mm}$$

Note that because U_{mf} is approximately proportional to d_p^2 for laminar flow conditions. l should not be very sensitive to particle diameters for smaller mean diameters. However, for particles with group D (Geldart 1972) characteristics, the gas flow conditions will be transitional to turbulent.

Negligible temperature gradients will be set up within small particles because the relevant Biot numbers will be small generally, they may be ignored if the particle Biot number is less than 1/10. Barile, Seth, and Williams (1970) have studied the transfer of heat between the gas in a rising chain of bubbles and the continuous fluidized phase. Measured temperature profiles indicated a rapid approach to bubble/dense phase equilibrium (within the order of length of one bubble diameter), but this could still be significant when a highly exothermic reaction is occurring.

McGaw (1976) analysed the problem of gas-to-particle heat transfer in a cross-flow moving packed bed heat exchanger which is relevant to the design of particle coolers and circulating bed regenerative systems. He gave both a rigorous analysis, taking into account internal particle resistance to heat transfer and a simplification for the situation where that resistance may be neglected.

9.3   BED-TO-SURFACE HEAT TRANSFER

Under the usual bubbling fluidized bed operating conditions, the heat transfer coefficient h between an immersed surface and a gas fluidized bed can be thought to consist approximately of three additive components (Botterill, 1975), namely:

(a)   The particle convective component h_{pc}, which is dependent upon heat transfer through particle exchange between the bulk of the bed and the region adjacent to the transfer surface

(b)   The interphase gas convective component h_{gc}, by which heat transfer between particle and surface is augmented by interphase gas convective heat transfer and

(c)   The radiant component of heat transfer h_r.

Thus:

$$h \quad = \quad h_{pc} \quad + \quad h_{gc} \quad + \quad h_r \qquad (9.5)$$

| approximate range of significance | 40 μm → 1 mm | > 800 μm and at higher static pressure | higher temperatures (> 900 K) and difference |

The particle convective component h_{pc} is particularly dependent on bed bubbling behaviour (Section 9.3.1) and, at higher fluidizing velocities when bubbles increasingly shroud the heat transfer surface, bed-to-surface heat transfer by this mechanism is reduced. Allowance can be made for this (e.g. Baskakov et al., 1973) by reducing the contribution of the particles convective component according to the fraction of the time that the transfer surface is shrouded by bubbles, but this is not generally known.

The situation, however, can be expected to be different when the operational range is extended into the 'turbulent' and 'fast fluidization' regimes (see Chapter 7). Much higher gas velocities are then employed. For particles of millimetre order diameter, the velocity for turbulent fluidization approaches that of the particle terminal velocity, and bed behaviour is characterized by a high degree of solids eddy motion superimposed on cocurrent and countercurrent solid/gas flow (Staub, 1979). For powders of group A, the operating velocities for fast fluidization exceed the particle terminal velocity and there is very extensive back mixing in the transporting solids (Yerushalmi and Cankurt, 1979). Under these conditions, it would seem profitable, as Staub (1979) suggested, to apply the correlation methods earlier developed for flowing dense solid/gas suspensions. This is briefly referred to below (Section 9.4).

9.3.1 Particle convective component h_{PC}

The particle convective component is responsible for the characteristic marked increase in the bed-to-surface heat transfer coefficient observed as the particle circulation develops when the fluidized bed begins to bubble (Fig. 9.1). With a continuing increase in the fluidizing gas flowrate, the coefficient

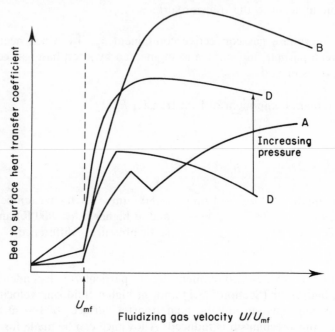

Figure 9.1 Effect of gas flowrate on bed-to-surface heat transfer for powders of groups A, B, and D characteristics when radiant transfer may be neglected.

passes through a maximum. With finer group A powders, which expand stably after the minimum fluidizing velocity is exceeded before bubbling begins, there is an additional minor peak in the rising portion of the curve (Khan, Richardson, and Shakiri, 1978).

It was Mickley and his coworkers who first appreciated the part played by the particle heat capacity in the process. The volumetric heat capacity of the particle is of the order of a thousandfold that of the gas at atmospheric pressure, so it is the particles which have the capacity to transport heat through the bulk of the bed. (Not until fluidizing beds of group D material is there significant heat transfer by gas convection and the interphase gas convective component h_{gc} becomes truly important.) Thus, particles in the bulk of the bed exchange heat with the fluidizing gas and, by conduction through the gas, with each other. Particles usually stay within the bulk of the bed long enough to come to the same temperature as their neighbours. Then a 'packet' of particles at the bulk bed temperature is swept into close proximity with the heat transfer surface under the bubble-induced circulation patterns occurring within the bed. When they first arrive close to the heat transfer surface there is a high local temperature difference between the bed material and the surface; consequently heat transfers rapidly. The longer the particles reside close to the surface, the more closely the surface and local bed temperatures approach. Highest mean temperature differences are therefore to be expected under those bubbling bed conditions which can generate a rapid exchange of material between the vicinity of the transfer surface and the bulk of the bed, i.e. with low particle residence time adjacent to the transfer surface as suggested by the basic Mickley and Fairbanks (1955) model which can be expressed in the form:

$$h = \sqrt{(k_{mf}\, \rho_{mf}\, C_p\, S)} \qquad\qquad (9.6)$$

The terms representing the thermal conductivity and density of the bed, k_{mf} and ρ_{mf}, are for the material at the condition of minimum fluidization representative of the condition of the 'packet material'; C_p is the heat capacity of the particles; and S is a 'stirring factor' representative of the frequency of

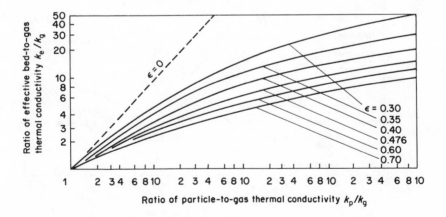

Figure 9.2 Diagram for estimation of effective continuous phase thermal conductivity (Baskakov, 1968). *Reproduced by permission of the author.*

material replacement at the surface. The bed thermal conductivity may be estimated using the plot given in Fig. 9.2 (following Baskakov, 1968) but the stirring factor S is not generally known (see later and Eq. 9.13). A limit to this unsteady state model, however, is reached with larger particles when their time constant for cooling t exceeds the particle replacement rate, and the heat transfer process more nearly approximates to a steady state process, i.e.:

$$t \sim \frac{1}{36} \frac{\rho_p C_p d_p^2}{k_g} > \tau \sim 1 \text{ s} \qquad (9.7)$$

Decker and Glicksman (1981) then suggest the use of the heat transfer resistances measured for packed beds in the estimation of rates of heat transfer.

The chief resistance to heat transfer lies in the low thermal conductivity of the fluidizing gas, and this is the limiting factor. Particle–particle and particle–surface contact areas are too small for significant heat transfer to

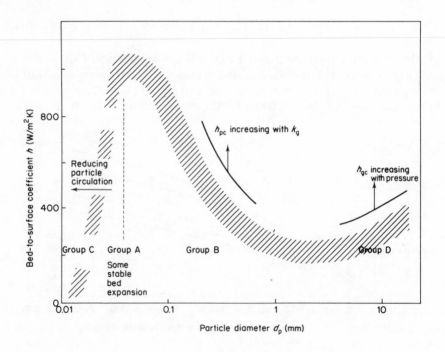

Figure 9.3   General range of bed-to-surface heat transfer coefficients.

occur through points of solid contact; heat must flow by conduction through the gas and there is evidence of additional resistance to heat transfer in the region of increased voidage at the transfer surface (Botterill, 1975). The strong inverse dependence of coefficient upon particle size over the range ~ 800 to 80 μm (Fig. 9.3) is a consequence of the increase in the percentage area of transfer surface through which the heat can flow to the particles by short transfer paths. With very fine powders, however, bed circulation is inhibited by interparticle forces, and this is responsible for the marked fall in coefficient as particle size is further reduced and the bed has the fluidization characteristics of a group C system. Bed-to-surface coefficients increase with an increase in the gas thermal conductivity upon operating beds at higher temperatures (see below, Fig. 9.5). It is not until particle size and/or static operating pressure are sufficiently increased for the fluidizing gas flow condition through the emulsion phase to enter the transitional or turbulent flow regime that significant interphase convective transfer occurs through the gas (h_{gc}, mechanism (b), Section 9.3), and not until then does pressure have a significant effect on bed-to-surface heat transfer apart from through its minor influence on overall bed behaviour.

A simple model (Botterill, 1975) suggests that particle packing close to the transfer surface will be important. The voidage there will be highest with closely size bed materials and will be affected by the material size distribution. Thus, bed-to-surface coefficients were observed to fall as predicted for a reduction in particle packing density at the surface when a heat transfer element was 'stirred' through a stably expanded bed to give low effective particle/surface residence times:

$$h_{\text{pc}} \propto \left(\frac{H_{\text{mf}}}{H} \right)^{2/3} \tag{9.8}$$

(Botterill *et al.*, 1966). With increasing gas flowrates, for either group A or group B materials, there comes a point where the effect of reducing the particle residence time caused by the increasing bubble-induced mixing is countered by the increased bed expansion and the blanketing effect of the bubble flow across the transfer surface. The resultant maximum in the bed-to-surface coefficient (Fig. 9.1) occurs at velocities relatively closer to that for minimum fluidization as mean particle size increases. Todes (1965) has given the approximate correlation for Re_{opt} which is used on Fig. 9.4.

For powders falling within Geldart's group A, the minor peak commented on above (Fig. 9.1) is a consequence of some initial restricted particle circulation by diffusive mixing as the bed fluidizes. Additional continuous phase expansion leads to a reduction in the bed-to-surface coefficient. With a further increase in gas velocity, the bed reaches the minimum bubbling

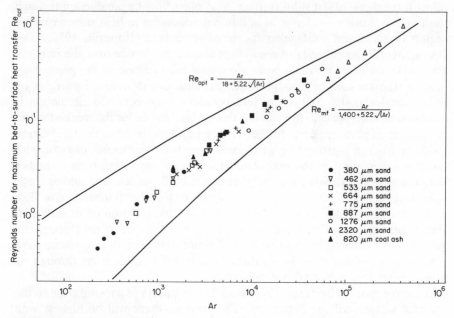

Figure 9.4 Reynolds number, Re_{opt}, for maximum bed-to-surface heat transfer as a function of Archimedes number Ar and the correlation of Todes (1965) for Re_{opt}.

condition and the continuous phase collapses back to a less expanded condition. Particle packing is then denser close to the surface and the bed-to-surface coefficient increases under the action of the bubble-generated mixing to reach the principal peak. Because of the greater degree of continuous phase expansion obtained with group A materials, however, the coefficient is less than the small particle size might first have led one to expect.

The design of the distributor and any internals will obviously have a strong influence on the bubble-induced solids convection patterns established within the bed (see Chapters 4 and 5; also Kunii and Levenspiel, 1969 ; Davidson and Harrison, 1971; Botterill, 1975).

Work reported by de Groot (1967) illustrates the pronounced change in bed behaviour consequent upon changing bubble development patterns which can result from change in the scale of the equipment. It also illustrates the great effect that a change in mean particle size or, rather, of fines concentration of bed material can have through its effect on maximum stable bubble size. As stressed in the introduction to this chapter, workers are warned about the limitations of published correlations. They reflect the particular behaviour of the experimental beds used, and because these have usually been of small scale in which bubble size is limited by the bed diameter, definite, simple solids circulation patterns develop unlike those often occurring in large scale equipment. Bubbles rise up the centre of the bed and there is the return of solids 'stick/slip' flow down the column wall. The tests have also most usually only been carried out at comparatively low temperatures, although the results are presented in generalized correlations. Van Heerden (1952) tested the then-available correlations with four hypothetical systems. For one of these systems there was greater than a four hundred-fold range in the predicted coefficient! The article by Gel'perin and Einstein (1971), in which they propose a generalized correlation, further illustrates the scatter in the reported measurements, and casts doubt on the value of such correlations at all. Nevertheless, Zadbrodsky's (1966) approximate correlation for the maximum bed-to-surface coefficient neglecting radiation effects is to be recommended for powders of Geldart's group B for which the particle convective component predominates and with which there is negligible stable bed expansion — generally within the mean size range of 100 to 800 μm mean diameter:

$$h_{max} = 35.8 \; \rho_p^{0.2} \; k_g^{0.6} \; d_p^{-0.36} \qquad (9.9)$$

This dimensional correlation was determined by following the response of small spherical calorimetric heat probes which quickly approached the bed temperature on immersion in the bed. It contains the particle density ρ_p, which is related to its heat capacity and the other two key variables: particle diameter d_p and gas thermal conductivity k_g. It allows well for the effect of changing gas conductivity with a change in the operating temperature until, beyond 600°C, radiant heat transfer becomes increasingly significant (see Example 9.2 and Fig. 9.5; see also Botterill, Teoman, and Yüregir, 1981). In Fig. 9.5 the differences between the values obtained with an ordinary oxidised probe and one of fine gold, and hence of low emissivity, in tests with a bed of 380 μm sand directly indicate the size of the radiant transfer component at higher temperatures.

Because of the limiting effect of the gas thermal conductivity, the absolute temperatures of the bed and surface influence the obtainable bed-to-surface heat transfer coefficient. Thus, higher heat transfer coefficients were obtained when a hotter calorimetric probe was plunged into a bed of sand at 350°C rather than when the probe was colder than the bed before insertion. If bed-to-surface heat transfer coefficients are measured using a calorimetric spherical probe of high Biot number so that it takes minutes rather than seconds after immersion to reach the bed temperature, it is possible to measure the effect of the changing surface temperature on the transfer coefficient. Typically, for a bed of 270 μm alumina operating at 500°C the coefficient increased from 410 W/m^2 K with the sphere surface at 75°C to 580 W/m^2 K when it had nearly heated up to bed temperature (Botterill, Teoman, and Yüregir, 1984a).

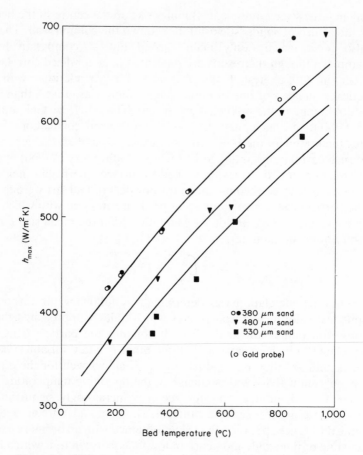

Figure 9.5 Variation in maximum bed-to-surface heat transfer coefficient with temperature for beds of sand of different particles sizes and predictions of Zabrodsky's equation (Eq. 9.9) (Botterill, Teoman, and Yüregir, 1981).

One can conservatively expect to obtain values of 70 per cent. of the predicted maximum to a horizontal cooling tube immersed in a hot bed under good operating conditions when k_g is based on the bed temperature. This reduction is a consequence of two effects: the disturbance caused to the local fluidization condition by immersing the tube in the bed and the decrease in gas thermal conductivity adjacent to a surface which is maintained consistently at a lower temperature than that of the bed (see Section 9.4 below).

Absolute pressure has little effect on bed-to-surface heat transfer coefficients except when fluidizing larger denser particles when the interphase gas convective component of heat transfer (Section 9.3.2 below) becomes significant. The value of $Re_{mf} \sim 12.5$ at $Ar \sim 26,000$ marks the approximate boundary between the dominance of the convective and interphase gas convective components (Botterill, Teoman, and Yüregir, 1981, 1984b).

Example 9.2

Predict the change in bed-to surface heat transfer coefficient for a 10 mm diameter immersed sphere as a function of operating conditions (data of Fig. 9.5). The bed is heated by combustion of propane and the thermal conductivity of the fluidizing gas mixture can be predicted following Mason and Saxena (1958). The bed material is sand of close size distribution:

$$d_p = 480 \ \mu m, \qquad \rho_p = 2,667 \ kg/m^3$$

For air at 170°C, $k_g \sim 3.55 \times 10^{-2}$ W/m K and using Eq. (9.9):

$$h_{max} = 35.8 \ (2,667)^{0.2} \ (3.55 \times 10^{-2})^{0.6} \ (480 \times 10^{-6})^{-0.36}$$
$$= 366 \ W/m^2 \ K$$

Other predicted values are compared with the measured values in Table 9.1.

Table 9.1 Correspondence between predictions of the Zabrodsky correlation (Eq. 9.9) and measured values for a bed of 480 μm sand fluidized by air

Bed temperature, °C	Excess air, %	Estimated gas conductivity, W/m K	Estimated coefficient h_{max}, W/m² K	Maximum measured coefficient, W/m² K
170	Negligible proportion of combustion products	3.55×10^{-2}	366	370
360	300	$4.6 \ \times 10^{-2}$	428	430
540	90	$5.8 \ \times 10^{-2}$	492	500
620	70	$6.2 \ \times 10^{-2}$	512	510
800	40	7.45×10^{-2}	572	610
960	20	8.45×10^{-2}	617	690

For the case of smaller, less dense powders, Khan, Richardson, and Shakiri (1978) give the correlation:

$$Nu_{max} = 0.157 \ Ar^{0.475} \qquad (9.10)$$

from experiments with group A powders of mean diameters between 40 and 96 μm. This has not been tested over a range of operating temperatures and it could be expected that changes in gas physical properties resulting from changing temperature would affect the fluidization conditions and influence the bed behaviour, so influencing bed-to-surface heat transfer coefficients (Botterill, Teoman, and Yüregir, 1981). It is also a matter of experience that the behaviour of large reactors can be sensitive to changes in the size distribution of group A powders consequent upon the effect of changing fines concentration through their generation by attrition or loss by elutriation. Bed-to-surface heat transfer coefficients have been observed to vary by as much as a factor of 2 in the course of operation. This is presumably a consequence of change in bed behaviour adjacent to vertical heat transfer surfaces so that the local stability of the bed can change, thereby affecting the exchange of material between the surface region and bulk of the bed.

Many fundamental models have been proposed to describe the particle convective heat transfer process (see Botterill, 1975) but these cannot generally be applied for predictive purposes because the conditions and thermal properties of the bed directly adjacent to the heat transfer surface are unknown. An extreme situation is that for a slugging bed — the operational condition of some high aspect ratio reactors. Donsi et al. (1984) analyse this situation successfully on the basis of the slug spacing being the characteristic length of relevance for the heat transfer process and including allowance for additional solids transfer in the wake of the slug by analogy with the behaviour of a slug in a liquid system. A simple model containing the principal features of the fluidized bed condition is the packet replacement model (Mickley and Fiarbanks, 1955). However, it predicts excessively high coefficients for very short particle residence times adjacent to the transfer surface. Accordingly, it has been suggested that perhaps the particles do not come directly into contact with the surface and that there is an effective gap between them. The replacement model was therefore modified to incorporate a resistance to heat transfer in the region of increased voidage close to the transfer surface. Gel'Perin, Einstein, and Zakovski (1966) give its solution in the form:

$$h_{pc} = \frac{1}{R_a} \left[1 - \frac{R_w}{2R_a} \ln \left(1 + \frac{2R_a}{R_w} \right) \right] \qquad (9.11)$$

where R_a is the resistance to heat transfer within the packet phase and R_w is the additional wall region resistance. Ra may be estimated from development of Eq. (9.6) as:

$$R_a = \left(\frac{\tau \pi}{4 \, k_{mf} \, \rho_{mf} \, C_p} \right)^{1/2} \qquad (9.12)$$

τ being the packet residence time adjacent to the surface. An approximate correlation for the prediction of particle residence time adjacent to a surface τ and the time that the surface is shrouded by bubbles f_o for vertical cylindrical surfaces of 15 to 30 mm diameter immersed in beds of mean particle size between 120 and 650 μm and fluidized with carbon dioxide and helium at temperatures between ambient and 550°C has been given by Baskakov *et al.* (1973). That for the particle residence time takes the form:

$$\tau = 0.44 \left[\frac{d_p\, g}{U_{mf}^2\, (U/U_{mf} - A)^2} \right]^{0.14} \frac{d_p}{D_o} \qquad (9.13)$$

and the fraction of time the surface is shrouded by bubbles is given by:

$$f_o = 0.33 \left[\frac{U_{mf}^2\, (U/U_{mf} - A)^2}{d_p\, g} \right]^{0.14} \qquad (9.14)$$

where the value of the constant A varied with the bed material and the tube diameter. However, packet residence time is usually unknown *a priori*, although developments in the understanding of mixing processes (Chapter 5) may eventually permit a realistic estimate of it and of the fraction of time bubbles shroud the transfer surface. The resistance in the wall region R_w is independent of time to a first approximation. More recently, this resistance has been associated with an effective gap between particle and surface originating from their inherent roughness (Decker and Glicksman, 1981). However, if the resistance is considered to be generally the region of increased voidage which extends to about half a particle diameter from the surface and the heat transfer through this region is solely by steady state conduction, R_w can be represented by:

$$R_w = \frac{d_p}{2k_w} \qquad (9.15)$$

where k_w is the effective thermal conductivity in the near wall region and can be estimated following Kunii and Smith (1960).

Very good correspondence between observation and prediction was obtained on this basis using Eq. (9.11) to predict the coefficients for flowing packed bed experiments in which the particle residence time at the heat transfer surface was known (Denloye and Botterill, 1977). Fortuitously, the effect of a resistance represented by a 'gas gap' of one-tenth of a particle diameter was tested in earlier theoretical studies (Botterill, 1975) and found

234 GAS FLUIDIZATION TECHNOLOGY

to correspond well in its predictions with observed experimental observations. It corresponds closely with predictions based on the more complicated Kunii and Smith model and is recommended for approximate estimates. R_w would be greater and more complex in form if conditions had developed such that there was a return flow of solids descending in 'stick/slip' flow at the heat transfer surface which was occasionally disturbed by bubbles penetrating to the surface. The basic forms of this latter model have been developed by Yoshida, Kunii, and Levenspiel (1969) but, as stressed above, it cannot usually be applied because the physical properties and condition of the particulate film adjacent to the transfer surface are not generally known. However, Bock and Molerus (1980) have demonstrated the basic validity of this type of model in experiments on heat transfer to vertical tubes in beds up to 1 m in diameter in which simultaneous measurements of local bubble behaviour and time-averaged local heat transfer were made. The claims of general applicability often made for specific mechanistic models, however, are very questionable. Where bubbling processes are explicity modelled, it is salutary to test out the implications of that element in the model and see if the bubble flow predicted is credible (Botterill, 1986). In any case, no simple correlation incorporates the change in mixing consequent upon the transition from group D to group B type behaviour (see Section 9.3.2 below). It is often necessary for the models, too, to fall back on limited experimental tests to evaluate a 'constant' which appears in the model (e.g. Heyde and Klocke, 1979; Martin, 1980).

Example 9.3

Estimate heat transfer coefficients as a function of packet residence time for sand of mean particle diameter 200 μm fluidized in air. Assumed physical properties are $\rho_p = 2{,}600$ kg/m^3, $k_p = 18.9$ W/mK, $C_p = 800$ J/kg K; $\rho_g = 1.2$ kg/m^3, $k_g = 0.026$ W/mK, $C_g = 1{,}050$ J/kg K; bed voidage, $\epsilon = 0.4$.

Solution

Using Eq. (9.11):

$$h_{pc} = \frac{1}{R_a} \left[1 - \frac{R_w}{2R_a} \ln \left(1 + \frac{2R_a}{R_w} \right) \right]$$

and Eq. (9.12):

$$R_a = \left(\frac{\tau \pi}{4 \, k_{mf} \, C_p \, \rho_{mf}} \right)^{1/2}$$

k_{mf} may be estimated from Fig. 9.2 as $10k_g$ (k_p/k_g = 727) and $\rho_{mf} = \rho_p$ $(1 - \epsilon)$ = 2,600 × 0.6 kg/m^3.

R_a estimated for τ = 40 s as $R_a = \left(\dfrac{40\pi}{4 \times 0.26 \times 800 \times 1{,}560} \right)^{1/2}$

$$= 9.8 \times 10^{-3} \text{ m}^2 \text{ K/W}$$

R_w estimated as $0.1 \dfrac{d_p}{k_g} = \dfrac{0.1 \times 200 \times 10^{-6}}{0.026} = 7.7 \times 10^{-4} \text{ m}^2 \text{ K/W}$

h_{pc} for τ = 40 s $= \dfrac{1}{9.8 \times 10^{-3}} \left[1 - \dfrac{7.7 \times 10^{-4}}{2 \times 9.8 \times 10^{-3}} \ln \left(1 + \dfrac{2 \times 8.8 \times 10^{-3}}{7.7 \times 10^{-4}} \right) \right]$

$$= 89 \text{ W/m}^2\text{K}$$

and for shorter residence times as follows:

Residence time τ, s	R_a, (m^2 K/W) × 10^3	R_w, (m^2 K/W) × 10^4	h_{pc}, W/m^2 K
40	9.8	7.7	89
20	7.0	7.7	120
10	4.9	7.7	163
5	3.5	7.7	213
1	1.6	7.7	379
0.5	1.1	7.7	418
0.1	0.51	7.7	713

Note, for comparison purposes, that Eq. (9.9) predicts a maximum value of h_{pc} of 414 W/m^{2}k, which seems to be a reasonable value for a freely fluidized system. Following Kunii and Smith (1960), the alternative estimate for R_w obtained is 7.1×10^{-4} m^2 K/W.

9.3.2 Interphase Gas Convective Component, h_{gc}

The interphase gas convective component becomes significant when working with solids of large mean particle diameter (> 800 μm) and at higher static operating pressures — Geldart's group D materials (Figs 9.1 and 9.3). The gas flow condition through the bed is then turbulent or at least in the transitional flow regime.

A theoretical model of some complexity for this component has been proposed by Adams and Welty (1979). Experimentally, Baskakov and Suprun (1972) estimated its size by analogy from mass transfer measurements and recommended the following correlations:

$$Nu_{gc} = 0.0175 Ar^{0.46} Pr^{0.33} \text{ for } U > U_m \qquad (9.16)$$

and

$$Nu_{gc} = 0.0175 Ar^{0.46} Pr^{0.33} \left(\frac{U}{U_m}\right)^{0.3}$$
$$\text{for } U_{mf} < U < U_m \qquad (9.17)$$

where U is the fluidizing gas velocity and U_m the corresponding value for the maximum bed-to-surface heat transfer. An alternative empirical correlation for the interphase gas convective component has been derived by Botterill and Denloye (1978) as:

$$\frac{h_{gc} d_p^{1/2}}{k_g} = 0.86 Ar^{0.39} \quad 10^3 < Ar < 2 \times 10^6 \qquad (9.18)$$

which has dimensions of $m^{-1/2}$. The assumption was made that it is measured by the heat transfer coefficient at the point of fluidization. This is not strictly true because, although no particle convection will occur with the bed in its quiescent state, there will be conductive heat transfer through the gas between the particle and surface close to the 'point of contact' to augment the interphase gas convective component when the particle thermal time constant is larger than its residence time close to the surface (Decker and Glicksman, 1981). The relative significance of this conductive contribution compared with the gas convective component reduces as particle size and operating pressure increase, so the assumption is most in question for particles < 1 mm mean diameter fluidized at atmospheric pressure at elevated temperatures. The corresponding maximum particle convective component (assuming the two components to be additive) was correlated against the Archimedes number by the following dimensionless equation:

$$\frac{h_{pc_{max}} d_p}{k_g} = 0.84 Ar^{0.15} \qquad (9.19)$$

Experimental conditions covered operation up to static pressures of 9 atm with air, carbon dioxide, and argon as the fluidizing gases, with freon at atmospheric pressure and with mean particle sizes between 160 and 2,370 μm. The bed materials included sand, ash, and copper shot and were closely sized. Experimental values for the bed-to-surface coefficient components for air fluidized beds of sand operated at atmostpheric pressure and 6 atm as a function of particles size are shown in Fig. 9.6.

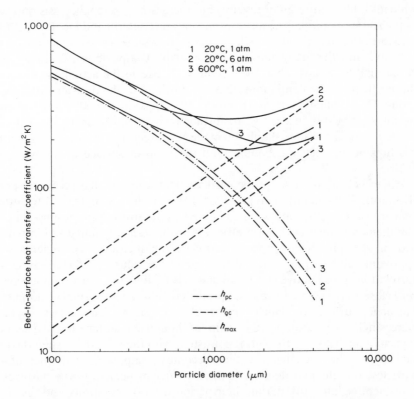

Figure 9.6 Effect of particle size on maximum bed-to-surface heat transfer coefficient h_{max} at atmospheric and 6 atm pressure operation. Interphase gas convective component h_{gc} taken as that for quiescent bed; particle convective component h_{pc} estimated by difference (Botterill, Teoman, and Yüregin, 1981).

Before using the above correlations (Eqs 9.16 to 9.19) it should be noted that there is some evidence that fines in the size distribution can have a large effect. Thus, contrary to the predictions of Eqs 9.16 to 9.19, Golan et al. (1979) came to the conclusion that particle size had a negligible effect on the bed-to-surface heat transfer with four beds of mean particle diameter between 800 and 2,700 μm with top fractions ranging from 1,410 to 5,600 μm, but it is significant that the fines distribution (1,410 μm downwards) was very similar in all four cases. Thus it is possible that the fines present may have determined what was happening and affected the turbulence in the gas, so reducing the contribution of the interphase gas convective component.

The Baskakov and Suprun (1972) predictions corresponded well with the experimental coefficients obtained for quiescent beds of large particles at ambient temperature and pressure, but increasing discrepancy was observed as the static operating pressure was increased (Botterill and Denloye, 1978)]. The form of the component will be relatively insensitive to an increase in gas velocity, and this has been substantiated by Catipovic et al. (1980). This is reasonable because there will be little change in the actual gas flow conditions through the continuous phase or adjacent to the transfer surface with group D bubbling beds and, indeed, the maximum bed-to-surface coefficient is relatively insensitive to further increase in fluidizing gas velocity, though some allowance should be made for direct heat transfer between the gas flowing through the bubbles and the transfer surface (Catipovic et al, 1980).

An effect of relative probe/particle heat capacity has been observed (Botterill, Teoman and Yüregir, 1984a). When, for example, bed-to-surface heat transfer coefficients were estimated by following the response of a 10 mm diameter spherical calorimetric test probe suddenly immersed into a hot bed of 2.3 mm diameter sand particles, the measured coefficients were approximately 30 per cent. higher than the values predicted using the correlations given above (Eqs 9.18 and 9.19). However, correspondence was very close between measured and predicted values when using a calorimetric exchange surface of much higher heat capacity. The effect becomes increasingly significant as the ratio of immersed surface-to-particle heat capacity falls below $\sim$ 10. This is consistent with the view that, when the heat capacity of the exchange element becomes comparable with that of the particles, the element does not extract sufficient heat from the particles to cool them significantly during their period of close proximity and the time-averaged particle-to-surface temperature difference is higher, resulting in higher effective values of the heat transfer coefficients. This effect could be expected, for example, to increase the heat transfer rate to crushed coal as it is charged to a fluidized bed combustor above that predicted by the simple correlation.

Heat transfer coefficients are further enhanced when the object is free to move (Rios and Gibert, 1984). Polchenock and Tamarin (1983) have developed the following empirical equation for the prediction of the transfer coefficient between fluidized bed particles and freely circulating objects:

$$\text{Nu}_{max} = 0.41 \text{Ar}^{0.3} \left(\frac{d_p}{D_o}\right)^{0.2} \left(\frac{\rho_p}{\rho_o}\right)^{-0.07} \psi^{0.66} \tag{9.20}$$

where the subscript p refers to the bed particle and o to the object being fed to the bed, and ψ is the bed particle shape factor. The correlation was derived from tests with bed particles of different materials of densities equal to or less than that of sand and of diameters up to 6.3 mm. They predominantly had group D characteristics but included some group B materials. The objects were spherical or cylindrical.

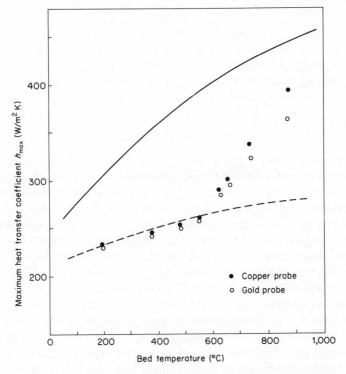

Figure 9.7 Variation in maximum bed-to-surface heat transfer coefficient with temperature for 1,150 μm alumina compared with predictions of Zabrodsky's equation, (Eq. 9.9) and the Botterill and Denloye equations (Eqs 9.18 and 9.19).

Figure 9.7 illustrates the close correspondence between the predictions of the simple correlations (Eqs 9.18 and 9.19) and measured values for heat transfer between a bed of closely sized 1,148 μm alumina and surface of high heat capacity until, at about 600°C, there is a sharp change in the temperature dependence of the heat transfer coefficient. The transition is consistent with a change in the dominant transfer mechanism consequent upon the change in gas physical properties with increasing temperature; Re_{mf} has fallen to ~12.5 and the corresponding Archimedes number is about 26,000 (Botterill, Teoman, and Yüregir, 1981, 1984b). At higher temperatures, the bed is behaving with the characteristics of a group B material and the particle convective mode of heat transfer becomes dominant. (The increasing difference between coefficients measured to an oxidized copper and a fine gold surface at the higher operating temperature also illustrates the increasing contribution of radiant heat transfer). It seems probable that the interstitial gas flow conditions change at the transition and that there is a change from 'slow' to 'fast' (see Chapter 4) bubbling behaviour, with a consequent increase in the effectiveness of solids mixing as bed operating temperature is increased. It is conceivable that bed operating conditions exist under which it is possible to have effectively slow bubbling behaviour in the lower bed regions but fast bubbling higher within the bed with consequent implications for bed-to-surface heat transfer.

9.3.3 Radiative Component h_r

For temperatures above which radiant heat transfer is increasingly significant (generally above 600° C; see the difference between the results for gold and copper surfaces, Figs 9.5 and 9.7), local particle packing and the gas conductivity decrease somewhat in importance (despite the fact that gas conductivity increases with temperature leading to an increase in the particle convective component of heat transfer in particular). Baskakov et al. (1973) illustrate the increasing fraction of the total heat transfer coefficient attributable to radiation from tests at temperatures up to 850°C (Fig. 9.8). It can be seen that the effect is also of greater importance with beds of larger particles. As suggested above (Section 9.3), the three components contributing to the heat transfer process are only approximately additive. It seems that the radiative process transfers some heat at the expense of the particle convective mode. Thus, the heat transfer surface receives radiant energy from the whole particle surface area visible to it and, with larger particles at high temperatures, relatively less heat flows by conduction through the particle to pass to the transfer surface through the shortest gas conduction paths (Botterill, 1975).

In high temperature systems, the bed material is likely to be of a refractory nature and these materials generally have low emissivities. However, it has been pointed out by Zabrodsky (1973) that a particulate bed will have a different overall emissivity to that of its constituent material. Pikashov et al. (1969) reported the measured effective emissivities reproduced in Table 9.2,

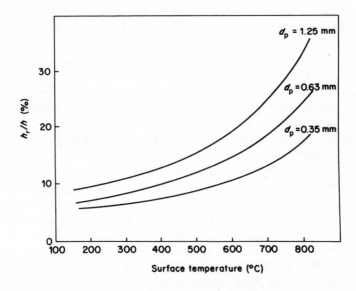

Figure 9.8 Fractional component of total heat transfer coefficient attributable to radiation, bed temperature 850°C, and for three mean particle sizes (Baskakov *et al.* 1973). *Reproduced by permission of the author.*

Table 9.2 Measured effective emissivities (Pikashov *et al.*, 1969), probably underestimated by 10 to 20 per cent (Makhorin, Pikashov, and Kuchin. 1978)

Material of bed	d_p mm	Gas velocity U/U_{mf}	Bulk bed temperature, K	Emissivity of particle	Effective bed emissivity
Fused MgO and SiO$_2$ (irregularly shaped particles)	1 – 1.15	1.2 – 3.0	770 – 1,470	–	0.95
River sand (rounded particles)	1 – 1.5	1.2 – 3.0	770 – 1,370	0.60	0.85
Chamotte (irregularly shaped particles)	1 – 1.5	1.2 – 3.0	720 – 1,370	0.60	0.80
Zirconium dioxide ZrO$_2$ (rounded particles)	0.21 – 1	1.2 – 4.0	870 – 1,320	0.23	0.59
Alumina Al$_2$O$_3$ (rounded particles)	1.5 – 2	1.2 – 3.0	1,070 – 1,720	0.27	0.59

which Makhorin, Pikashov, and Kuchin (1978) subsequently suggested, were about 10 to 20 per cent. too low after further refining their measurement technique. Trace quantities of transition metal oxides will also increase the emissivity of refractory oxides markedly. The 'effective emissivity' of an isothermal bed will be higher than that of the actual individual constituent particles because of the effect of surface reflections. However, even allowing for variation in emissivity with temperature, which they also measured, in practical heat transfer situations there also remains the influence of an immersed surface at a temperature different from that of the bulk of the bed on the temperature of the bed material directly adjacent to it. (It was the effect of this reducing the local radiant flux which was responsible for the lower emissivity values first reported; Table 9.2). For a bed of particles of 250 to 500 μm mean particle diameter operating at 1,100°C, Makhorin, Pikashov, and Kuchin (1978) reported that the layer of particles directly adjacent to a surface maintained at 100°C cooled at an initial rate of between 450 and 860°C/s so that their temperature very rapidly fell 50 to 100°C. Baskakov et al. (1973) have attempted to deal with this problem on the basis of an 'apparent emissivity' giving estimates according to the difference between the bed and surface temperatures (Fig. 9.9). It may be anticipated

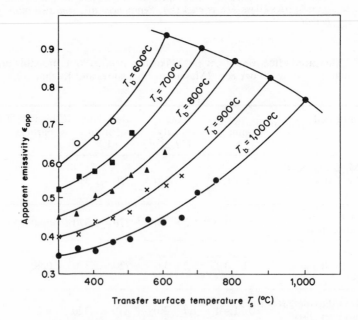

Figure 9.9 Variation in apparent emissivity with transfer surface and bed temperature for a bed of alumina (Baskakov et al., 1973). *Reproduced by permission of the author.*

that the total bed-to-surface heat transfer coefficient h will be of the order of 600 W/m^2 K for an immersed tube in a bed operating at 800°C which is of similar magnitude to the likely steam side coefficient for a unit raising steam at 300°C and 20 atm pressure. Under such circumstances the tube surface temperature may approximate to 550 to 600°C, leading to an apparent emissivity of about 0.6, which seems reasonable.

In contrast to lower temperature systems, maximum heat transfer coefficients with group B bed materials in particular are likely to be less sensitive to the gas flowrate. Thus, when the bed is bubbling vigorously while bubbles will increasingly shroud the heat transfer surface, they will also open up the bed so that the heat transfer surface can see deeper into it where there has been less influence of the cooled immersed transfer surface and the bed is behaving more like a black body radiator (Botterill, Teoman, and Yüregir, 1984a). Similarly, the apparent emissivity of the free surface of the bed will be higher. For rule of thumb estimates, the radiative component can be estimated using absolute temperatures and an adaptation of the Stefan–Boltzman equation in the form:

$$h_r = \frac{5.673 \times 10^{-8} \, \epsilon_r \, (T_b^4 - T_s^4)}{T_b - T_s} \qquad (9.21)$$

where ϵ_r is a reduced emissivity to take into account the different emissive properties of the surface ϵ_s and bed ϵ_B, given by:

$$\epsilon_r = \frac{1}{(1/\epsilon_s + 1/\epsilon_B) - 1} \qquad (9.22)$$

As indicated above, because of the effect of a cooling surface on the immediately adjacent particles, the bed temperature there will be lower than its bulk temperature; allowance can be made for this when using the bulk bed temperature in Eq. (9.21) by using an apparent emissivity ϵ_{app} in place of ϵ_r (Fig. 9.9) following Baskakov et al. (1973). A reasonable value for alumina for ϵ_{app} would be 0.6 to 0.7 and somewhat lower for sand. ϵ_r may be as low as 0.1 with a shiny metal surface. An alternative formula recommended by Panov et al. (1978) for approximate estimates is:

$$h_{rad} = 7.3 \, \sigma \epsilon_p \, \epsilon_s \, T_s^3 \qquad (9.23)$$

9.4 USE OF IMMERSED TUBES TO INCREASE THE AVAILABLE HEAT TRANSFER SURFACE AREA

If the walls of the bed provide insufficient heat transfer surface, the available transfer area may be increased by the immersion of tubes within the bed.

Indeed, particularly with highly exothermic catalystic reactions employing group A type materials, the provision of this extra heat transfer surface often dominates the reactor design.

Insertion of the tube, of course, affects the general bed fluidization behaviour because of the influence of the inserts on bubble development. This, in turn, influences the solids circulation patterns and the particle convective component of heat transfer. However, because of the influence of the tubes on bed behaviour, beds with such inserts will be less sensitive to the overall effects of changing scale than will unobstructed beds (see Section 9.3.1 above). Indeed, the use of vertical tubular inserts has been advocated as a means of controlling bubble growth and hence as a means of reducing the effect of scale on bed behaviour. The beneficial effect of horizontal tubes in breaking up slugs has also been reported. However, although the behaviour of beds containing inserts is less sensitive to change in scale, there is much variation in their behaviour according to details of their configuration, and predictions between the published correlations vary by about a factor of 2. The final choice between horizontal or vertical tube arrangements would seem to depend largely upon constructional convenience — how best to introduce the tubes into the bed, support them, and allow for thermal expansion. Thus, vertical tubes are predominantly used with the deep catalyst beds in high aspect ratio catalytic reactors and horizontal tubes in low aspect ratio fluidized bed combustors. Erosion can be severe if jetting bed material impinges on a tube. There is also danger that tubes exposed within the freeboard may be prone to erosion damage, although one means of obtaining turndown with a steam-raising combustion bed is to arrange to vary the operating air flowrate in order to control the bed expansion and so vary the number of horizontal tubes immersed in the bed (Temmink and Meulink, 1983). Results, for example, from a pilot plant facility burning < 25 mm coal show an approximately linear decrease for the bed-to-tub coefficient from a value of ~ 330 W/m^2 K for an in-bed tube to 150 W/m^2 K at a tube height of 300 mm above the expanded bed height.

When horizontal tubes are immersed within a bed, much heat transfer surface is exposed to the cross-flow of solids, which is advantageous from the point of view of heat transfer. Counter to this there is also a tendency for bubbles to shroud parts of the surface (one would expect this to be the downward facing surface), and for particles to defluidize and form a stagnant cap on the top of the tube. There is evidence from tests that this is the general pattern (see Botterill, 1975; Saxena et al, 1978), but the region at which the maximum coefficient occurs moves to higher positions round the tube as the fluidizing gas velocity is increased (Fig. 9.10). Newby and Keairns (1978) report measurements where the maximum coefficient even occurs at the top surface. When tubes are mounted in a staggered array there is a greater

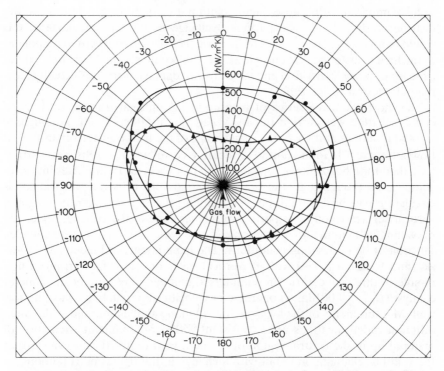

Figure 9.10 Polar diagram showing variation in local heat transfer coefficient round a 35 mm diameter tube immersed in a bed of 0.37 mm alumina operating at 500°C; excess fluidizing gas velocities of 0.89 m/s (●) and 0.23 m/s (▲), corresponding value of U_{mf} 0.11 m/s (Botterill, Teoman, and Yüregir, 1986a).

tendency for the rising bubbles to displace the stagnant particles from the top of the tubes periodically. Although very close tube spacing seriously affects the particle circulation (in the extreme, the bed may even rise above the tubes so that they become the effective gas distributor if they are so closely spaced), McLaren and Williams (1969) found no drastic reduction in heat transfer coefficients as the tube spacing was progressively reduced in low temperature tests. Their heat transfer results for both in-line and staggered arrays fell on a single curve if the results were plotted in terms of the narrowest gap between the tubes. Thus, coefficients fell by about 25 per cent. as the gap reduced from 282 to 15 mm in tests which fluidized ash of wide size distribution and mean particle diameters between 400 and 500 μm but including small fractions of greater than 1.78 mm diameter. They commented on obtaining coefficients some 20 per cent. lower when the tube bundle was located close to the distributor, a region strongly influenced by the functioning of the distributor and where the bubbles generating particle circulation are still small compared with those higher within the bed. If gas jetting at the distributor should increase particle exchange at the transfer surface, this would, conversely, be expected to increase the particle convective component of the heat transfer coefficient. However, this would also be likely to be at the expense of particle attrition and tube erosion damage if the jets actually blow the particles on to the tubes.

The most extensive series of measurements on heat transfer to a single immersed tube was carried out by Vreedenberg (1958, 1960), although he worked predominantly over the rising range of the heat transfer versus the gas flowrate curve. He found it more difficult to correlate the results for a vertical tube than for a horizontal one, which is to be expected. In the latter instance, complicated though it is, the prevailing conditions of strong vertical solids circulation past horizontal tubes is a simpler process than that obtaining at a vertical surface when local particle residence time at the surface, under the prevailing bed circulation conditions, can be expected to be a more complex function of bed operating conditions. The experiments of Piepers, Wiewiorski, and Rietema (1984) with 19 vertical tubes of 2 m length arranged in a bundle of three concentric rings immersed in a bed of 66 μm mean diameter catalyst ($U_{mf} = 0.0018$ m/s) operated at velocities up to 0.4 m/s illustrate this well. Measurements of the local coefficient along a tube mounted at the three different radial locations were consistent with a predominant circulation of material up in the central region of the bed, with return flow downwards at the wall but changing in detail with a change in the operating conditions.

To characterize his material, Vreedenberg used a particle Reynolds group which included both the particle density and diameter, and he did not, therefore, include the gas density in it. However, he did not use this group directly in his correlations but used, instead, the tube diameter in a modified Reynolds group. For a vertical tube he also included the bed diameter in his correlations, which is somewhat surprising. For a horizontal tube, Vreedenberg's correlation (1958) takes the form:

$$\left(\frac{hD_T}{k_g}\right)\left(\frac{k_g}{C_p\,\mu}\right)^{0.3} = 0.66\left(\frac{U\,D_T\,\rho_p\,(1-\epsilon)}{\mu\epsilon}\right)^{0.44} \tag{9.24}$$

$$\text{for } \frac{U\,d_p\,\rho_p}{\mu} < 2{,}050$$

and

$$\left(\frac{hD_T}{k_g}\right)\left(\frac{k_g}{C_p\,\mu}\right)^{0.3} = 420\left(\frac{U\,D_T\,\rho_p}{\mu}\,\frac{\mu^2}{d_p^3\,\rho_p g}\right)^{0.3} \tag{9.25}$$

$$\text{for } \frac{U\,d_p\,\rho_p}{\mu} > 2{,}550$$

The form of the correlations for vertical tubular inserts was considerably more complicated (Vreedenberg, 1960).

Andeen and Glicksman (1976) introduced the factor $(1 - \epsilon)$ into the correlation (Eq. 9.25) so that it should be able to follow the observed decrease in the bed-to-tube coefficient with increasing gas flowrate beyond the maximum. Andeen, Glicksman, and Bowman (1978) subsequently reported that the modified form of correlation would also fit results for horizontal flattened tubes if the constant is adjusted appropriately, but their experimental results show considerable scatter within themselves.

The very complete review by Saxena *et al.* (1978) of work on heat transfer to immersed tubes at temperatures below which radiant transfer is significant draws attention to the wide range of results reported. In various circumstances, the different correlations lead to unrealistic predictions with the Vreendenberg correlations best following the observed temperature dependence of the coefficient (Botterill, Teoman, and Yüregir 1984a, 1984b). Of seven other correlations also tested, those by Grewal and Saxena (1981) and Ternovskaya and Korenberg (1971) tended to predict the more extreme values (Fig. 9.11), but none predicted the right overall temperature dependence. As reliable for horizontal tubes, and easier to use, are the simple correlations recommended in Sections 9.3.1 and 9.3.2 above, namely: for materials of group B, characteristics to use 70 per cent. of $h_{pc,max}$ as predicted by Eq. (9.9) to allow for the effect of the immersed tube when the hot bed is being cooled (Fig. 9.11a) and about 75 per cent. of the sum of the components predicted by Eqs (9.18) and (9.19) for beds of group D characteristics, immersion of the tube not then affecting the coefficient so much. The boundary between the two classes of behaviour is at $Re_{mf} \sim 12.5$ and $Ar \sim 26,000$, as outlined above (Botterill, Teoman and Yüregir, 1984a, 1984b). This occurs at about 550°C for the 1.15 mm alumina bed (Fig. 9.11b). The detail of the results obtained by Piepers, Wiewiorski, and Rietema (1984) for heat transfer to vertical tubes would suggest that it is not so easy to estimate an average bed-to-vertical tube coefficient. Some tests on vertical (Bock and Molerus, 1980) and horizontal arrays (Catipovic *et al.*, 1980) showed little difference between the maximum coefficients obtainable to a tube within an array and that to a single isolated tube.

Staub and Canada (1978) reported interesting tests of large particle bed models ($d_p = 600$ and 2,600 μm fluidized at high velocities and pressures up to 10 atm in the so-called 'turbulent' fluidized bed regime. They reported results which show a maximum heat transfer coefficient at gas velocities one-quarter of the particle terminal velocity, typical values being of about 230 W/m^2 K for 550 μm particles fluidized under atmospheric pressure conditions; they also found a dependence of the coefficient on the 0.2 to 0.3 power of the gas density depending on the particle size and tube geometry.

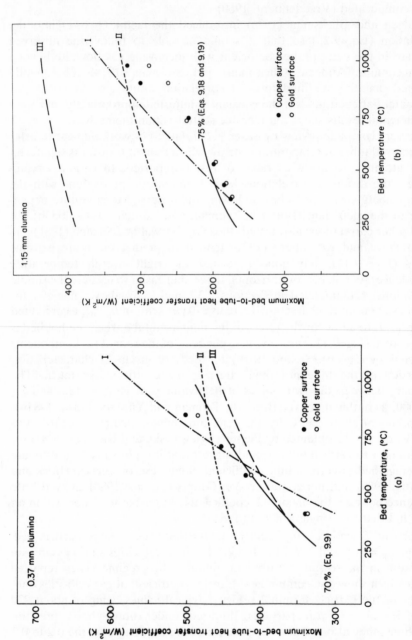

Figure 9.11   Comparison between predictions of representative correlations by (I) Vreedenberg (1958), (II) Grewal and Saxena (1981), and (III) Ternovskaya and Korenberg (1971), and results for heat transfer to a single 35 mm o.d. tube; particle mean diameter (a) 0.37 mm and (b) 1.15 mm alumina (Botterill, Teoman, and Yüregir, 1984b).

This compares with powers of 0.39 and 0.15 for the gas and particle convective components, as indicated by the correlations of Eqs (9.18) and (9.19), but one can hardly expect the correlations for ordinary bubbling bed flow to be applicable. Staub (1979) has developed a credible heat transfer model based on the correlations established for flowing dense solids/gas suspensions. This uses gas and solid superficial velocities pertinent to the upflow and downflow situations as a prediction within his flow model. The simplified form of his general relation predicts the ratio of Nusselt numbers (based on the tube diameter) for the conditions with the bed present Nu and without particles Nu_{gas} in the form:

$$\frac{Nu}{Nu_{gas}} = \left[1 + \left(\frac{150}{d_p \times 10^6} \right)^{0.73} \left(\frac{0.42 \rho_p (1 - \epsilon_{av}) \Delta Z_m^{0.4}}{\rho_g U} \right) \right]^{0.45} \quad (9.26)$$

for 20 μm $< d_p <$ 1,000 μm; for 1,000 μm $< d_p <$ 3,000 μm a value of $d_p = 10^{-3}$ m should be used in Eq. (9.26). ΔZ_m in the expression is a mixing length which is the average distance between tube centres in this model. The average bed hold-up $(1 - \epsilon_{av})$ is taken from bed expansion data or the correlation data given by Staub and Canada, (1978). Wood, Kuwata, and Staub (1980) have extended this model to heat transfer to tubes in the splash zone of a fluidized bed of large particles.

Because of the reported range of behaviour, it is only prudent to carry out tests on a model of as large a scale as possible. Preferably, when concerned with a multi-unit system, this should be a 'unitcell' of the proposed system, i.e. incorporating all the features of one full-scale complete unit of the proposed design. This would be the area of bed fed from one fuel distribution point, say, in the design of a combustor. However, because bed behaviour can change markedly with operating temperature (Botterill and Teoman, 1980; Botterill, Teoman, and Yüregir, 1981), the results of cold tests should be interpreted with caution.

For a bed of given cross-sectional area, the depth of bed required in order to immerse adequately the tubes providing the necessary heat transfer surface area will be dependent on the chosen tube diameter and pitch. The available surface area of tube within a given volume of bed increases both as the tube diameter is increased and as the pitch is decreased. Higher fluidizing gas pumping power cost in order to overcome the pressure drop across the bed is the penalty to be paid for deeper beds. Thus, for example, the choice can be important in the design of atmospheric pressure fluidized bed combustors for steam-raising plant. At higher superficial velocities, a larger immersed surface area may be necessary in order to be able to hold the required combustion bed temperature, and the cost of the bed depth through increased pumping power charges must be balanced against the increase containment costs of a bed of greater cross-sectional area (Elliott, Healey, and Roberts, 1971). Staub and Canada (1978) point out that finned tubes, if sufficiently effective in increasing available heat transfer surface area, can be of value in reducing the necessary bed height.

In applications where it is required to remove more or less heat from a bed operating as a heat source, the generally high bed-to-surface coefficient may well pose problems when the heat demand is small. As noted earlier in this section, one way to overcome this and provide adequate turn down is by arranging more or less heat transfer surface to be immersed within the bed by varying the fluidizing gas flowrate and therefore varying the degree of overall bed expansion. This may not be operationally satisfactory for the maintenance of efficient fuel combustion, and there is again danger of increased attrition damage to the particles and erosion damage to horizontal tubes exposed in the free board while the bed is being operated in a less expanded state, although the effect of the immersed tubes is to reduce the violence of bubble eruptions at the bed surface. Alternatively, it is possible to regulate the rate of heat removal through tubes permanently immersed within a section of the bed in smaller units by varying the degree of fluidization adjacent to the tubes. Thus, if there is a section of distributor with a separate gas supply for that region, the degree of fluidity of bed around the tubes can be regulated, so controlling the heat transport within it. Because horizontal fluidbed 'conductivities' are only a few per cent. of those in the vertical direction, it is also necessary to ensure that adequate bed material circulation will occur from the bulk of the bed to the heat transfer section. A design to do this is a feature of the combustor described by Virr (1976).

9.5   FINNED TUBE TRANSFER UNITS

The use of finned tube heat transfer elements has been advocated for applications where the bed-side coefficient may be limiting. Generally, in deep bed applications, bed-to-surface coefficients based on the total exposed surface area are considerably reduced. Thus, Staub and Canada (1978) give values showing something like a 30 per cent. reduction after allowing a correction for the fin efficiency, but this obviously depends on the geometry of the system (see also Saxena et al., 1978). Not only will particle access within the fins depend on the relative size of particle and fin spacing but it will also depend on the relative location of neighbouring tubes. Nevertheless, rough calculations indicate that a tube with 2.1 fins per centimetre length and with a fin structure that will not exceed a maximum fin tip temperature of 500°C in a 900°C combustion bed, offers a threefold improvement in heat transfer capability compared with a bare tube bank (Staub and Canada, 1978).

Shallow fluidized beds have an advantage through their low pressure drop, and the most efficient gas/solid contacting also occurs in the region of the distributor, where bubbles are growing rapidly by coalescence (see Chapter 4). They also afford the basis for very efficient waste heat recovery units (Virr, 1976). In this latter application, finned tubes are immersed in shallow fluidized beds and the regions between the fins behave virtually as separate, dilute fluidized beds. Bubbles remain restricted in size by the closeness of the fin spacing, which may be as small as 3 to 5 mm or 15 to 20 times the particle diameter. In operation, bed expansion as high as 400 per cent. is usual while still obtaining high bed-to-surface coefficients, as good as those to a bare tube, because of the very short particle residence times obtaining at the heat transfer surface; particle velocities are possibly an order of magnitude higher than in a deep bed and this more than compensates for the greatly reduced bed density. Based on the outside area of the tube, obtainable coefficients are in the range of 1 to 4 kW/m^2 K. Because the particles in the bed present a very high surface area to the gas, very good heat recovery can be achieved from a hot fluidizing gas. Typically a unit can be operated with an overall pressure drop as low as 50 mm water gauge from the suction developed by an exhaust fan which is used to suck the hot flue gases through the bed. For space heating, 150 kW of hot water has been recovered using a small unit to cool flue gases from 600 to 180°C. Another unit is producing 3.5 tonne/h of steam from the waste gases from a 1,200 h.p. diesel engine in a 32,000 tonne dead weight oil tanker.

9.6 FLUIDIZED SOLIDS AS A HEAT TRANSFER MEDIUM

Reference was made in the introduction to the present limited applications of fluidized solids as a basic heat transfer medium. these have mainly been in metallurgical heat treatment processes (Baskakov, 1968; and Virr, 1976). In one application requiring a reducing atmosphere, for example, a rich fuel/air mixture was introduced through the distributor. Higher within the bed, above the zone where the specimens for treament were placed, additional air was injected to complete the combustion of the fuel and provide the required heat release to maintain the overall bed temperature. Natural bed circulation caused the hot solids to flow downwards through the bed from the principal combustion zone to the heat treatment zone, replacing those carried upwards towards the upper combustion region by rising bubbles.

Work described by Newby and Keairns (1978) formed part of a design study for the development of a fluidized bed stream-raising heat exchanger in a liquid metal cooled nuclear reactor system. In this scheme, heat would be transferred between the lower horizontal tube array through which the liquid metal circulated to an upper bank of steam-raising tubes by the fluidized bed. Thus they thought to use the fluidized solids as an inert heat transfer medium, so obviating the danger of direct sodium/steam contact in the event of tube failure.

There is much potential for the use of beds of particles in heat storage associated with generation and recovery systems. There is a possible advantage in separating the heat generation from steam-raising sections of the boiler plant by interposing a large store between the combustion bed and the steam generating plant (Botterill and Elliott, 1964). For an electricity generating power station, the combustion equipment could then be rated at the average load over 24 hours, whereas only the heat exchangers and electricity generating equipment would have to be rated for the peak capacity. A cube of side 40 m would provide adequate store for a 500 MW unit, and this is no bigger but much cheaper than a larger boiler. In the defluidized state heat losses from the store would be low. The system offers improved reliability against combustor failure through the period of grace afforded by the energy in store. By controlling the rate at which material is fluidized and withdrawn from the store, the turndown problem occurring when tubes are immersed directly within the combustion bed (referred to in Section 9.4 above) is overcome; such a system posseses very quick startup capabilities, limited only by the rate at which the turbine can accept load.

Many high temperature processes produce dirty exhaust gases at varying rates. In the steel industry, for example, the BOS unit operates intermittently, producing exhaust gases at 1,600°C, and the order of energy to be disposed of is typically about 8 GJ/t of steel production. Currently, this is wasted and is expensive to dispose of. It could be transferred to a store using a falling cloud heat exchanger and recovered from there using fluidbed techniques for transport and the heat transfer operation. A solids flow rate of 3 kg/s is readily capable of transporting energy at a rate of 2 MW. Basic studies on the feasibility of this and solar energy storage utilizing fluidized bed tehcniques are outlined by Bergougnou et al. (1981).

9.7 GENERAL POSITION

A more detailed review of work on fluidized bed heat transfer up to 1984 has been undertaken by Botterill (1986). Gas-to-particle heat transfer is not usually a limiting factor in fluidized bed operations because of the large surface area presented by particulate solids. The correlation (Eq. 9.1) may be used to estimate the heat transfer coefficient, although precise measurements, as outlined in Section 9.2 above, are difficult to make. This remains a problem area.

The three basic mechanisms involved in bubbling bed-to-surface heat transfer have been identified as outlined in Section 9.3 above. To a first approximation they are additive. It is the capacity of the circulating particles to transport heat which is the significant feature of the particle convective component h_{pc} (Section 9.3.1). In this mechanism, gas conductivity is the limiting factor and the level of operating temperature is therefore important because of its influence on it. Particle packing and residence time close to the transfer surface are usually the controlling factors; the former is dependent on particle size distribution and bed fluidizing conditions, and the latter is also much influenced by the general design of the bed. For particles of Geldart's group B between about 100 and 800 μm, Zabrodsky's dimensional correlation (Eq. 9.9) is to be recommended. For group A materials where there can be significant bed expansion there is the correlation (Eq. 9.10), but this has only been tested at temperatures close to ambient.

The interphase gas convective component h_{gc} (Section 9.3.2) becomes significant when gas flow conditions through the continuous phase cease to be laminar. Gas heat capacity and hence static operating pressure are important. A correlation for its estimation is given in Eq. (9.18) and, for the corresponding maximum particle convective component, in Eq. (9.19). Transition from group D to group B type behaviour occurs as Re_{mf} falls below ~12.15 with Ar ~26,000 as operating temperatures increase (Fig. 9.7).

Progress has been made in modelling radiant heat transfer (Lindsay 1983), Rule of thumb estimates can be made using an apparent emissivity to allow for the influence of the immersed heat transfer element on the bed directly adjacent to it (Section 9.9.3). Thus, although the bed emissivity and that of its surface will be higher than for the individual particles forming the bed (e.g. see Table 9.2), an apparent emissivity for the bed adjacent to the transfer surface between 0.6 and 0.7 is more reasonable in many instances.

The range of coefficients to be expected if radiative transfer can be neglected is indicated in Fig. 9.3. Knowing the necessary conditions to obtain good bed-to-surface heat trasnfer, the challenge remains to engineer the system so that it can be obtained. As indicated in Section 9.4, the available correlations for the prediction of heat transfer to arrays of tubes predict a wide range of values. For ease of application and general reliability of estimate, the simpler correlations as outlined in Section 9.3 are to be preferred. Thus a conservative estimate for the bed-to-horizontal-tube coefficient for beds of group B type materials would be about $0.7 \times h_{pc,max}$, as predicted by the Zabrodsky equation (Eq. 9.9), and for the group D type materials, about 75 per cent. of the sum of the predictions for the correlations of Eqs (9.18) and (9.19). Different correlations will apply for the condition of turbulent fluidization (Eq. 9.26). It should be appreciated that most of the correlations for heat transfer to tubes have been developed from lower temperature experiments only. 'Unitcell' tests need to be carried out and particularly so because the effects of lateral mixing may become very important when there is a large number of tubes immersed within the bed. Operations with 'large' local heat sources should be tested if these are likely to occur in practice. Cold tests need to be interpreted with care because of the pronounced changes in bed behaviour that can occur with a change in temperature of operation (Botterill and Teoman, 1980; Botterill, Teoman, and Yüregir, 1981, 1984a, 1984b). The conditions of fluidization can be used to regulate heat removal from a bed, but this must also be determined according to the particular operational characteristics of the bed. As indicated in Sections 9.5 and 9.6, the potential of fluidized solids in basic heat transfer operations is only just beginning to be exploited.

9.8 NOMENCLATURE

A	constant in Eqs (9.13) and (9.14)	–
C_g	heat capacity of gas	J/kg K
C_p	heat capacity of particle	J/kg K
d_p	diameter of particle	m
D_o	diameter of immersed or freely circulating object	m
D_T	diameter of immersed tube	m
f_o	fraction of time surface is shrouded by bubbles	–
g	acceleration due to gravity	m/s^2
h	heat transfer coefficient	W/m^2 K
h_{gc}	interphase gas convective component of bed to surface heat trasnfer coefficient	W/m^2 K
h_{gp}	gas-to-particle heat trasnfer coefficient	W/m^2 K
h_{pc}	particle convective component of bed-to-surface heat trasnfer coefficient	W/m^2 K
h_r	radiative component of bed-to-surface heat transfer coefficient	W/m^2 K
H	bed height	m
H_{mf}	bed height at minimum fluidization	m
k_e	effective thermal conductivity of bed	W/m K
k_g	thermal conductivity of gas	W/m K
k_{mf}	thermal conductivity of bed at minimum fluidization	W/m K
k_p	thermal conductivity of particle	W/m K
k_w	effective thermal conductivity in near wall region	W/m K
R_a	resistance to heat transfer within packet phase	m^{2}K/W
R_w	resistance to heat transfer at wall	m^{2}K/W

S	stirring factor (Eq. 9.6)	–
S_B	surface area of solids/unit volume of bed	m
U	gas velocity	m/s
U_{mf}	gas velocity at minimum fluidization	m/s
t	cooling time constant	s
T_B	temperature of bed	K
T_g	temperature of gas	K
T_p	temperature of particle	K
T_s	temperature of surface	K
ΔZ_m	mixing length (Eq. 9.26)	m
ϵ	bed voidage	–
ϵ_{app}	apparent emissivity	–
ϵ_B	emissivity of bed	–
ϵ_{eff}	effective emissivity	–
ϵ_p	particle emissivity	–
ϵ_r	reduced emissivity (Eq. 9.22)	–
ϵ_s	emissivity of surface	–
μ	gas viscosity	N s/m^2
ρ_g	gas density	kg/m^3
ρ_{mf}	density of bed at minimum fluidization	kg/m^2
ρ_p	particle density	kg/m^3
σ	Stefan-Boltzman constant	W/m^2 K^4
τ	packet residence time	s
ψ	particle shape factor	–
Ar	Archimedes number, $gd_p^3\rho_g(\rho_p - \rho_g)/\mu^2$	
Nu$_{gc}$	interphase gas convective Nusselt number, $h_{gc}d_p/k_g$	
Nu$_{gp}$	gas-to-particle Nusselt number, $h_{gp}\, d_p/k_g$	
Pr	gas Prandtl number, $C_g\mu/k_g$	
Re$_p$	particle Reynolds number, $d_p\rho_gU/\mu$	
Re$_{mf}$	Reynolds number for minimum fluidization condition, $d_p\rho_gU_{mf}/\mu$	

9.9 REFERENCES

Adams, R.L., and Welty, J.R. (1979). *A.I.Ch.E. J.*, **25**, 395.
Andeen, B.R., and Glicksman, L.R. (1976). ASME-AIChE Heat Transfer Conference, Paper 76-HT-67.
Andeen, B.R., Glicksman, L.R., and Bowman, R. (1978). *In Fluidization — Proc of the Second Engineering Foundation Conference*, (Eds J.F. Davidson and D.L. Keairns) Cambridge University Press, p. 345.

256   GAS FLUIDIZATION TECHNOLOGY

Barile, R.G., Seth, H.K., and Williams, K.A. (1970). *Chem. Eng. J.*, **1**, 236.
Baskakov, A.P. (1968). *High Speed Non-oxidising Heating and Heat Treatment of Metals in a Fluidized Bed*, Metallurgy, USSR.
Baskakov, A.P., Berg, B.V., Virr, O.K., Phillippovsky, N.F., Kirakosyan, V.A., Goldobin, J.M., and Suprun, V.M. *Powder Technol.*, **8**, 273.
Baskakov, A.P., and Suprun, V.M. (1972). *Int. Chem. Eng.*, **12**, 119.
Bergougnou, M.A., Botterill, J.S.M., Howard, R.J., Newey, D.C., and Teoman, Y. (1981). *Third International Conference on Future Energy Concepts*, London, p. 61.
Bock, H.J., and Molerus, O. (1980). *Proceedings of the 1980 Fluidization Conference* (Eds J.R. Grace and J.M. Matsen), Plenum Publishing Corporation, New York, p. 217.
Botterill, J.S.M. (1975). *Fluid Bed heat Transfer*, Academic Press, London.
Botterill, J.S.M. (1986). "Advances in Fluidized Bed Heat Transfer" in *Transport Phenomena in Fluidising Systems*. (Eds. L.K. Doraiswamy and B.D. Kulkarni) Wiley Eastern/Wiley Halsted, *in press*.
Botterill, J.S.M., Brundrett, G.W., Cain, G.L., and Elliott, D.E. (1966). *Chem. Eng. Prog. Symp. Ser*, **62**, (62), 1.
Botterill J.S.M., and Denloye, A.O.O. (1978). *A.I.Ch.E. Symp. Ser.*, **74** (176), 194).
Botterill, J.S.M., and Elliott, D.E. 1964). *Engineering*, **198**, 146 (July 31).
Botterill, J.S.M., and Teoman, Y. (1980). *Proceedings of the 1980 Fluidization Conference* (Eds J.R. Grace and J.M. Matsen), Plenum Publishing Corporation, New York, p. 93.
Botterill, J.S.M., Teoman, Y., and Yüregir, K.R. (1981). *A.I.Ch.E. Symp. Ser.*, **77** (208), 330.
Botterill, J.S.M., Teoman, Y., and Yüregir, K.R., (1984a). *Powder Technol.*, **39**, 177.
Botterill, J.S.M., Teoman, Y., amd Yüregir, K.R., (1984b). XVI ICHMT Symposium, Dubrovnik, 5–7 Sept.
Catipovic, N.M., Jovanovic, G.N., Fitzgerald, T.J., and Levenspiel, O. (1980). *Proceedings of the 1980 Fluidization Conference* (Eds J.R. Grace and J. Matsen) Plenum Publishing Corporation, New York, p. 225.
Davidson, J.F., and Harrison, D. (1971). *Fluidization*, Adademic Press, London.
Decker, N., and Glicksman, L.R. (1981). A.I.Ch.E. Symp. Ser, **77** (208), 341.
de Groot, J.H. (1967). In *Proceedings of the International Symposium on Fluidization*, Ed. A.A.H. Drinkenburg Netherlands University Press, Amsterdam, p. 348.
Denloye, A.O.O., and Botterill, J.S.M. (1977). *Chem. Eng. Sci.*, **32**, 461.
Donsi, G., Lancia, A., Massimilla, L., and Volpicelli, G. (1984). In *Fluidization* (Eds D. Kunii and R. Toei), Engineering Foundation Conference, p. 347.
Elliott, D.E., Healey, E.M., and Roberts, A.G. (1971). Conference of the Institute of Fuel and L'Institut Francais des Combustibles et de L'Energie, Paris.
Geldart, D. (1972). *Powder Technol.*, **6**, ?85.
Gel'Perin, N.I., and Einstein, V.G. (1971). In *Fluidization* (Eds J.F. Davidson and D. Harrison), Academic Press, London, p. 471.
Gel'Perin, N.I., Einstein, V.G., and Zakovski, A.V. (1966). *Khim Prom.*, **6**, 418.
Golan, L.P., Cherrington, D.C., Diener, R., Scarborough, C.E., and Wiener, S.C. (1979). *Chem. Eng. Prog.*, **75**, 62 (July).
Grewal, N.S., and Saxena, S.C. (1981). *Ind. Eng. Chem. Proc. Des. & Dev.*, **20**, 108.
Heyde, M., and Klocke, H-J. (1979). *Chem. Ing. Technik*, **51**, 318.
Khan, A.R., Richardson, J.F., and Shakiri, K.J., (1978). In *Fluidization — Proceedings of the Second Engineering Foundation Conference* (Eds J.F. Davidson and D.L. Keairns), Cambridge University Press, p. 345.
Kunii, D., and Levenspiel, O. (1969). *Fluidization Engineering*, Wiley.

Kunii, D., and Smith, J.M. (1960). *A.I.Ch.E. J.*, **6**, 71.
Lindsay, J. (1983). Radiative Heat Trasnfer in Fluidized Beds, Ph.D. Thesis, University of Cambridge.
Littman, H., and Sliva, D.E. (1970). *Heat Transfer 1979*, Paper CE.1.4, Fourth International Heat Transfer Conference, Versailles, Elsevier Publishing Corp.
McGaw, D.R. (1976). *Powder Technol.*, **13**, 231.
McLaren, J., and Williams, D.F. (1969). *J. Inst. Fuel.*, **42**, 303.
Makhorin, K.E., Pikashov, V.S., and Kuchin, G.P. (1978). In *Fluidization — Proceedings of the Second Engineering Conference* (Eds J.F. Davidson and D.L. Keairns), Cambridge University Press, p. 39.
Martin, H. (1980). *Chem. Eng. Technik.*, **52**, 199.
Mason, E.A., and Saxena, S.C. (1958). *The Physics of Fluids*, **1**, 361.
Mickley, H.S., and Fairbanks, D.F. (1955). *A.I.Ch.E. J.*, **1**, 374.
Newby, R.A., and Keairns, D.L. (1978). In *Fluidization — Proceedings of the Second Engineering Foundation Conference* (Eds J.F. Davidson and D.L. Keairns), Cambridge University Press, p. 320.
Panov, O.M., Baskakov, A.P., Goldobin, Yu.M., Phillippovsky, N.F., and Mazur, Yu.S., (1978). *Inzh-Fiz. Zh.*, **36**, 409.
Piepers, H.W., Wiewiorski, P., and Rietma, K. (1984). In *Fluidization* (Eds D. Kunii and R. Toei), Engineering Foundation Conferences, p. 339.
Pikashov, V.S., Zabrodsky, S.S., Makhorin, K.E., and Il'Chenko, A.I. (1969). *Bull. B.S.S.R., Acad. Sci., Physical Energetics Ser.*, **100**, No. 2.
Polchenok, G.I., and Tamarin, A.I., (1983). *Inzh-Fiz.Zh.*, **45**, 427.
Richardson, J.F., and Ayers, P. (1959). *Trans. Instn. Chem. Engrs*, **37**, 314.
Rios, G.M., and Gibert, H. (1984). *Fluidization* (Eds D. Kunii and R. Toei) Foundation Conferences, p. 363.
Saxena, S.C., Grewal, N.S., Gabor, J.D., Zabrodsky, S.S., and Galershtein, D.M. (1978). *Advances in Heat Transfer* (Eds T.F. Irvine and J.P. Hartnett), Vol. 14, Academic Press, New York, p. 147.
Singh, A.N., and Ferron, J.R. (1978). *Chem. Eng. J.*, **15**, 169.
Staub, F.W. (1979). *J. Heat Transfer*, **101**, 391.
Staub, F.W., and Canada, G.S. (1978). In *Fluidization — Proceedings of the Second Engineering Foundation Conference* (Eds J.F. Davidson and D.L. Keairns), Cambridge University Press, p. 339.
Temmink, H.M.G., and Meulink, J. (1983). 'Operating experience with TNO 2m × 1m atmospheric fluid bed boiler facility', 3rd European Coal Utilization Conference, 11–13 Oct., Amsterdam.
Ternovskaya, A.N., and Korenberg, Yu.G. (1971). *Pyrite Klining in a Fluidized Bed*, Izd Khiminya, Moscow.
Todes, O.M. (1965). *Applications of Fluidized Beds in Chemical Industry*, Part II, Izd. Zanie, Leningrad, pp. 4–27.
van Heerden, C. (1952). *J. Appl. Chem.*, **2**, Supplement Issue No. 1, S7.
Virr, M.J. (1976). In *Proceedings of the Fourth International Conference on Fluidized Bed Combustion*, The Mitre Corporation, p. 631.
Vreedenberg, H.A. (1958). *Chem. Eng. Sci.*, **9**, 52.
Vreedenberg, H.A. (1960). *Chem. Eng. Sci.*, **11**, 274.
Wood, R.T., Kuwata, M., and Staub, F.W. (1980). *Proceedings of the 1980 Fluidization Conference* (Eds J.R. Grace and J. Matsen), Plenum Publishing Corporation, New York, p. 235.
Yerushalmi, J., and Cankurt, N.T. (1979). *Powder Technol.*, **24**, 187.
Yoshida, K., Kunii, D., and Levenspiel, O. (1969). *Int. J. Heat and Mass Transfer*, **12**, 529.

258   GAS FLUIDIZATION TECHNOLOGY

Zabrodsky, S.S. (1963). *Int. J. Heat and Mass Transfer,* **6**, 23.
Zabrodsky, S.S. (1966). *Hydrodynamics and Heat Transfer in Fluidized Beds,* The MIT Press.
Zabrodsky, S.S. (1973). *Int. J. Heat and Mass Transfer,* **15**, 241.

Gas Fluidization Technology
Edited by D. Geldart
Copyright © 1986 by John Wiley & Sons Ltd.

CHAPTER 10

Fluid Bed Drying

D. REAY

10.1 INTRODUCTION

Fluidizing with hot air is an attractive means for drying many moist powders and granular products. The technique has been used industrially since 1948, and today it enjoys widespread popularity for drying crushed minerals, sand, polymers, fertilizers, pharmaceuticals, crystalline materials, and many other industrial products.

The main reasons for this popularity are as follows:

(a) Efficient gas solids contacting leads to compact units and relatively low capital cost combined with relatively high thermal efficiency.

(b) The handling of the particles is quite gentle compared to some other types of dryer. This is important with fragile crystals.

(c) The lack of moving parts, other than feeding and discharge mechanisms, keeps reliability high and maintenance costs low.

The main limit on the applicability of fluid bed dryers is that the material being dried must be fluidizable. Some potential feedstocks are too wet to fluidize satisfactorily. This is usually due to an excessive amount of surface moisture on the particles, causing them to agglomerate. This problem may be overcome by flashing off the surface moisture in a pneumatic conveying dryer preceding the fluid bed dryer. Another limitation is encountered if the product has a very wide size distribution, so that at an air velocity high enough to fluidize the large particles there is an unacceptable loss of small particles from the bed. To some extent this limitation can be overcome nowadays by the use of a vibrating fluid bed (see below).

259

10.2 THE TECHNOLOGICAL OPTIONS

The potential user of fluid bed drying has a wide variety of equipment to choose from. The main categories are described briefly below. Within each category there are many variations offered by equipment vendors.

10.2.1 Batch dryers

The first distinction to be made is between batch and continuous fluid dryers. Batch dryers are normally used when the production scale is small and several different products have to be made on the same production line. Figure 10.1

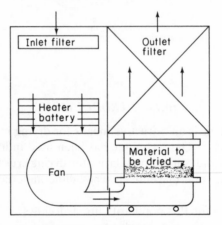

Figure 10.1 Batch fluid bed dryer.

shows a typical batch fluid bed dryer. The wet feed is loaded into the cabinet and clamped to the filter sock module. The cabinet doors are then closed and the blower started. An adjustable damper controls the degree of air recirculation. The circulating air may be heated by a steam tube battery or by gas firing. Batch fluid bed dryers are particularly popular in the pharmaceutical and dyestuffs industries.

10.2.2 Continuous 'well-mixed' dryers

The first continuous fluid bed dryer was the 'well-mixed' type, illustrated in Fig. 10.2, which was introduced in the United States in 1948. It is usually of circular cross-section, and takes its name from the fact that the particle residence time distribution approaches the perfect mixing law:

$$E(t) = \frac{1}{t_R} \exp \left(-\frac{t}{t_R} \right) \qquad (10.1)$$

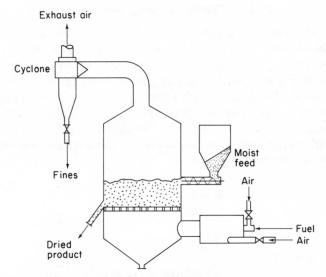

Figure 10.2   Continuous 'well-mixed' fluid bed dryer.

In this expression, $E(t)\mathrm{d}t$ is the fraction of particles with residence times between $t + \mathrm{d}t$, and t_R is the mean residence time. Because of the near-perfect mixing the bed has a nearly uniform composition and temperature equal to the composition and temperature of the outlet product stream. Hence, the moist feed falls into a bed of almost dry particles. For this reason this type of continuous fluid bed dryer can handle wetter feedstocks than can other types to be described later.

The main disadvantage of the 'well-mixed' type of fluid bed dryer is the wide particle residence time distribution, leading to a wide range of moisture content in the product particles. The average moisture content of the product may be acceptable, but 40 per cent. of the particles stay in the dryer for less than half the average residence time and 10 per cent. for less than one-tenth of the average (see Chapter 5, Section 5.6). Hence, some particles will emerge quite wet. For some products, particularly many polymers, this is unacceptable. Furthermore, the wide distribution makes it difficult to achieve a very low average product moisture content.

Despite this drawback, the 'well-mixed' bed is still the most popular type of continuous fluid bed dryer in North America. In Western Europe it has been superseded in many applications by the 'plug flow' bed or by the vibrated fluid bed.

10.2.3   Continuous 'plug flow' dryers

These are beds of shallow depth (typically 0.1 m) in which the particles flow along a channel whose length/width ratio is much greater than unity (typically in the range 4:1 to 30:1). The objective is a more or less close approach to plug flow of the particles, so that they all emerge with approximately the same moisture content. For length/width ratios up to about 10 the particle flow channel may be straight, but for higher ratios a reversing path or spiral path is more practicable as shown in Fig. 10.3. Straight channel designs are often provided with baffles normal to the direction of particle flow in an attempt to improve the approach to plug flow.

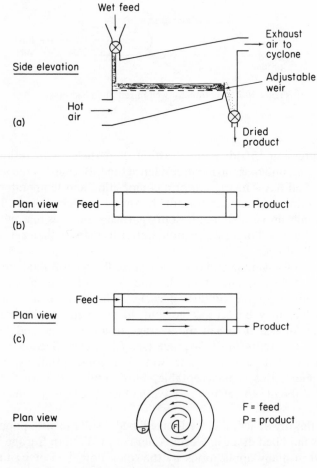

Figure 10.3 Continuous 'plug flow' fluid bed dryer: (a) straight path; (b) reversing path; (c) spiral path.

In addition to reducing the spread of product particle moisture contents, a plug flow bed will normally require a smaller bed volume than a well-mixed bed to achieve the same average product moisture content. However, the simple plug flow bed has some disadvantages. Firstly, the moist feed falls into an area where the particles are still comparatively wet. Consequently, there may be fluidization difficulties at the feed end with some feeds which could be handled quite satisfactorily in a well-mixed bed. Secondly, the hot air passing through the bed towards the discharge end does comparatively little drying and therefore does not give up much of its heat. Hence, the thermal efficiency of a simple plug flow bed is lower than that of a well-mixed bed. Thirdly, the temperature of the particles rises as they flow along the bed and towards the discharge end it approaches the inlet air temperature. With a heat-sensitive product this limits the inlet air temperature which can be used, thereby reducing the thermal efficiency still further. Variations on the simple plug flow bed have been developed to overcome these difficulties. The most important of these variations are the vibrated fluid bed and the multi-stage bed.

10.2.4 Vibrated fluid bed dryers

This is simply a straight channel plug flow bed with a vibrating distributor, as shown in Fig. 10.4. Alternatively, it may be viewed as a vibrating fluid bed

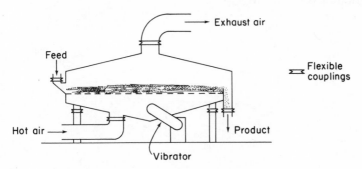

Figure 10.4   Vibrated fluid bed dryer.

conveyor which uses hot air as the fluidizing medium. Compared to the simple plug flow bed, it has the advantage that any agglomerates arising at the feed will be kept moving by the vibrations of the distributor, hopefully until they have dried sufficiently to break up.

Furthermore, feeds with a wide size distribution can be processed successfully in this type of bed. The air velocity can be set low enough to avoid excessive elutration of the smaller particles while the largest particles are kept moving by the vibration of the distributor.

Finally, these beds are often used to dry feeds consisting entirely of large particles with a minimum fluidization velocity of the order of 1 m/s. With a static distributor an air velocity considerably in excess of the minimum would be needed to ensure adequate fluidization and conveying, but this would be far more air than is required to satisfy mass and heat balance considerations. If the distributor is vibrated the air velocity can safely be kept in the vicinity of the minimum fluidization velocity, with consequent savings in capital and operating costs.

Present technology imposes two limitations on vibrated fluid bed dryers. Firstly, the materials of construction available for vibration mountings limit air inlet temperature to about 400°C. Secondly, if the bed is to vibrate at its natural frequency, which is the simplest and most economical way of operating, bed lengths are limited to about 8 m. Therefore, vibrated fluid bed dryers are limited to intermediate temperatures and intermediate throughputs. Finally, they have a higher maintenance requirement than static fluid beds.

10.2.5 Multi-stage beds

Multi-stage beds are often used for the following purposes:

(a) To accomplish drying and cooling in the same vessel, which is usually a straight channel plug flow unit with a divided chamber below the distributor (see Fig. 10.5a.)

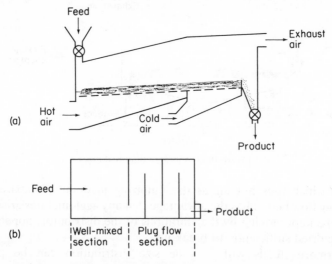

Figure 10.5 Multi-stage fluid bed dryers: (a) dryer plus cooler; (b) plug flow fluid bed following well-mixed fluid bed (plan view).

(b) To combine the ability of a 'well-mixed' bed to handle wet feeds with the ability of a 'plug flow' bed to achieve a comparatively uniform product moisture content, through having a 'well-mixed' section followed by a 'plug flow' section (see Fig. 10.5b)

Purposes (a) and (b) can be combined in a three-stage unit.

In case (b) the temperature of the air supplied to the 'well-mixed' stage can be higher than that supplied to the 'plug flow' stage because the higher evaporation rate in the first stage will help to keep the bed temperature down. A higher air velocity may also be used in the first stage to help in rapid dispersion of the wet feed.

10.2.6 Fluid bed dryers with internal heating

When a finely divided, heat-sensitive powder is being dried in a fluidized bed there are limitations on both the velocity and the temperature of the inlet air. Consequently, the rate of heat input from the air per unit distributor area may be quite low, and if the air is the only heat source the distributor area required to perform the drying duty may be large. It can be reduced by supplying part of the heat through steam tubes or heated baffles immersed in the bed. For example, if half the heat is supplied in this way the distributor area needed is halved. This can give substantial savings in capital and operating costs.

Heat transfer coefficients from immersed heating surfaces to the bed particles increase with decreasing particle size (see Chapter 9). The converse is true for gas-to-particle heat transfer, when account is taken of the enforced reduction in air velocity with decreasing particle size. Hence, the use of internal heating surfaces becomes more attractive the smaller the particle size.

When internal heating surfaces are used, the bed depth is usually determined by the need to keep the heat transfer surfaces immersed, rather than by consideration of drying kinetics.

10.2.7 Fluid bed granulation

Fluidized beds can also be used for making dry powder from a feed which is a slurry or solution. The feed is sprayed on to the bed, usually with a pneumatic atomizer to give a very fine atomization. The particles in the bed are continually growing, by one or both of two mechanisms. A drop may strike a particle and form a thin layer of liquid on the particle surface, whereupon the layer immediately dries and the particle grows by one layer. This mechanism gives a hard, dense particle structured like an onion. Alternatively, the particle may strike another particle before the layer has dried, in which case the liquid may act as a binder and hold the particles together. This gives a product consisting of agglomerates of finer particles. Which mechanism predominates depends very much on the materials and operating conditions, and must be ascertained by experiment.

In batch granulation the bed initially contains fine seed particles. The required amount of feed is then added at an appropriate rate. After feed addition is complete, there may be a further period during which the grown particles are thoroughly dried, followed finally by a cooling period when the fluidizing medium is changed from hot to cold air. Batch fluid bed granulation is very popular in the pharmaceutical industry.

In continuous fluid bed granulation, a stream of particles is continually withdrawn from the bed and classified into fines, size, and oversize. The fines and crushed oversize are returned to the bed for further growth, while product of the required size is taken off. Figure 10.6 shows the principal options.

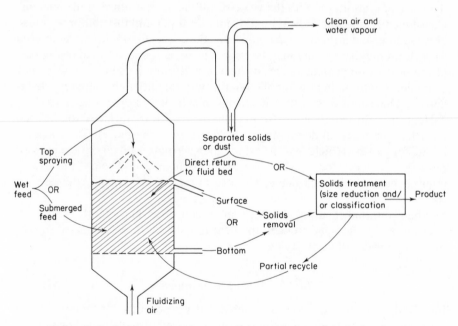

Figure 10.6 Continuous fluid bed granulation: variations of the basic process.

The main potential problem with fluid bed granulation is an operational one. If the ratio of liquid to solid in the bed becomes too high the particles will rapidly form large agglomerates the size of golf balls and fluidization will be lost. The plant must then be shut down and the bed dug out. The operator gets only a few minutes warning of the onset of this 'wet quenching' process.

10.3 DRYING KINETICS

10.3.1 Batch drying curves

If a batch of moist powder is dried in a fluidized bed with periodic withdrawal of samples for determination of moisture content, the resulting curve of powder moisture content X (defined as the weight of H_2O divided by the weight of dry solid) versus time t will probably look something like the curve of Fig. 10.7(a). This is conventionally divided into two portions, the first

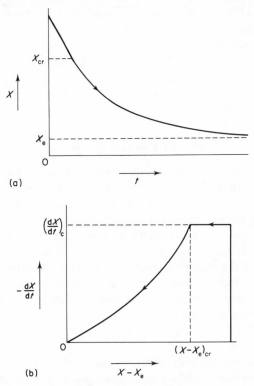

Figure 10.7 Batch drying curves.

called the constant rate period and the second called the falling rate period. The moisture content at the transition between the two periods is called the critical moisture content X_{cr}. If drying is continued for long enough, X will approach the equilibrium moisture content X_e. For a given material this is a function of relative humidity and temperature. At any point on the curve the amount of removable moisture remaining, $X - X_e$, is called the free moisture content. It should be noted that in reality X_{cr} rarely appears as a sharply defined point on the experimental drying rate curve. There is usually some curvature at the transition from constant rate to falling rate. X_{cr} is probably best defined as the point at which the forward extrapolation of the constant rate line intersects the backwards extrapolation of the falling rate curve.

The rate of drying $- \, dX/dt$ can be determined at any point by differentiating the X versus t curve. A graph of $- \, dX/dt$ versus the free moisture content $X - X_e$ is an alternative way of representing the drying characteristics of a material (see Fig. 10.7b).

As a rough approximation, the constant rate period may be regarded as corresponding to the removal of surface moisture from the particles, while the falling rate period corresponds to the removal of internal moisture. Since most materials will not fluidize satisfactorily if there is substantial surface moisture, the constant rate period in fluidized bed drying may be so short as to be unobservable, except under very mild drying conditions. On the other hand, a non-porous material such as sand may show virtually no falling rate period.

10.3.2   Constant rate period

In the constant rate period the surface of the particle is wet enough for the layer of air adjacent to the surface to be saturated. Hence, the drying rate is determined by the rate at which vaporized moisture can be transported across the boundary layer surrounding a particle. During this period the temperature of the particle surface remains constant at the wet bulb temperature T_{wb} of the air. If p_{wb} is the partial pressure of vapour at the wet bulb temperature, p is the partial pressure in the bulk air stream, and K_p is the mass transfer coefficient based on partial pressure, the rate of moisture removal N_c per unit particle surface area (called the drying flux) in the constant rate period is given by:

$$N_c = K_p \, (p_{wb} - p) \tag{10.2}$$

Since the particle temperature does not rise during this period, all the heat transferred across the boundary layer from gas to particle must be used for evaporation. Hence, an alternative formulation for Eq. (10.2) is:

$$N_c = \frac{h_{gp}}{\lambda} \, (T - T_{wb}) \tag{10.3}$$

where h_{gp} is the gas-to-particle heat transfer coefficient, λ is the latent heat of vaporization and T is the bulk gas temperature.

For the air–water system the wet bulb temperature of the air is almost equal to the adiabatic saturation temperature, which is readily obtained from psychrometric charts. For organic solvents in air or an inert gas, T_{wb} can be calculated if the psychrometric ratio of the solvent–gas system is known (see, for example, Keey, 1978). Hence, if either K_p or h_{gp} is known, N_c can be calculated for any drying conditions.

There is a plethora of data on gas-to-particle heat transfer coefficients in fluidized beds of small particles of diameter less about 1 mm. This subject is covered in Chapter 9. One of the most frequently used correlations is that proposed by Kothari and discussed in Kunii and Levenspiel (1969):

$$\text{Nu} = 0.03\text{Re}_p^{1.3} \qquad (10.4)$$

This suggests that the heat transfer coefficient, and hence the drying rate in the constant rate period, should be proportional to the gas velocity U raised to the power 1.3. This was confirmed by Mostafa (1977) in experiments on the drying of small particles of silica gel, molecular sieve, and vermiculite in a fluidized bed.

Many fluid bed dryers, particularly those with vibrating distributors, operate on particles of diameter greater than 1 mm. Many less data are available for this size range, and what there are suggest a much lower dependence of drying rate on gas velocity. No generally applicable correlation is yet available. Zabeschek (1977) found that with a fluidized bed of aluminium silicate particles the particle-to-gas mass transfer coefficient was roughly proportional to $U^{0.5}$ when the particle diameter was 2.76 mm, and was almost independent of U when the particle diameter was 4.30 mm. Subramanian, Martin, and Schlünder (1977) also concluded, from an analysis of other investigators' results, that transfer rates are effectively independent of gas velocity when the particle diameter is much greater than 1 mm.

Equations (10.2), (10.2), and (10.4) have a number of other practical uses. They enable one to predict the effects of bulk gas temperature, bulk gas humidity, and small particle diameter on the drying rate in the constant rate period.

10.3.3 Falling rate period

In the falling rate period the rate of moisture migration to the surface of a particle is insufficient to keep the layer of air adjacent to the particle surface saturated. Hence, the drying rate is no longer determined solely by conditions in the boundary layer. It also depends on the pore structure of the material and on the mechanism of moisture migration. There may, in fact, be several simultaneous mechanisms. These include capillary action, vapour diffusion, diffusion along internal surfaces, and, in the case of cellular materials, diffusion across cell walls. The balance between these mechanisms may change as drying proceeds. For example, capillary motion may predominate during the early part of the falling rate period when the pores are relatively full, while vapour diffusion may dominate towards the end when only small pockets of moisture remain in the solid structure. In general, the falling rate curve cannot be predicted *a priori* and must be determined by experiment.

However, having determined the falling rate curve experimentally at one set of operating conditions, it is often possible to predict approximately how it will change when the operating conditions are changed. This is useful in that it enables a range of process options to be explored with only a small amount of experimental data. To do this we make use of the concept of the characteristic drying curve, first proposed by van Meel (1958), elaborated by Keey (1978), and shown by experiment to be valid for several widely different materials (see, for example, Zabeschek, 1977, and Gummel and Schlünder, 1977).

Essentially, the characteristic drying curve concept is based on the assumption that at any point on a batch drying curve the drying flux N is given by:

$$N = N_c f \left(\frac{X - X_e}{X_{cr} - X_e} \right) = N_c V \qquad (10.5)$$

N_c is the drying flux which would be observed at the prevailing external conditions if the material was wet enough to be in the constant rate period of drying. Above the critical moisture content V takes the value unity; below the critical moisture content it is assumed to be a function only of the free moisture content remaining in the material, expressed in dimensionless form by using the critical free moisture content as a reference quantity. N_c can be estimated theoretically by the methods outlined above, while the form of V can be estimated by fitting Eq. (10.5) to a single measured batch drying curve (see Fig. 10.8); V is assumed to be unaffected by changes in external conditions. For analytical purposes Eq. (10.5) is sometimes simplified by assuming that:

$$V = \frac{X - X_e}{X_{cr} - X_e}$$

that is:

$$N = N_c \left(\frac{X - X_e}{X_{cr} - X_e} \right)$$

For many materials this may be an adequate approximation if only a rough answer is required; it is equivalent to representing the curve on Fig. 10.7(b) by a straight line joining the points (N_c, X_{cr}) and $(0, 0)$. Greater accuracy can be achieved by approximating the curve either by a series of straight line portions or by an empirical power function.

The concept cannot be expected to hold exactly, since the critical moisture content tends to increase slowly as drying conditions are made more severe. However, it has been shown to hold approximately for materials as diverse as molecular sieve, paper, textiles, ion exchange resin, iron ore, gelatin, and

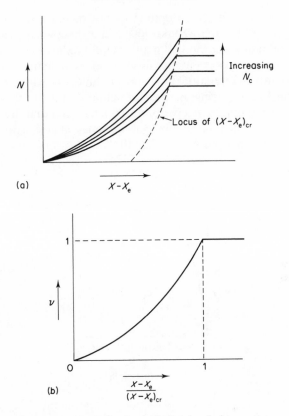

Figure 10.8 Normalization of batch drying curves:
(a) batch drying curves at different external conditions;
(b) normalized batch drying curve.

crystalline agglomerates. Its virtue is that it enables batch drying curves to be transformed from one set of operating conditions to another without any fundamental knowledge of the structure of the material or of the mechanism of moisture movement within it. It is unlikely to give acceptable accuracy when applied to conditions giving an order of magnitude difference in the drying rate, because then the change in critical moisture content will be too great.

10.3.4   *Effect of bed depth on drying rate*

The available experimental evidence indicates that for materials which dry relatively quickly, $- \mathrm{d}X/\mathrm{d}t$ is almost inversely proportional to bed depth (Mostafa, 1977; Reay and Allen, 1982). Therefore, the drying rate $- W_B \, \mathrm{d}X/\mathrm{d}t$, where W_B is the dry bed weight, is effectively independent of bed depth, and increasing the bed depth does not increase the drying rate. It appears that most of the drying takes place in a shallow layer just above the distributor, and increasing the bed depth merely reduces the frequency with which an individual particle enters this layer. If we assume that the gas flow can be divided into an interstitial flow and a bubble flow with most of the heat and mass transfer occurring in the former and not much gas exchange occurring between them, then a plausible explanation is that the mass transfer rate is so rapid that the interstitial gas becomes almost saturated a short distance above the distributor (see Fig. 10.9a).

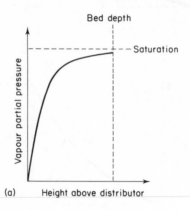

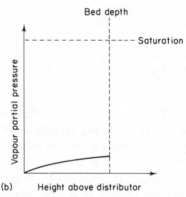

Figure 10.9 Postulated increase of vapour partial pressure with height above the distributor in the interstitial gas flow: (a) fast drying; (b) slow drying.

On the other hand, with a material such as wheat which dries very slowly, $- dX/dt$ is almost independent of bed depth (Reay and Allen, 1982). Therefore, the drying rate $- W_B \, dX/dt$ is directly proportional to bed depth. It may be surmised that in this case the interstitial gas is still far from saturated at the bed surface (see Fig. 10.9b). If relatively quick drying materials have an appreciable falling rate period they also tend towards this type of behaviour when the drying rate has declined to a very low value.

PROCESS DESIGN OF CONTINUOUS FLUID BED DRYERS

10.4.1   Mass and heat balances

Let F = solid mass flowrate (dry basis)

G = gas mass flowrate (dry basis)

X = solid moisture content (kg liquid/kg dry solid)

Y = gas moisture content (kg vapour/kg dry gas)

Subscripts:

.   i = in   o = out

A moisture mass balance over the dryer yields:

$$F(X_i - X_o) = G (Y_o - Y_i) \tag{10.6}$$

Let H_F = enthalpy of 1 kg of dry solid + associated liquid

H_g = enthalpy of 1 kg of dry gas + associated vapour

Q_W = rate of heat loss from walls

Q_h = rate of heat input from heater immersed in bed

A heat balance over the dryer yields:

$$FH_{Fi} + GH_{gi} + Q_h = FH_{Fo} + GH_{go} + Q_W \tag{10.7}$$

Let C_s, C_l, C_g and C_v be specific heats of solid, liquid, gas and vapour respectively. Let T_F be the temperature of the solid and associated liquid, and T_g the temperature of the gas and associated vapour. Let λ be the latent heat of vaporization at 0°C (2,500 kJ/kg for water). Then, referring to an enthalpy base of 0°C, we have:

$$H_{Fi} = (C_s + X_iC_l) T_{Fi} \tag{10.8}$$

$$H_{Fo} = (C_s + X_0C_l) T_{Fo} \tag{10.9}$$

$$H_{gi} = (C_g + Y_iC_v) T_{gi} + Y_i\lambda \tag{10.10}$$

$$H_{go} = (C_g + Y_oC_v) T_{go} + Y_o \lambda \tag{10.11}$$

Substitution in Eq. (10.7) and rearrangement yields:

$$G (C_g + Y_iC_v) T_{gi} - (C_g + Y_oC_v) T_{go} - \lambda(Y_o - Y_i) \quad + Q_h$$
$$= F (C_s + X_oC_l) T_{Fo} - (C_s + {}_iC_l) T_{Fi} \quad + Q_W \tag{10.12}$$

If a psychrometric chart is available for the gas-vapour system in question, H_{gi} and H_{go} would normally be read off directly from it. Equation (10.12) would still be used for computerized calculations, however, and is essential for many organic vapour–inert gas systems for which no psychrometric chart is available.

10.4.2   Calculation of required mean particle residence time t_R

The starting point of this calculation is always a batch drying curve measured in laboratory equipment at a suitable bed depth z and gas velocity U. The latter is chosen to be high enough to ensure good fluidization but low enough to avoid excessive elutriation. The measured batch drying curve can be transformed to other bed depths and gas velocities by means of the rules presented earlier.

For simulation of a plug flow fluid bed dryer the batch drying curve should be measured with the air inlet temperature T_{gi} constant at the desired value. This will usually be dictated by material temperature limitations, with an appropriate safety margin. If the deviation from plug flow is small, the temperature and moisture content histories of the material as it flows along the bed will correspond to its temperature and moisture content histories on the batch drying curve, so the required mean particle residence time t_m can be read directly from the appropriate curve.

The calculation is more complicated for a bed with a well-mixed particle residence time distribution. This case has been analysed in detail by Vanecek, Markvart, and Drbohlav (1966). In this case, in order to simulate the passage of particles through the continuous bed the batch drying curve should refer to a constant bed temperature rather than a constant air inlet temperature. This means that during the course of the laboratory experiment the air inlet temperature must be progressively reduced as the drying rate declines in order to prevent the bed temperature rising. A good control system is needed.

In the continuous operation let X_t be the moisture content of a particle after residence time t in the bed, and let X_o be the average moisture content of the product. If $E(t)$ is the residence time distribution function of the product we have:

$$X_o = \int_0^\infty X_t \, E(t) \, dt \tag{10.13}$$

and using Eq. (10.1) for the residence time distribution in a perfectly mixed vessel we obtain:

$$X_o = \frac{1}{t_R} \int_0^\infty X_t \exp\left(-\frac{t}{t_R}\right) dt \tag{10.14}$$

In this expression the function X_t is given by the batch drying curve measured at constant bed temperature.

In the general case, Eq. (10.14) must be integrated numerically with an arbitrarily chosen value of t_R, and by trial and error the value of t_R found which gives the desired X_o. This is not difficult with a computer or a programmable calculator.

In certain special cases an analytical solution may be possible. For example, suppose the batch drying curve can be approximated by a linear falling rate curve with no constant rate period, i.e.:

$$-\frac{dX}{dt} = k \, (X - X_e) \qquad \text{where } k \text{ is a constant}$$

Integrating between $(X_i, 0)$ and (X_t, t) we obtain:

$$\frac{X_t - X_e}{X_i - X_e} = \exp \, (-kt)$$

Therefore:

$$X_t = X_e + (X_i - X_e) \exp \, (-kt)$$

Substituting for X_t in Eq. (10.14) and integrating yields:

$$X_o = X_e + \frac{X_i - X_e}{1 + kt_R}$$

Therefore:

$$t_R = 1/k \, \left(\frac{X_i - X_e}{X_o - X_e} - 1\right) \qquad (10.15)$$

10.4.3 Calculation of the bed dimensions

Let the bed area be A, the bed depth z, and the bed density ρ_B (easily measured in the laboratory for a given gas velocity U). Then:

Bed hold-up $= Ft_R = \rho_B z A$

and

$$A = \frac{Ft_R}{\rho_B z} \qquad (10.16)$$

Since t_R is calculated using a batch drying curve corresponding to a selected bed depth z, A can be calculated from Eq. (10.16).

The calculation of t_R, and hence of A, may be repeated for other bed depths using the batch drying curve transformation rules. However, if the material is one which dries relatively quickly so that $- \mathrm{d}X/\mathrm{d}t$ is approximately inversely proportional to bed depth, then A is likely to be sensibly independent of bed depth. On the other hand, if the material dries very slowly so that $- \mathrm{d}X/\mathrm{d}t$ is almost independent of bed depth, A is likely to be inversely proportional to bed depth.

Having calculated A, the corresponding gas mass flowrate G at the selected gas velocity can be calculated from:

$$G = \rho_g U A \qquad (10.17)$$

The gas exit humidity Y_o can now be calculated from the mass balance Eq. (10.6). If reasonable values are assumed for Q_h and Q_W, the gas exit temperature T_{go} can then be estimated from the heat balance Eq. (10.12). The psychrometric chart should then be used to ascertain how far T_{go} is above the dew point of the exit gas. A difference of at least 10°C is usually considered necessary to avoid the risk of condensation in the cyclones. If they are too close, additional bed area may be required to increase G and thereby reduce Y_{go}.

If desired, the entire calculation procedure can be repeated with batch drying curves corresponding to different gas velocities until an optimal design is achieved.

Example 10.1: Design of a continuous perfectly mixed fluid bed dryer

Let 6,000 kg/h of a wet porous solid entering at 20°C be dried from a water content of 20 per cent. by weight (dry basis) to an average water content of 4 per cent. by weight. The bed depth is to be 200 mm, the bed temperature 80°C, and the superficial mass velocity $\rho_g U$ of the inlet air 0.70 kg/m²s. Under these conditions the bed density is 500 kg/m³. The humidity of the inlet air is 0.005 kg H_2O/kg dry air. The specific heat of the dry solid is 1.0 kJ/kg °C and the specific heat of liquid water is 4.2 kJ/kg °C. There is no internal heater in the bed. Heat loss from the walls may be taken as 5 per cent. of the heat content of the inlet air. At the selected operating conditions the constant bed temperature batch drying curve for this material consists entirely of a linear falling rate curve obeying the equation $- \mathrm{d}X/\mathrm{d}t = 0.0005X \text{ s}^{-1}$. Calculate the required mean residence time, bed area, mass flowrate of air, and air inlet temperature. Check that the exhaust air is far enough above its dew point to avoid condensation problems in the cyclone, and estimate the thermal efficiency of the dryer.

Solution

$$X_i = 0.20, \qquad X_o = 0.04, \qquad X_e = 0$$

$$F = \frac{6,000}{1.20 \times 3,600} = 1.39 \text{ kg/s dry solid}$$

From Eq. (10.15): $t_R = \dfrac{1}{0.005}\left(\dfrac{0.20}{0.04} - 1\right) = 800 \text{ s}$

From Eq. (10.16): $A = \dfrac{1.39 \times 800}{500 \times 0.200} = 11.12 \text{ m}^2$

From Eq. (10.17): $G = 0.70 \times 11.12 = 7.78 \text{ kg/s}$

From the mass balance Eq. (10.6):

$$Y_o = \frac{F}{G}(X_i - X_o) + Y_i$$

$$= \frac{1.39}{7.78}(0.20 - 0.04) + 0.005$$

$$= 0.034 \text{ kg } H_2O/\text{kg dry air}$$

Reference to a psychrometric chart shows that air of this humidity has a dew point of 34°C. Therefore, since the air leaves the bed at 80°C there is no risk of condensation.

$$T_{Fi} = 20°C, \qquad C_s = 1.0 \text{ kJ/kg°C}, \qquad C_l = 4.2 \text{ kJ/kg°C}$$

Therefore, from Eq. (10.8):

$$H_{Fi} = (1.0 + 0.20 \times 4.2)\, 20 = 36.8 \text{ kJ/kg}$$

and from Eq. (10.9):

$$H_{Fo} = (1.0 + 0.04 \times 4.2)\, 80 = 69.6 \text{ kJ/kg}$$

From a psychrometric chart, air at 80°C with a humidity of 0.034 has an enthalpy $H_{go} = 148 \text{ kJ/kg}$.

$$Q_h = 0, \qquad Q_w = 0.05\, H_{gi}$$

Rearranging the heat balance Eq. (10.7) yields:

$$H_{gi} - \frac{0.05\, GH_{gi}}{G} = \frac{F}{G}(H_{Fo} - H_{Fi}) + H_{go}$$

Therefore:

$$H_{gi} = \left\{ \frac{F}{G}(H_{Fo} - H_{Fi}) + H_{go} \right\} \frac{1}{0.95}$$

$$= \left\{ \frac{1.39}{7.78}(69.6 - 36.8) + 148 \right\} \frac{1}{0.95}$$

$$= 162 \text{ kJ/kg}$$

Reference to a psychrometric chart shows that air with an enthalpy of 162 kJ/kg and a humidity of 0.005 has a temperature of 147°. This is the required air inlet temperature.

Heat content of the inlet air $= H_{gi}$

$$= 7.78 \times 162$$

$$= 1{,}260 \text{ kJ/s}$$

For water at 80°C, the latent heat of vaporization = 2,300 kJ/s. Therefore:

heat used for evaporation $= F (X_i - X_o)$

$$= 2{,}300 \times 1.39 \ (0.20 - 0.04)$$

$$= 507 \text{ kJ/s}$$

thermal efficiency $= \dfrac{507}{1{,}260} \times 100 = 40\%$

10.5 PARTICLE RESIDENCE TIME DISTRIBUTION IN 'PLUG FLOW' BEDS

Perfect plug flow of particles is an idealization which is never achieved in a real 'plug flow' fluid bed dryer. The extent of the deviation from the ideal will depend mainly on the geometry of the bed, the mean residence time of the particles, the gas velocity, and the size and density of the particles. In many applications it is important to control this deviation within specified limits.

Particle mixing in a horizontal plane, which is what determines the particle residence time distribution (RTD) in shallow beds, has been shown to obey Fick's second law of diffusion for bed depths up to 10 cm (Reay, 1978):

$$\frac{\partial c}{\partial t} = D_p \frac{\partial^2 c}{\partial x^2} \tag{10.18}$$

where c is the concentration of labelled particles monitored at a distance x along the bed, t is time, and D_p is the diffusivity of the particles. If the particles have a net flow velocity v_s along the bed an additional term must be added to the right-hand side of Eq. (10.18), which becomes:

$$\frac{\partial c}{\partial t} = D_p \frac{\partial^2 c}{\partial x^2} - v_s \frac{\partial c}{\partial x} \tag{10.19}$$

The net flow velocity $v_s = L/t_R$, where L is the bed length. In terms of the dimensionless variables $\theta = t/t_R$ and $\xi = x/L$, Eq. (10.19) becomes:

$$\frac{\partial c}{\partial \theta} = \frac{D_p t_R}{L^2} \frac{\partial^2 c}{\partial \xi^2} - \frac{\partial c}{\partial \xi} \tag{10.20}$$

The dimensionless group $B = D_p t_R / L^2$, known as the axial dispersion number, is a measure of the degree of particle backmixing along the length of the bed, and hence of the degree of deviation from plug flow. For perfect plug flow $B = 0$ and for perfect mixing $B = \infty$.

From experiments on a variety of materials in beds of depth ranging up to 10 cm, Reay (1978) derived the following correlation for particle diffusivity:

$$D_p = 3.71 \times 10^{-4} \frac{(U - U_{mf})}{U_{mf}^{1/3}} \quad m^2/s \qquad (10.21)$$

where the gas velocity U and the minimum fluidization velocity U_{mf} are in metres per second. The term $(U - U_{mf})$ in the numerator is clearly a measure of the extent of bubbling in the bed, but the denominator term is completely empirical.

Having established that particle diffusion in a horizontal direction in a shallow fluidized bed obeys the same basic equation as molecular diffusion, we can make use of the body of theory developed for deviation from plug flow of fluids through process vessels. A good account of the latter is given in Chapter 9 of the book by Levenspiel (1972). For our present purpose the most important concepts and results are the following:

(a) The RTD can be represented by function $E_\theta = f(B, \theta)$, where $E_\theta d\theta$ is the fraction of particles which have dimensionless residence times between θ and $\theta + d\theta$.

(b) $\int_0^\theta E_\theta \, d\theta$ = fraction of particles with dimensionless residence times less than θ.

(c) For small deviations from plug flow, i.e. for small values of B, the solution to Eq. (10.20) yields the symmetrical curve:

$$E_\theta = \frac{1}{2 \sqrt{(\pi B)}} \exp \left\{ -\frac{(1-\theta)^2}{4B} \right\} \qquad (10.22)$$

The mean is at $\theta = 1$ and the dimensionless standard deviation of the RTD about the mean is given by:

$$\sigma_\theta = \frac{\sigma}{t_m} = \sqrt{(2B)} \qquad (10.23)$$

This approximation is reasonable up to $B = 0.1$, which covers most practical cases of true 'plug flow' fluid bed dryers.

(d) An alternative way of representing deviations from plug flow is to consider the bed as consisting of a number n of perfectly mixed beds in series. For perfect plug flow $n = \infty$. With this model the dimensionless standard deviation $\sigma_\theta = \sqrt{(1/n)}$. Hence, comparing the two models we have $n = 1/(2B)$.

From the above analysis the designer can select the maximum value of B which can be tolerated if the deviation from plug flow is to be kept within the specified limits. D_p can be estimated from Eq. (10.21) and the required t_R estimated from the batch drying curve. Hence, the minimum acceptable bed length L can be calculated. The process design calculations outlined earlier yield the required bed area A, so the required bed width W can be found. In the case of a bed with reversing or a spiral flow channel, W and L should be interpreted as the width and total length of the particle flow channel.

It should be emphasized that the validity of the above analysis has only been confirmed for bed depths up to 10 cm. There is some evidence in the literature that D_p may be an order of magnitude larger in much deeper beds (Mori and Nakamura, 1965; Highley and Merrick, 1971). this is thought to be due to the establishment of regular particle circulation cells in deeper beds, in contrast to the more random particle motion in shallow beds. This large increase in D_p with deeper beds suggests the use of the shallowest practicable bed depth when the objective is a close approach to plug flow.

Example 10.2: Calculation of the residence time distribution in a 'plug flow' fluid bed dryer

Bed length L	$= 5$ m
Mean particle residence time t_R	$= 1,000$ s
Gas velocity U	$= 0.4$ m/s
Minimum fluidiization velocity U_{mf}	$= 0.1$ m/s

Calculate the equivalent number of perfectly mixed stages in series.

Solution

From Eq. (10.21):
$$D_p = 3.7 \times 10^{-4} \frac{0.4 - 0.}{0.1^{1/3}}$$
$$= 2.4 \times 10^{-4} \text{ m}^2/\text{s}$$

Therefore:
$$B = \frac{D_p t_R}{L^2} = \frac{2.4 \times 10^{-4} \times 1,000}{25} = 0.0096$$

and thus:
$$n = \frac{1}{2B} = 52$$

10.6 NOMENCLATURE

A	bed area	m^2
B	axial dispersion number, $D_p t_R / L^2$	—
c	concentration of labelled particles	—
c_g	specific heat of gas	kJ/kg K
c_l	specific heat of liquid	kJ/kg K
c_s	specific heat of solids	kJ/kg K
c_v	specific heat of vapour	kJ/kg K
d_p	particle diameter	m
D_p	particle diffusivity	m^2/s
E_t	residence time distribution (RTD) function	—
E_θ	dimensionless RTD function	—
$f(\ )$	function of parameter in parenthesis	—
F	solids feed rate (dry basis)	kg/s
G	gas mass flowrate (dry basis)	kg/s
h_{gp}	gas-to-particle heat transfer coefficient	$W/m^2 K$
H_{Fi}	enthalpy of inlet solids	kJ/kg
H_{FO}	enthalpy of outlet solids	kJ/kg
H_{gi}	enthalpy of inlet gas	kJ/kg
H_{go}	enthalpy of outlet gas	kJ/kg
k	proportionality constant	—
k_g	thermal conductivity of gas	W/m K
k_p	mass transfer coeffcient	$kg/m^2 s$ (N/m^2)
L	dryer length	m
n	number of perfectly mixed stages	—
N	drying rate	$kg/m^2 s$
N_c	drying rate in constant rate period	$kg/m^2 s$
Nu	Nusselt number, $h_{gp} d_p / k_g$	—
p	partial pressure of vapour	N/m^2
p_{wb}	partial pressure of vapour at the wet bulb temperature	N/m^2
Pr	Prandtl number. $C_g \mu / k_g$	—
Q_h	rate of heat supply from immersed heater	kJ/s
Q_w	rate of heat loss from dryer	kJ/s
Re_p	Reynolds number, $U d_p \rho_g / \mu$	—

T	bulk temperature of gas	K
T_{Fi}	solids inlet temperature	K
T_{Fo}	solids outlet temperature	K
T_{gi}	gas inlet temperature	K
T_{go}	gas outlet temperature	K
T_{wb}	wet bulb temperature	K
t	time	s
t_R	mean residence time	s
U	gas velocity	m/s
U_{mf}	minimum fluidization velocity	m/s
v_s	net flow velocity of particles	m/s
W	Bed width	m
W_B	weight of solids in bed (dry basis)	kg
x	distance along bed	m
X	moisture content (dry basis)	kg H_2O/kg solids
X_{cr}	critical moisture content	kg H_O/kg solids
X_e	equilibrium moisture content	kg H_2O/kg solids
X_i	inlet moisture content	kg H_2O/kg solids
X_o	outlet moisture content	kg H_2O/kg solids
$\bar{X}_o$	mean outlet moisture content from a well-mixed bed	kg H_2O/kh solids
X_t	moisture content at time t	kf H_2O/kg solids
Y_i	gas inlet humidity	kg H_2O/kg dry air
Y_o	gas outlet humidity	kg H_2O/kg dry air
z	bed height	m
θ	dimensionless time, t/t_R	—
λ	latent heat of vaporization	kJ/kg
μ	viscosity of gas	Ns/m^2
V	defined in Eq. (10.5)	—
ξ	dimensionless distance, x/L	—
ρ_B	bulk density of bed	kg/m^3
ρ_g	density of gas	kg/m^3
σ	standard deviation of RTD function	—
σ_θ	standard deviation of dimensionless RTD function	—

10.7 REFERENCES

Gummel, P., and Schlünder, E.U. (1977). *Verfahrenstechnik,* **II** (12), 743.
Highley, J., and Merrick, D. (1971). *Chem. Eng. Prog. Symp. Ser.,* **67** (116), 219.
Keey, R.B. (1978). *Introduction to Industrial Drying Operations*, Pergamon Press.
Kunii, D., and Levenspiel, O. (1969). *Fluidization Engineering*, John Wiley and Sons, Chapter 7.
Levenspiel, O. (1972). *Chemical Reaction Engineering*, 2nd ed., John Wiley and Sons, Chapter 9.
Mori, Y., and Nakamura, K. (1965). *Kagaku Kogaku,* **29**, 868.
Mostafa, I. (1977). 'Studies of fluidized bed drying' Ph.D. Thesis, Imperial College of Science and Technology, University of London.
Reay, D. (1978). *Proceedings of the First International Symposium of Drying*, McGill University, Montreal, Canada, published by Science Press, p. 136.
Reay, D. and Allen, R.W.K. (1982). Proceedings of the Third International Drying Symposium, Birmingham, **2**, p. 130–140.
Subramanian, D., Martin, H., and Schlünder, E.U. (1977). *Verfahrenstechnik,* **II** (12), 3.
van Meel, D.A. (1958). *Chem. Eng. Sci.,* **9**, 36.
Vanecek, V., Markvart, M., and Drbohlav, R. (1966). *Fluidised Bed Drying,* Leonard Hill.
Zabeschek, G. (1977). Ph.D. Dissertaion (in German), Universität Karlsruhe, Germany.

Gas Fluidization Technology
Edited by D. Geldart
Copyright © 1986 John Wiley & Sons Ltd.

Chapter 11

Fluid Beds as Chemical Reactors

J.R. GRACE

11.1 INTRODUCTION

The earliest applications of fluidization were for the purpose of carrying out chemical reactions (see Chapter 1). Since that time there have been many successful chemical processes involving fluid bed reactors. The principal applications are listed in Table 11.1 under three headings:

Table 11.1 Some applications of fluidized beds as chemical reactors

---

Catalytic gas phase reactions

Hydrocarbon cracking
Catalytic reforming
Phthalic anhydride manufacture
Acrylonitrile production
Aniline production
Synthesis of high-density and low density polyethylene
Fischer-Tropsch synthesis
Chlorination or bromination of hydrocarbons
Oxidation of SO_2 to SO_3
Methanol to gasoline process

Non-catalytic gas phase reactions

Hydrogenation of ethylene
Thermal cracking, e.g. to give ethylene or for fluid coking

Gas-solid reactions

Roasting of sulphide and sulphate ores
Calcination of limestone, phosphates etc.
Incineration of waste liquors and solids refuse
Combustion of coal, coke, peat, biomass
Gasification of coal, peat, wood wastes
Pyrolysis of coal, silane
Catalyst regeneration
Fluorination of UO_2 pellets
Chlorination of rutile, ilmenite
Hydrogen reduction of ores

(a) Catalytic gas phase reactions. These are cases in which the solid particles serve as catalysts, undergoing no chemical changes and limited physical changes during the course of the reaction. The catalyst materials are manufactured or selected to combine favourable activity and selectivity characteristics with suitable physical properties — particle size and density, resistance to attrition, etc. The most important application historically is the catalytic cracking of petroleum.

(b) Non-catalytic gas phase rections. In this case the particles are inert. Their purpose is to provide a uniform heat transfer medium or nuclei upon which the product material can be deposited (as in fluid coking) or both. Applications in this category are relatively rare.

(c) Gas-solid reactions. Reactions of this type involve the fluidized particles themselves. The useful product may be the solid, a gas, thermal energy, or some combination of these. In reactions of this type there is generally less control over the properties of the solids than can be achieved in catalytic reactions. Moreover, the solids undergo changes in their chemical composition and their physical characteristics during the course of the reaction. The most important applications of fluidization for gas-solid reactions are in roasting of ores, calcination reactions, regeneration of catalysts, and processing of coal and biomass.

The principal *advantages* of fluidized beds which make them attractive for many chemical processes relative to fixed beds and other types of gas-solid reactors are:

(a) Temperature uniformity. 'Hot spots' are avoided.

(b) Favourable rates of heat transfer to immersed tubes or to the walls of the column.

(c) Ease of solids handling. Solids can be continuously added and removed from the system. In catalytic reactors the deactivated solids may be fed to a parallel regenerator bed and then recycled to the main reactor.

(d) Scale of operation. Successful operations have been achieved with columns as small as 0.05 m and as large as 30 m in diameter.

(e) Turndown. The gas flowrate can be varied over a wide range.

(f) Pressure drop. The pressure drop across a fluidized bed of solids is less than for the same bed at the same superficial gas velocity under fixed bed conditions, especially for fine particles.

On the other hand, fluidized beds have certain limitations and *disadvantages* which must be clearly recognized when considering what type of reactor is best suited to a given process:

(a) Backmixing of both solids and gas is generally substantial, resulting in lower conversions than with most competing types of reactor.

(b) By-passing of gas via bubbles or jets causes gas-solids contacting to be unfavourable. This leads to a further lowering of conversions, and may also contribute to poor reactor selectivity.

(c) Entrainment of solids may lead to loss of expensive materials or product as well as to pollution of the atmosphere and the need for pollution control equipment.

(d) Attrition, erosion, and agglomeration may cause serious operational problems.

(e) Scale-up and design of fluidized bed reactors is an uncertain undertaking because of the complexity of fluidized beds and limitations in the extent to which their behaviour can be predicted and modelled.

(f) There are some limitations on the size of particles that can be treated. Generally speaking, gas fluidized beds are not well suited for particles of mean size less than about 30 μm or greater than about 3 mm.

Some of these limitations can be lessened by modifications in design. For example, particles too small to fluidize properly can often be treated by stirring or vibrating the bed. Backmixing of solids can be greatly reduced by using stages in series. By-passing of gas can be lessened by means of properly designed baffles. However, all of these modifications involve considerable extra cost and operational complexity which may make the fluid bed process uncompetitive.

In the remainder of this chapter, we consider methods of modelling fluid bed reactors for purposes of design, scale-up, simulation, and control. Considerable effort has gone into devising models, but the results have been disappointing on the whole. Nevertheless, the models help to give a good understanding of the phenomena involved and of the factors and variables which influence the performance of fluidized bed reactors. While it is impossible to predict the conversion or selectivity of fluidized beds with complete certainty, the models do provide a rational basis for scaling-up fluidized beds and for simulating existing reactors. Because the contacting between solids and gas is quite different in three different regions — the region just above the gas distributor (which we will call the grid region), the bed proper, and the freeboard region above the bed surface — these regions are considered separately.

11.2   REACTION KINETICS: CATALYTIC REACTIONS

Fluidized beds are far too complex hydrodynamically to be used for deriving reaction kinetics. Instead, kinetic parameters should be obtained in reactors where the hydrodynamics can be described with confidence. Fixed beds are well characterized by dispersed plug flow models (see Levenspiel, 1972; Wen and Fan, 1975), but temperature gradients may be severe. Consequently, it may be better to use a spinning basket reactor (Tajbl, Simons, and Carberry,

1966) or other suitable laboratory reactors (e.g. Carberry, 1964) to derive reaction kinetics.

Kinetic data should be obtained for a range of temperatures spanning the range in which operation is to take place. It is also essential that kinetic data be obtained under much higher conversion conditions than are anticipated for the fluidized bed as a whole. This is because of the two-phase nature of fluidized beds: Most reaction occurs in a dense phase region where the gas is highly converted, the final composition being achieved by mixing of this gas with poorly reacted dilute phase gas.

Catalytic reactions occurring on the interior surface of porous catalyst particles involve the following sequence of physical and chemical steps:

(a) Mass transfer of reactants from the bulk gas to the outside of the catalyst particle.

(b) Diffusion of reactant into the catalyst pores.

(c) Chemisorption of reactant species

(d) Chemical reactions on the surface

(e) Desorption of product molecules

(f) Diffusion of products back to the exterior surface of the particle

(g) Mass transfer of products from the exterior surface into the bulk

Catalyst particles used in most fluidized bed processes are quite small, typically 50 to 120 μm in mean diameter, small enough for external mass transfer and internal diffusion resistances to be minor. Temperature gradients inside the catalyst particles can also usually be neglected. For many purposes it is permissible to use a single apparent or overall volumetric rate constant k_n, defined as:

$$r_A = \frac{-1}{V_p} \frac{dN_A}{dt} = k_n C_A^n \qquad (11.1)$$

where n is the order of reaction and C_A is the local bulk concentration of gaseous reactant A. We will assume rate expressions of this type for catalytic reactions throughout this chapter. The rate constant k_n is based on unit particle volume. Where diffusional resistances are significant, the k_n thus defined includes allowance for the effectiveness of the catalyst pores. If the kinetics follow more complex rate expressions (e.g. Smith, 1970), these expressions may be readily incorporated within the reactor models described below, but solution will then generally require numerical integration.

In addition to the order of reaction and rate constant as a function of temperature, laboratory kinetic tests should yield the stoichiometry of the desired reaction and indications of side reactions or other difficulties. The heat of reaction and equilibrium conversion can be calculated from thermodynamic considerations. In this chapter we consider only cases in which gas volume changes due to reaction can be neglected. Allowance for volume changes accompanying reaction is complicated due to the fact that most reaction takes place in the dense phase of fluidized bed reactors, while most of the added volumetric flow probably finds its way into the bubble phase.

11.3 BUBBLING REGIME: GAS MOTION IN AND AROUND BUBBLES

Bubbles in fluidized beds are responsible for by-passing of gas and for limiting gas-solid contacting. While there are important similarities between bubbles in fluidized beds and bubbles in liquids, the gas motion in the two cases is fundamentally different. In gas-liquid systems, the gas must circulate entirely within the bubble. In fluidized systems, on the other hand, the bubble boundary is permeable, and gas circulates through the bubble, entering at the bottom and leaving through the upper surface. This factor, commonly referred to as 'throughflow', has important consequences for gas-solid contacting and for interphase (bubble to dense phase) mass transfer in aggregative fluidized beds.

Consider a bubble rising at steady velocity u_b. Whether or not elements of gas leaving the top of the bubble recirculate through the dense phase and reenter the bottom of the same bubble depends on the ratio

$$\alpha = \frac{\text{bubble velocity}}{\text{remote interstitial velocity}} = \frac{u_b \, \epsilon_{mf}}{U_{mf}} \qquad (11.2)$$

For $\alpha > 1$ recirculation does occur. The region surrounding the bubble in which gas recirculates (see Fig. 11.1) is called the 'cloud'. According to the

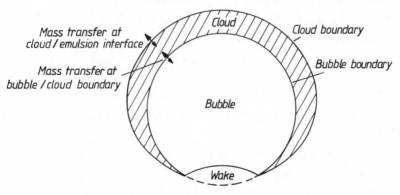

Figure 11.1 Shape of cloud surrounding a two-dimensional gas bubble in a fluidized bed, as observed by Rowe, Partridge and Lyall (1964) for $\alpha = 2.5$. Resistances to mass transfer may be accounted for at the bubble/cloud boundary and/or at the cloud/emulsion interface.

theory of Davidson and Harrison (1963), the cloud for a spherical bubble also
has a spherical boundary with the same centre and with volume:*

$$V_c = V_b \left(\frac{3}{\alpha - 1} \right) = \frac{\pi}{2} \frac{d_{eq}^3}{(\alpha - 1)} \qquad (11.3)$$

An alternative analysis (Murray, 1965) leads to a non-spherical cloud which is
smaller and which has its centroid above the centre of the (assumed) spherical
bubble. An empirical equation (Partridge and Rowe, 1966) which has been
fitted to the results of this analysis gives

$$V_c = V_b \left(\frac{1.17}{\alpha - 1} \right) = \frac{0.195 \, \pi \, d_{eq}^3}{\alpha - 1} \qquad (11.4)$$

Experimental results (Rowe, Partridge, and Lyall, 1964; Anwer and Pyle,
1974) indicate closer agreement with the Murray analysis, but both theories
have been used as the bases for reactor models, as we will see below. For a
review of the fluid mechanics of fluidization and analyses of gas motion
around bubbles, see Jackson (1971).

As shown by Eqs (11.2) to (11.4), the cloud volume is strongly related to α
and hence to particle size. Typical cloud volumes are estimated in the
following sample problem.

Example 11.1: Cloud volumes

Estimate the ratio of cloud volume to bubble volume for a bubble of volume
equivalent diameter 0.10 m rising through an air-fluidized bed of:

(a) Cracking catalyst particles: $d_p = 80 \, \mu m$, $U_{mf} = 0.005$ m/s, $\epsilon_{mf} = 0.50$

(b) Silica sand: $d_p = 250 \, \mu m$, $U_{mf} = 0.056$ m/s, $\epsilon_{mf} = 0.45$

(c) Silica sand: $d_p = 450 \, \mu m$, $U_{mf} = 0.16$ m/s. $\epsilon_{mf} = 0.45$

(d) Dolomite: $d_p \doteq 700 \, \mu m$, $U_{mf} = 0.32$ m/s, $\epsilon_{mf} = 0.48$

(e) Dolomite: $d_p = 1.2$ mm, $U_{mf} = 0.54$ m/s, $\epsilon_{mf} = 0.48$

Solution

The rise velocity of the bubble relative to the dense phase in each case is
(from Eq. 4.9):

$$u_b = 0.71 \, (g d_{eq})^{1/2} = 0.71 \, (9.8 \times 0.10)^{1/2} \text{ m/s} = 0.70 \text{ m/s}$$

---

*The terms cloud and cloud volume refer in this book only to the outer dense phase mantle
(particle and interstitial gas) surrounding the bubble in which gas recirculates. Occasionally
(e.g. Partridge and Rowe, 1966) the terms are used to include also the gas contained in the
bubble.

The values of U_{mf} and ϵ_{mf} together with values of α calculated from Eq. (11.2) and V_c/V_b from Eqs (11.3) and (11.4) appear below:

Case	U_{mf}, m/s	ϵ_{mf}	α	V_c/V_b from Eq. (11.3)	V_c/V_b from Eq. (11.4)
a	0.005	0.50	70	0.043	0.017
b	0.056	0.45	5.6	0.65	0.25
c	0.16	0.45	2.0	3.1	1.2
d	0.32	0.48	1.05	54	21
e	0.54	0.48	0.62	No clouds since $\alpha < 1$	

Clouds form the basis for a number of the reactor models to be discussed below. For $\alpha < 1$, gas emerging from the upper surface of bubbles does not recirculate. Instead the gas short circuits through the bubbles on its way through the bed. Hence cloud models are not applicable to large particle systems (often called 'slow bubble' systems) where $\alpha < 1$. A qualitative explanation of why α plays such a role may be helpful.

All of the theories suggest that the velocity of the throughflow gas at the bubble roof is proportional to the interstitial gas velocity, U_{mf}/ϵ_{mf}. (Davidson's theory gives a proportionality constant which is three times greater than Murray's, and this accounts for the larger predicted clouds.) The gas emerging into the dense phase from the bubble surface encounters particles which are moving tangentially around the bubble surface with velocities of order u_b and decreasing with increasing radial distance from the bubble surface. This situation is shown in Fig. 11.2. If the interstitial velocity U_{mf}/ϵ_{mf} is much less than u_b (i.e. for large α), the particles near the bubble drag the gas around the bubble surface in a tight trajectory and a thin cloud is formed. As U_{mf}/ϵ_{mf} increases (i.e. as α decreases), the gas elements penetrate further into the dense phase before the drag caused by the solid particles is able to reverse the direction. A larger cloud therefore results. For a sufficiently large U_{mf}/ϵ_{mf} where $\alpha < 1$, the gas is able to penetrate so far that it never recirculates.

Pyle and Rose (1965) predicted gas streamlines inside rising bubbles. The most notable feature of their predictions was that there were regions at all values of α in which gas would circulate entirely within the bubble, without ever making contact with particles. Total by-passing of this gas would therefore occur. In practice, molecular diffusion, bubble dilations, and raining through of solids are expected to cause rapid mixing of the gas within each bubble. Hence almost all subsequent workers have assumed perfect mixing within each bubble unit, and we will limit our attention to models of this type.

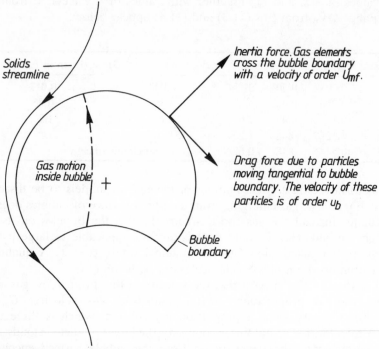

Figure 11.2 Forces acting on an element of gas emerging from a bubble into the dense phase. The frame of reference is moving with the bubble.

11.4 BUBBLING REGIME: INTERPHASE MASS TRANSFER

The performance of fluidized bed reactors is greatly influenced by the rate of mass transfer between the dilute phase (bubble phase or bubble plus cloud phase) and the remaining dense phase. As shown in Fig. 11.1, two resistances to mass transfer may be considered for cases in which cloud formation occurs:

(a) Resistance at bubble boundary. Mass transfer at the outer surface of the bubble occurs due to two parallel mechanisms: convective flow (also called throughflow) and diffusion.

(b) Resistance at cloud boundary. Mass transfer at the cloud boundary occurs due to molecular diffusion, due to gas adsorption on (or capture by) particles moving through the cloud region, and due to shedding of elements of the cloud during wake shedding, bubble shape dilations, or as a result of coalescence.

Most of the chemical reactor models assume that one of these resistances is much larger than the other, and this allows the clouds to be lumped with either the dense phase or bubbles so that two-phase models, rather than three-phase models, can be adopted. An exception is the model of Kunii and Levenspiel (1969) which allows for resistances in series at the bubble and cloud boundaries. If the resistance at the cloud boundary is ignored, then no distinction is made between gas elements in the cloud and those in the dense phase at large at the same level, and the models consider a bubble phase and a dense phase comprising all (or essentially all) particles and interstitial gas. If, on the other hand, the resistance to mass transfer at the bubble boundary is ignored, the clouds are lumped with the bubbles as one phase. The remainder of the bed, often called the emulsion, then comprises the other phase.

Interphase mass transfer rates have been measured or deduced in a number of studies. Almost all of the work has been for single isolated bubbles. If due allowance is made for end effects and for the tendency of bubbles to grow when injected into a bed fluidized at superficial velocities somewhat in excess of U_{mf}, Sit and Grace (1978) have shown that the transfer rate can be expressed as the sum of two terms. For isolated three-dimensional bubbles the interphase mass transfer coefficient is then given by:

$$k_q = \frac{U_{mf}}{4} + \left(\frac{4 \mathcal{D} \, \epsilon_{mf} \, u_b}{\pi \, d_{eq}} \right)^{1/2} \tag{11.5}$$

The first term represents the throughflow component, as given by the analysis of Murray (1965). (In a number of reactor models the corresponding result from Davidson's analysis (Davidson and Harrison, 1963) is adopted, leading to the first term being larger by a factor of 3.) The second term in Eq. (11.5) describes diffusion according to the penetration theory. The bulk flow or convective term is dominant for large particles, but the diffusion term becomes more important for small particles, as shown in the following worked example.

Example 11.2: Interphase mass transfer rates

Determine the two components of the interphase mass transfer coefficient and the total coefficient for the transfer of hydrogen in air at 20°C for the five cases of Example 11.1. The molecular diffusivity is 6.9×10^{-5} m/s.

Solution

From Example 11.1, $d_{eq} = 0.10$ m and $u_b = 0.70$ m/s. We are given $\mathcal{D} = 6.9 \times 10^{-5}$ m²/s. Inserting these values into Eq. (11.5) with the appropriate values of U_{mf} and ϵ_{mf} from Example 11.1, we obtain the following results:

Case	Particles	Convection term, mm/s	Diffusion term, mm/s	k_q, mm/s	Bulk flow/k_q
a	80 μm catalyst	1.3	17.5	18.8	0.07
b	250 μm sand	14.0	16.6	30.6	0.46
c	450 μm sand	40.0	16.6	56.6	0.71
d	700 μm dolomite	80.0	17.2	97.2	0.82
e	1.2 mm dolomite	135.0	17.2	152	0.89

In addition to the influence of decreasing particle size, the diffusion term also becomes more important as the molecular diffusivity increases and as the bubble size decreases.

The fact that Eq. (11.5) gives a good representation of transfer rates from the bubble phase implies that the principal resistance to mass transfer resides at the bubble boundary, not at the cloud boundary. Theories for mass transfer at the cloud boundary based on diffusional transfer alone (Partridge and Rowe, 1966; Kunii and Levenspiel, 1969) predict much lower mass transfer coefficients than are measured or inferred in practice. This implies that neglected mechnisms of transfer at the cloud surface (in particular due to adsorption by particles moving through the cloud region, due to unsteady bubble motion, or due to shedding of cloud/or wake elements) are sufficient to cause rapid dispersion of gas in the dense phase once gas elements are transferred from a bubble. See Walker (1975), Chavarie and Grace (1976), Lignola, Donsi and Massimilla (1983), and Grace (1984) for further discussion and evidence on this point.

For bubbles undergoing interaction and coalescence, there is evidence of enhanced interphase mass transfer (Toei *et al.*, 1969; Pereira, 1977; Sit and Grace, 1978, 1981) compared with isolated bubbles. Indications are that the throughflow term in Eq. (11.5) is augmented by 20 to 30 per cent. for a freely bubbling bed when averaged over time, while the diffusion term is unaffected (Sit and Grace, 1981). For freely bubbling beds the following relationship should be used (Sit and Grace, 1981):

$$k_q = \frac{U_{mf}}{3} + \left(\frac{4 \, \mathscr{D} \, \epsilon_{mf} \, \bar{u}_b}{\pi \, \bar{d}_{eq}} \right)^{1/2} \tag{11.6}$$

The enhancement of transfer is therefore more important for large particles, where the bulk flow term is dominant, than for fine particles, where the diffusion term plays the major role. Equation (11.6) is used in reactor modelling below.

Interphase mass transfer rates can also be inferred from residence time distribution data, if an appropriate unsteady state two-phase model is used. This possibility is covered in Section 11.9 below.

11.5 SINGLE-PHASE REACTOR MODELS

Early workers tended to treat fluidized beds as if the gas and solids were intimately mixed without segregation into dilute and dense phases. This led to simple 'single-phase models' in which it was assumed that reactor performance could be determined from the residence time distribution of gas elements. In practice, however, reactor performance is likely to be affected less by the residence time distribution than by the 'contact time distribution', i.e. by the distribution of times that gas elements spend in the vicinity of, and in effective contact with, solid particles. Gas elements in the bubbles are in poor contact with solids, while those in the dense phase are in intimate contact. Hence single-phase models are unable to account for the true flow and mixing behaviour of aggregative fluidized beds, and we must therefore look at more complex models in which two or more phases are considered.

Single-phase models do, however, provide limits for the two-phase models considered below. If we consider a first-order reaction ($n = 1$), the limiting results are as follows:

Plug flow reactor
$$\frac{C_{\text{Aout}}}{C_{\text{Ain}}} = \exp\left(-k_1'\right) \tag{11.7}$$

Perfectly mixed reactor
$$\frac{C_{\text{Aout}}}{C_{\text{Ain}}} = \frac{1}{1 + k_1'} \tag{11.8}$$

where k_1' is a dimensionless rate constant defined by:

$$k_1' = \frac{k_1 \times \text{total particle volume}}{\text{volumetric fluid flowrate}} \tag{11.9}$$

One or other of these results (depending on the assumed dense phase gas mixing pattern) will be approached as limiting cases for the simpler two-phase models considered below, as mass transfer of reactants between phases becomes very rapid or as the proportion of gas passing through the dilute phase approaches zero.

11.6 TWO- AND THREE-PHASE REACTOR MODELS: BASIC ASSUMPTIONS

In modelling any chemical reactor system, a primary objective must be to represent accurately the key physical or hydrodynamic features of the system before inserting chemical kinetics and other chemical parameters. At the same time, it is important that models be sufficiently straightforward that their application does not demand excessive computation. The many models of fluid bed reactors that have been proposed and described differ in complexity and in their view of what constitute the principal physical or hydrodynamic features of the fluidized bed reactor.

The models considered in this section are based upon parallel one-dimensional flow in two or three phases, as illustrated in Fig. 11.3. The word

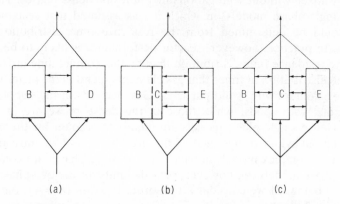

Figure 11.3 Schematic diagrams showing phases chosen in alternative fluid bed reactor models: B = bubbles; C = clouds; D = dense phase; E = emulsion. (a) Two-phase models with all solids assigned to the dense phase; (b) two-phase models with clouds included with bubbles; (c) three-phase models.

'phase' in this context refers to a region which may include both gas and solid particles. The phases are distinguished from one another in terms of the volume fraction of solids, by physical appearance, and/or through their flow characteristics. In one group of models, shown in Fig. 11.3(a), the dilute phase is either free of particles or assumed to contain only widely dispersed solids; in these models no distinction is made between the cloud and other interstitial gas elements. In a second group of models, Fig. 11.3(b), the cloud gas and solids (with or without bubble wakes) are lumped with the bubbles to give a combined bubble/cloud dilute phase. Gas within the cloud is then assumed to have the same composition as gas within the bubble surrounded by the cloud. In a third group the cloud is treated separately from both bubbles and the remaining dense phase (often called the 'emulsion'), so that there are three phases. This third group of models is illustrated schematically in Fig. 11.3(c).

Having decided upon the general physical nature of the phases the model builder must then decide on the distribution of gas flow between the phases, on the extent of interphase mass transfer, and on the degree of axial dispersion in each phase. Since each of these decisions is controversial, the number of alternative assumptions which can be made is large, and the number of combinations of assumptions even larger. Table 11.2 outlines the assumptions upon which models have been based in the critical areas of phase composition, flow distribution, axial dispersion and interphase mass transfer. Table 11.3 lists some of the more popular and more recent models, and shows

Table 11.2 Competing assumptions for two- and three-phase reactor models

---

A. Nature of dilute phase:

1. Bubble phase completely free of particles.
2. Bubbles containing some widely dispersed solids.
3. Bubble–cloud phase, i.e. clouds are included.

B. Division of gas between phases:

1. Governed by two-phase theory of fluidization (Chapter 4).
2. All gas carried by bubbles.
3. Some downflow of gas in the dense phase permitted.
4. Other or fitted parameter.

C. Axial dispersion in dilute phase:

1. Plug flow.
2. Dispersed plug flow.

D. Axial dispersion in dense phase:

1. Plug flow.
2. Dispersed plug flow.
3. Stagnant.
4. Well-mixed tanks in series.
5. Perfect mixing.
6. Downflow.
7. Bubble-induced turbulent fluctuations.

E. Mass transfer between phases:

1. Obtained from independent gas mixing or mass transfer studies.
2. Fitted parameter for case under study.
3. Empirical correlation from previous or pilot plant data.
4. Bubble to dense phase transfer obtained from experimental or theoretical single bubble studies.
6. Enhancement due to bubble interaction included.

F. Cloud size (if applicable):

1. As given by Davidson theory.
2. Murray or modified Murray.
3. Wake not specifically included or assumed negligible.
4. Wake added to cloud.
5. Recognized but assumed negligible.

G. Bubble size:

1. Not specifically included.
2. Single representative size for entire bed.
3. Allowed to increase with height.
4. Obtained from separate measurements, correlation, or estimated.
5. Kept as fitting parameter.

Table 11.3　Assumptions or approaches embodied in some of the principal two-phase and three-phase reactor models; letters and numbers in the table refer to Table 11.2

Authors	Assumptions						
	A	B	C	D	E	F	G
(a)　Two-phase models:							
Shen and Johnstone (1955)	1	1	1	1 or 5	3	NA	1
Lewis, Gilliland, and Glass (1959)	2	2	1	1 or 5	3	NA	1
May (1959)	1	1	1	2	1	NA	1
Van Deemter 1961	1	4	1	2	1	NA	1
Orcutt, Davidson, and Pigford (1962)	1	1	1	1 or 5	4	NA	2,5
Partridge and Rowe (1966)	3	1	1	1	5	2,4	4
Mireur and Bischoff (1967)	1	1	1	2	1,3	NA	1
Kato and Wen (1969)	3	2	1	4	1	1,3	3,4
Bywater (1978)	1	4	1	7	5	5	2,4
Darton (1979)	1	1	1	5	4,5	5	3,4
Werther (1980)	1	1	1	1	3	NA	3,4
Grace (1984)	2	2	1	3	4,6	NA	2,4
(b)　Three phase models:							
Kunii and Levenspiel (1969)	2,1	1 or 2	1	3	4,5	1,4	2,4
Fryer and Potter (1972a)	1	3	1	6	4,5	4,5	2,4
Fan, Fan, and Miyanami (1977)	2	1	2	2	4,5	2,3	3,4

the assumptions which have been adopted in these models, using the categories given in Table 11.2. This list of models is far from exhaustive. An expanded listing of pre-1970 models appears in the article by Grace (1971). Except for assumptions regarding axial dispersion in the dilute phase, where all models except one assume plug flow, we find that there is no unanimity in the assumptions adopted by different authors. Even in cases where the same number appears in a given column in Table 11.3, competing models may employ different approaches, For example, there are a number of expressions which may be used to predict interphase mass transfer rates, as discussed in Section 11.4.

Not surprisingly, models whose assumptions differ as radically as those listed in Table 11.3 can give very different predictions of chemical conversions and selectivities. For simulation of existing reactors, operated over limited ranges of variables, it may be possible to use a number of different models, especially if fitted constants are introduced. On the other hand, for reactor scale-up or for prediction of reactor performance from first principles, it is extremely important to choose a suitable model from the many alternatives. Experimental testing of a number of competing models was carried out by

Chavarie and Grace (1975). Reactant concentration profiles in both the bubbles and dense phase were shown to give a much more discriminating test of the models than has been achieved with overall conversion determinations. A qualitative comparison of the key features of the primary models tested appears in Table 11.4. Generally speaking, all of the models tested were

Table 11.4 Comparison of key assumptions underlying models investigated by Chavarie and Grace (1975)

Model of	Interphase mass transfer	Solids assigned to bubble phase	Gas flow through bubble phase	Axial dispersion in dense phase
Orcutt, Davidson, and Pigford	High	None	Two-phase theory	None
Orcutt, Davidson, and Pigford	High	None	Two-phase theory	Perfect mixing
Rowe (1964)	None	Davidson cloud	$(U-U_{mf})A$ + gas carried in clouds	None
Partridge and Rowe (1966)	Low	Murray cloud	$(U-U_{mf})A$ + gas carried in clouds	None
Kato and Wen (1969)	Intermediate	Davidson cloud (later include wake)	All	Well-mixed units in series
Kunii and Levenspiel (1969)	Two resistances in series (one low, one high)	Little or none	All	None

deficient in certain respects, although some gave better representations than others. From a statistical point of view none of the models was acceptable, a conclusion which was also reached by Shaw, Hoffman, and Reilly (1974) using a very different experimental technique in which selectivities, rather than concentration profiles, were the basis for model evaluation.

In the next two sections two of the two-phase models are presented in greater detail. The first of these, one of the Orcutt, Davidson, and Pigford (1962) models, is presented because it is simple, widely used, and illustrates many of the features of the two-phase models. In Section 11.8 another model, the two-phase bubbling bed model, is presented. This is nearly as simple to apply as the Orcutt model and appears to give a better representation of the performance of bubbling fluidized bed reactors for most practical conditions.

The models treated in this chapter all assume negligible heat and mass transfer resistances between gas and particles within the dense phase. The entire bed is also treated as if it were isothermal. If these simplifying assumptions are not adopted, energy balances must be written in addition to mole balances, and separate accounting must be made of the particles. This complicates the models greatly and is beyond the scope of this book. For a discussion of non-isothermal reactor models and of the significance of heat and mass transfer resistance within the dense phase, see Bukur, Caram, and Amundson (1977).

In Sections 11.7 and 11.8 we consider only gas phase solid catalysed reactions. Application of these models to gas–solid heterogeneous reactions is treated in Section 11.12 below.

11.7 ORCUTT MODEL WITH PERFECT MIXING IN DENSE PHASE

The key assumptions of this model are that no solids are associated with the bubble phase while gas is assumed to be in plug flow in the bubble phase and perfectly mixed in the dense phase. We generalize the model somewhat by letting the fraction of gas which flows through the bubble phase at any height be β, and by letting the interphase mass transfer coefficient (volumetric rate of transfer per unit bubble surface area) be k_q. In the model proposed by Orcutt, Davidson, and Pigford (1962), reproduced by Davidson and Harrison (1963), the following expressions for β and k_q were assumed:

$$\beta = \frac{U - U_{mf}}{U} \tag{11.10}$$

$$k_q = 0.75\, U_{mf} + \frac{0.975 g^{0.25}\, \mathscr{D}^{0.5}}{d_{eq}^{0.25}} \tag{11.11}$$

Other values of the flow distribution parameter β and of the interphase mass transfer coefficient k_q can be used instead, based on more recent experimental evidence (see Chapter 4 and Section 11.4, respectively).

With the assumption that there are no particles in the bubble phase, the change in the molar flux of species A at any height in the bubble phase must be accounted for by interphase transfer, i.e.:

$$\beta U \, dC_{Ab} = k_q \, (C_{Ad} - C_{Ab})\, a_b \epsilon_b dz \tag{11.12}$$

where a_b is the interfacial bubble area per unit bubble volume. A mole balance over the entire dense phase yields:

$$(1-\beta)U(C_{Ain} - C_{Ad}) + \int_0^H k_q(C_{Ab} - C_{Ad})a_b\epsilon_b dz = (1-\epsilon_b)(1-\epsilon_{mf})Hk_n C_{Ad}^n \tag{11.13}$$

where k_n is the nth-order rate constant as defined in Eq. (11.1). Here mass transfer resistance within the dense phase is ignored so that the gas composition is treated as being the same in the particles and local interstitial gas. This assumption is valid unless the particles are large or the reaction very rapid (Bukur, Caram, and Amundson, 1977). The boundary condition on C_{Ab} is $C_{Ab} = C_{Ain}$ at $z = 0$. C_{Ad} is a constant throughout the column for this model in view of the assumption of perfect gas mixing in the dense phase. Integration of Eq. (11.12) yields:

$$C_{Ab} = C_{Ad} + (C_{Ain} - C_{Ad}) \exp \left\{ \frac{-k_q a_b \epsilon_b z}{\beta U} \right\} \tag{11.14}$$

Substitution of this expression into Eq. (11.13) gives an equation for C_{Ad} which can be solved for certain specific values of the reaction order n. The final exit concentration of A is then obtained from a mole balance at $z = H$, i.e.:

$$C_{Aout} = \beta \, [C_{Ab}]_{z=H} + (1-\beta)C_{Ad} \qquad (11.15)$$

Predictions of this model are given in Table 11.5 for irreversible reactions of

Table 11.5 Outlet concentrations derived from generalization of the model proposed by Orcutt, Davidson, and Pigford (1962) for some simple cases

Reaction	Dimensionless outlet concentration
0-order irreversible $(r_A = k_0)$	$\dfrac{C_{Aout}}{C_{Ain}} = 1 - k_0'$
Half-order irreversible $(r_A = k_{0.5} \, C_A^{0.5})$	$\dfrac{C_{Aout}}{C_{Ain}} = 1 + \dfrac{(k_{0.5}')^2}{2(1-\beta e^{-X^2})} \left\{ 1 - \sqrt{} \left\{ 1 + \dfrac{4(1-\beta^{-X})^2}{(k_{0.5}')^2} \right\} \right\}$
First-order irreversible: $A \to B$ $(r_A = k_1 C_A)$	$\dfrac{C_{Aout}}{C_{Ain}} = \dfrac{1 - \beta e^{-X} + \beta k_1' e^{-X}}{1 - \beta e^{-X} + k_1'}$
Second-order irreversible $(r_A = k_2 C_A^2$	$\dfrac{C_{Aout}}{C_{Ain}} = \beta e^{-X} + \dfrac{(1-\beta e^{-X})^2}{2k_2'} \left[\sqrt{} \left\{ 1 + \dfrac{4k_2'}{1-\beta e^{-X}} \right\} - 1 \right]$
First-order reversible: $A \rightleftharpoons B$, $(r_A = k_1(C_A - C_B/K_e))$	$\dfrac{C_{Aout}}{C_{Ain}} = \beta e^{-X} + \dfrac{(1-\beta e^{-X})^2}{1-\beta e^{-x} + k_1' - (k_1')^2/\{k_1' + K_e(1-\beta e^{-X})\}}$
Consecutive first-order: $A \to B \to C$; no B in feed $r_A = k_A C_A$; $r_B = k_B C_B$	$\dfrac{C_{Aout}}{C_{Ain}} =$ as for first-order irreversible case above $\dfrac{C_{Bout}}{C_{Ain}} = \dfrac{k_A'}{1+k_B' - \beta e^{-X}} \, \dfrac{(1-\beta e^{-X})^2}{1-\beta e^{-X}+k_A'}$

order 0. 0.5, 1, and 2, for a reversible first-order reaction, and for a consecutive first-order reaction. All of these reactions are assumed to be carried out under isothermal conditions with negligible volume change occurring due to reaction. All of the dimensionless outlet concentrations are given as functions of β and two other dimensionless groups, a dimensionless rate constant:

$$k_n' = \frac{k_n \, H_{mf}(1-\epsilon_{mf})C_{Ain}^{(n-1)}}{U} \tag{11.16}$$

and a dimensionless interphase mass transfer group:

$$X = \frac{k_q a_b \epsilon_b H}{\beta U} \tag{11.17}$$

X can also be thought of as the number of mass transfer units or the number of times a bubble is flushed out during passage through the bed.

The dimensionless outlet concentration is shown in Fig. 11.4 for a first-order

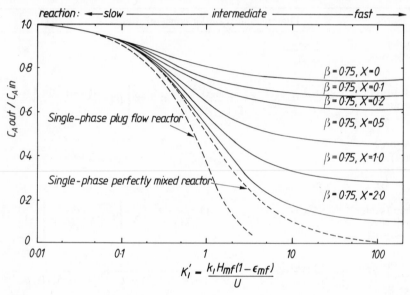

Figure 11.4 Outlet concentration for a first-order chemical reaction according to the Orcutt model for $\beta = 0.75$ and several different values of the interphase mass transfer group X.

reaction ($n = 1$) and for a particular value of β, $\beta = 0.75$, i.e. with 75 per cent. of the gas passing through the bubble phase. The limiting results for single-phase plug flow and perfectly mixed reactors, Eqs (11.7) and (11.8), respectively, are also shown in the figure as broken lines. As $X \to \infty$ or as $\beta \to 0$, the conversion, $1-C_{Aout}/C_{Ain}$, predicted by the two-phase theory approaches Eq. (11.8), i.e. that of the single-phase perfectly mixed reactor. For this particular two-phase model, conversions are always less favourable for finite X than for the perfectly mixed case. For slow reactions ($k_1' < 0.1$), the outlet concentration is controlled by chemical kinetics and virtually independent of the hydrodynamics. However, for intermediate and fast reactions, interphase transfer plays a very significant role, since, in the absence of particles in the bubble phase, gas must be transferred to the dense phase before reaction can take place. The expressions given in Table 11.5 could be used to construct curves for other values of β or for reactions of other orders. Generally speaking, the predictions become more sensitive to the dimensionless hydrodynamic parameters β and X with increasing order of reaction. An example showing the use of this model to estimate chemical conversions is given below in Example 11.3.

Although this model is useful in discussing the effects of different parameters and simple enough to lead to a number of analytic results for conversions and selectivities, as shown in Table 11.5, results are generally disappointing. The predicted gas concentration profiles are not observed in practice (Chavarie and Grace, 1975; Fryer and Potter, 1976). Moreover, the lack of any solids associated with the bubble phase appears to lead to underestimation of conversion for fast reactions. We turn therefore to a more realistic two-phase model.

11.8 TWO-PHASE BUBBLING BED REACTOR MODEL

The two-phase bubbling bed model (Grace, 1984) can be regarded as a simplification of the three-phase bubbling bed model of Kunii and Levenspiel (1969) and of the countercurrent backmixing model of Fryer and Potter (1972a). Under many circumstances it is possible to lump the cloud region with one of the other phases to give a simpler two-phase representation (Chavarie and Grace, 1975). The assumption of no net vertical gas flow in the dense phase introduced by Kunii and Levenspiel is adopted. The accuracy of the model is hardly affected for most practical purposes by these simplifications, providing that the parameters of the model are chosen carefully. Since the net vertical flow of gas is assumed to be entirely through the bubble phase, reflecting the tendency for downflowing solids to drag interstitial gas downwards with them, the β parameter of Section 11.7 is set equal to 1 and does not appear. On the other hand, we require two new parameters, ϕ_b and ϕ_d, to describe the fraction of the bed volume occupied by solids which are part of the bubble phase and dense phase, respectively. Values of ϕ_b and ϕ_d are discussed below.

Consider two parallel phases as shown in Fig. 11.3(a). The reaction is assumed to be isothermal and to involve negligible volume change. A mole balance on component A in the dilute or bubble phase yields:

$$U \frac{dC_{Ab}}{dz} + k_q a_b \epsilon_b (C_{Ab} - C_{Ad}) + k_n \phi_b C_{Ab}^n = 0 \qquad (11.18)$$

where ϕ_b is the fraction of the bed volume occupied by bubble phase solids. For the dense phase, a mole balance on component A gives:

$$k_q a_b \epsilon_b (C_{Ad} - C_{Ab}) + k_n \phi_d C_{Ad}^n = 0 \qquad (11.19)$$

where ϕ_d is the fraction of the bed volume occupied by solids assigned to the dense phase.

These equations can be integrated in a straightforward manner with the boundary conditions $C_{Ab} = C_{Ad} = C_{Ain}$ at $z = 0$ for several simple cases. The results for these cases are given in Table 11.6. For cases where the kinetics or

Table 11.6   Outlet concentrations derived from the two-phase bubbling bed reactor model for some simple cases

Reaction	Dimensionless outlet concentration
Zero-order irreversible $(r_A = k_0)$	$\dfrac{C_{Aout}}{C_{Ain}} = 1 - k_0^* (\phi_b + \phi_d)$
First-order irreversible: $A \rightarrow B \ (r_A = k_1 C_A)$	$\dfrac{C_{Aout}}{C_{Ain}} = \exp \left\{ \dfrac{-k_1^* \{ X(\phi_b + \phi_d) + k_1^* \phi_b \phi_d \}}{X + k_1^* \phi_d} \right\}$
Consecutive first-order: $A \rightarrow B \rightarrow C$; no B in feed; $r_A = k_A C_A$; $r_B = k_B C_B$	$\dfrac{C_{Aout}}{C_{Ain}} = \exp(-F_A)$ $\dfrac{C_{Bout}}{C_{Air}} = \dfrac{G\{ exp(-F_A) - \exp(-F_B) \}}{F_B - F_A}$ where $F_A = \dfrac{k_A^* \{ X(\phi_b + \phi_d) + k_A^* \phi_b \phi_d \}}{X + k_A^* \phi_d}$ $F_B = \dfrac{k_B^* \{ X(\phi_b + \phi_d) + k_B^* \phi_{\phi d} \}}{X + k_B^* \phi_d}$ $G = k_A^* \phi_b + \dfrac{X^2 k_A^* \phi_d}{(X + k_A^* \phi_d)(X + k_B^* \phi_d)}$

boundary conditions are more complex, numerical integration is required, and is generally straightforward. The results for first-order irreversible reaction in Table 11.6 can also be obtained by simplification of the Kunii and Levenspiel (1969) expression for the case where the mass transfer resistance between the cloud and emulsion.is negligible relative to that at the bubble interface. In the limiting cases where the dimensionless interphase mass transfer group, X, approaches infinity, the results for this model approach those for a single-phase plug flow reactor, e.g. Eq. (11.7) for a first-order reaction.

The dimensionless groups which determine chemical conversions and selectivities from this model are similar to, but not identical to, those which appeared in the model considered in the previous section. The mass transfer group is the same as given by Eq. (11.17), but with $\beta = 1$; i.e.:

$$X = \frac{k_q a_b \epsilon_b H}{U} \qquad (11.20)$$

where k_q should be obtained from Eq. (11.6), while:

$$a_b = \frac{6}{\bar{d}_{eq}} \qquad (11.21)$$

The mean bubble diameter $\bar{d}_{eq}$ can be obtained from one of the bubble growth correlations outlined in Chapter 4, evaluated at height $0.4H$, as advocated by Fryer and Potter (1972b). Similarly, ϵ_b and H can be estimated from relationships given in Chapter 4. For an nth-order reaction, the dimensionless rate constant group is:

$$k_n^* = \frac{k_n H C_{Ain}^{n-1}}{U} \qquad (11.22)$$

Let us now consider the variables ϕ_b and ϕ_d, giving the fraction of bed volume occupied by solids associated with the bubble and dense phase, respectively. Kunii and Levenspiel (1969) defined a variable $\gamma_b = \phi_b/\epsilon_b$, which is the volume fraction of bubbles occupied by dispersed or raining solids, and cited experimental evidence suggesting that $0.001 \leq \gamma_b \leq 0.01$. Hence the lower and upper limits on ϕ_b may be taken as:

$$0.001\epsilon_b \leq \phi_b \leq 0.01\epsilon_b \qquad (11.23)$$

The value of ϕ_d can be approximated by including all of the solids in the clouds, wakes, and emulsion in the dense phase and assuming the dense phase voidage to be ϵ_{mf}. Hence, we may write:

$$\phi_d = (1-\epsilon_b)(1-\epsilon_{mf}) \qquad (11.24)$$

The actual conversion should be bounded between that predicted using the upper and lower bounds for ϕ_b given in Eq. (11.23). For slow reactions the solids assigned to the bubble phase have little influence and the two limits are indistinguishable. However, for fast reactions, ϕ_b plays a significant role. The limits proposed for ϕ_b are conservative in that all cloud and wake solids are assigned to the dense phase, rather than to the bubble phase. In view of this fact, the model can also be used for the so-called 'slow bubble' or cloudless bubble case ($\alpha < 1$).

The dimensionless outlet concentration for an irreversible first-order reaction, as predicted by the model for $\phi_d = 0.4$, is plotted in Fig. 11.5. Two

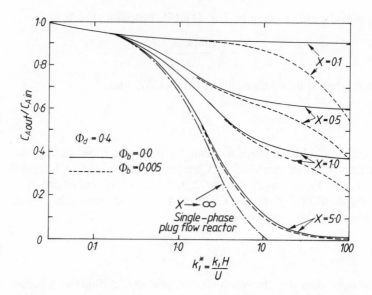

Figure 11.5 Outlet concentration for a first-order chemical reaction according to the two-phase bubbling bed reactor model with $\phi_d = 0.4$, $\phi_b = 0.0$ and 0.005, and different values of the interphase mass transfer group X.

values of ϕ_b, $\phi_b = 0$ and $\phi_b = 0.005$, are chosen to demonstrate the influence of this variable. Note that ϕ_b has little effect for slow reactions or for cases where interphase mass transfer is rapid, but that even a small fraction of solids in the bubble phase can lead to a significant increase in conversion for fast reactions and slow mass transfer (high k_1^* and low X). As in Fig. 11.4, all curves in Fig. 11.5 converge where the reaction becomes kinetically rate controlled, i.e at small values of the dimensionless rate constant. Conversion is again seen to depend strongly on interphase mass transfer for intermediate and fast reactions. In general, the conversion becomes somewhat more sensitive to the parameters X, ϕ_b, and ϕ_d as the order of reaction increases.

The following sample problem illustrates the use of the two-phase bubbling bed model and of the Orcutt model outlined in the previous section.

Example 11.3:   Chemical conversion for catalytic gas phase reaction

A first-order irreversible gas phase chemical reaction is to be carried out in a fluidized bed of diameter 1.30 m. The expanded bed height is 2.60 m. The catalyst particles have a superficial velocity and voidage at minimum fluidization of 0.020 m/s and 0.48, respectively, and an activity (based on particle volume) at the bed temperature of 12.8 s^{-1}. The operating superficial gas velocity is 0.50 m/s. The molecular diffusivity of the reacting species is 3.3 $\times$ 10^{-5} m^2/s. Use both the Orcutt model and the two-phase bubbling bed model to estimate the conversion if: (a) no baffles are present so that unrestrained bubble growth occurs and (b) baffles are present limiting the bubble size to 0.15 m.

Solution

We are given $D = 1.30$ m, $H = 2.60$ m, $U_{mf} = 0.020$ m/s, $\epsilon_{mf} = 0.48, k_1 = 12.8$ s^{-1}, $U = 0.50$ m/s, $\mathcal{D} = 3.3 \times 10^{-5}$ m^2/s.
Hence $U - U_{mf} = 0.48$ m/s.

Case (a) No baffles present

The mean bubble diameter, $\bar{d}_{eq}$, can be obtained from the Mori and Wen correlation (see Chapter 4), evaluated at a height of $0.4H = 1.04$ m above the distributor. Assuming bubbles form at the distributor with an initial diameter of 10 mm, we calculate $\bar{d}_{eq} = 0.300$ m. The corresponding value of u_A, the absolute rise velocity of the bubbles, is found using Eq. (4.41):

$$u_A = 0.71 \sqrt{(g\bar{d}_{eq})} + (U-U_{mf}) = 1.70 \text{ m/s}$$

Hence, from Eq. (4.42):

$$\epsilon_b \simeq \frac{U-U_{mf}}{u_A} = \frac{0.48}{1.70} = 0.282$$

Also:

$$a_b \simeq \frac{6}{\bar{d}_{eq}} = 20.0 \text{ m}^{-1}$$

and

$$H_{mf} = H(1-\epsilon_b) = 1.87 \text{ m}$$

Orcutt model:

$$\beta = \frac{U-U_{mf}}{U} = 0.96$$

From Eq. (11.11) the interphase mass transfer coefficient is:

$$k_q = 0.75 \times 0.020 + 0.975 \times \left[\frac{9.8}{0.300}\right]^{0.25} (3.3 \times 10^{-5})^{0.5} \text{ m/s}$$

$$= 0.0284 \text{ m/s}$$

Hence, from Eq. (11.17), the dimensionless mass transfer group is:

$$X = \frac{0.0284 \times 20.0 \times 0.282 \times 2.60}{0.96 \times 0.50} = 0.867$$

and

$$e^{-X} = 0.420$$

The dimensionless kinetic rate constant is obtained from Eq. (11.16):

$$k_1' = \frac{12.8 \times 1.87 \times (1-0.48)}{0.50} = 24.9$$

We may now evaluate the outlet concentration of the reacting species from Table 11.5; i.e.:

$$\frac{C_{Aout}}{C_{Ain}} = \frac{1 + 0.96 \times 0.420\,(24.9-1)}{1 - 0.96 \times 0.420 + 24.9} = 0.417$$

The estimated conversion is then simply $(1-C_{Aout}/C_{Ain})$ or 58 per cent.

Two-phase bubbling bed model:

The interphase mass transfer coefficient is now calculated from Eq. (11.6):

$$k_q = \frac{0.020}{3} + \left\{ \frac{4 \times 3.3 \times 10^{-5} \times 0.48 \times 1.70}{\pi \times 0.300} \right\}^{0.5} \text{ m/s}$$

$$= 0.0174 \text{ m/s}$$

Thus:

$$X = \frac{0.0174 \times 20.0 \times 0.282 \times 2.60}{0.50} = 0.509$$

while:

$$k_1^* = \frac{12.8 \times 2.60}{0.50} = 66.6$$

The fraction of the bed occupied by bubble phase solids is expected to lie in the range $0.001\epsilon_b$ to $0.01\epsilon_b$; therefore:

$$0.00028 \leqslant \phi_b \leqslant 0.00282$$

while the fraction of the bed occupied by dense phase solids is:

$$\phi_d = (1-0.282) \times (1-0.48) = 0.373$$

The conversion may now be calculated using the first-order reaction relationship in Table 11.6. For $\phi_b = 0.00028$, we calculate:

$$\frac{C_{Aout}}{C_{Ain}} = \exp\left\{\frac{-66.6\ \{0.509(0.00028 + 0.373) + 66.6 \times 0.00028 \times 0.373\}}{0.509 + 66.6 \times 0.373}\right\}$$

$$= 0.596$$

Similarly, for $\phi_b = 0.00282$, we obtain:

$$\frac{C_{Aout}}{C_{Ain}} = 0.503$$

The conversion is therefore predicted to lie in the 40 to 50 per cent. range.

Case (b) With baffles present

We are given $d_{eq} = 0.15$ m. Hence we may proceed with the calculations as before, yielding $u_A = 1.342$ m/s, $\epsilon_b = 0.358$, $a_b = 40.0$ m^{-1}, $H_{mf} = 1.67$ m. For the Orcutt model, $\beta = 0.96$ as before, while $k_q = 0.0309$ m/s, $X = 2.40$, and $k_1' = 22.2$ are new calculated values. We proceed as before, giving $C_{Aout}/C_{Ain} = 0.123$. Hence the new estimated conversion is 88 per cent. For the two-phase bubbling bed model, $k_1^* = 66.6$ as before, while we now obtain $k_q = 0.0201$ m/s, $X = 1.50$, $0.00036 \leqslant \phi_b \leqslant 0.00358$, and $\phi_d = 0.334$. We calculate limiting values of C_{Aout}/C_{Ain} as before, giving new lower and upper bounds for the conversion of 76 to 81 per cent.

Note the marked increase in conversion which can occur when baffles are introduced to keep bubbles relatively small, under conditions where hydrodynamic conditions are rate controlling. Note also that there can be significant differences between the predictions of alternative reactor models. The primary reason for the difference in the present case is the considerably larger interphase mass transfer group in the case of the Orcutt model.

The two-phase reactor models can be used to show the effect of reactor scale on conversion. As a process is scaled up from the laboratory scale to the industrial scale, it is a common experience that the conversion tends to decrease. This decrease in conversion is related to the fact that bubbles tend to grow larger in columns of larger scale (see Chapter 4). The larger bubbles, in turn, lead to: shorter gas residence time, since $u_b \propto d_{eq}^{-0.5}$; decreased interphase mass transfer ($X \propto d_{eq}^{-1.5 \text{ to } -1.75}$). An additional factor, in many cases, is that it is difficult to achieve as uniform a distribution of incoming gas at the grid for columns of larger scale. Separate models should be used for reactors which are small enough for slug flow to occur, (see Section 11.14). It is then possible to model the influence of reactor scale on conversion over the range of conditions in which experimental data are available. As illustrated in the worked example above, introduction of internal baffles or tubes can be used to keep bubbles small, thereby counteracting the adverse effect of increased reactor scale.

The expressions in Table 11.6 for consecutive first-order reactions of the type A → B → C can be used to predict the selectivity to an intermediate product B. Similar expressions are given in Table 11.5 for the Orcutt model. An important feature of fluidized bed reactors is that the selectivity of an intermediate for a given conversion of A is always less favourable for consecutive reactions than in a single-phase isothermal reactor. This is due to the fact that the reaction in a fluidized bed takes place primarily in the dense phase where the local conversion is high, whereas the flow is predominantly in the bubble phase where conversion is relatively low. This gives more opportunity for C to form in the dense phase than in the case where the same amount of A is converted in a homogeneous single-phase reactor, e.g. a packed bed.

11.9   BUBBLING REGIME: AXIAL AND RADIAL GAS MIXING

For packed beds, flow in the axial direction generally does not deviate greatly from plug flow, and deviations can be accounted for in terms of the dispersed plug flow model. This model assumes that the mixing steps are both small in scale and random in nature.

In gas fluidized beds, on the other hand, most of the axial mixing is due to phenomena associated with bubbles. Some causes of the spread in gas residence time distribution from plug flow are as follows:

(a) In general, bubble gas moves at a different velocity than interstitial gas. At $\alpha > 1$ elements associated with bubbles and their clouds tend to spend less time in the bed than gas elements in the dense phase. For $\alpha < 1$, they spend longer.

(b) Different bubbles rise at different velocities due to having different sizes, different coalescence histories, etc.

(c) Net solids circulation patterns become established, causing gas to move upwards in some regions more than in others. At $U/U_{mf} \gg 1$, gas in the dense phase can actually have a downwards absolute velocity (Latham, Hamilton, and Potter, 1968).

Although a number of workers have inferred overall axial dispersion from gas residence time distribution data, the dispersed plug flow model is inappropriate in this case, since the bubbles which induce mixing are neither small in scale nor random in nature. Hence, overall axial dispersion coefficients have no physical significance, and their correlation in terms of Peclet numbers should be avoided.

In general, measured residence time distributions tend to follow neither plug flow nor perfect mixing. (Plug flow may, however, be approached for very tall slugging beds or for a multi-stage unit.) The gas flow under transient conditions can be best described by extending the two-phase models discussed in the previous section to incorporate unsteady terms. May (1959) used a two-phase model in which plug flow is assumed in the bubble phase and dispersed plug flow is assumed in the dense phase* with crossflow (i.e. interphase mass transfer) between the two phases. Based on data available in the literature, Mireur and Bischoff (1967) have correlated the dense phase axial dispersion coefficient and crossflow in terms of H/D and U/U_{mf}, respectively.

For a transient case in which chemical reaction is ignored, the Orcutt model with perfect mixing in the dense phase, discussed in Section 11.7, may be written:

Bubble phase: $\quad \epsilon_b \dfrac{\partial C_b}{\partial t} + \beta U \dfrac{\partial C_b}{\partial z} = k_q a_b \epsilon_b \, (C_d - C_b) \qquad (11.25)$

Dense phase: $\quad (1-\epsilon_b)[\epsilon_{mf} + \epsilon_p(1-\epsilon_{mf})] \dfrac{dC_d}{dt} + (1-\beta) \dfrac{U}{H} (C_d - C_{in})$

$$= \frac{k_q a_b \epsilon_b}{H} \int_0^H (C_b - C_d) dz \qquad (11.26)$$

Here we have made allowance for ϵ_p, the internal porosity of the particles, and have assumed rapid equalization of concentration between the outer surface and inner volume of particle, usually a good approximation for the small particles used in fluidized bed processes. Transfer functions have been derived for the case where $\beta = 1$ by Heimlich and Gruet (1966) and Barnstone and Harriott (1967), and good matching between residence time distribution data and the model were demonstrated for reasonable values of the dimensionless interphase transfer parameter X for beds of 0.076 and 1.50 m diameter. This technique has been shown to be a possible means of estimating interphase mass transfer rates by Fontaine and Harriott (1972), who cycled the input concentration of different tracer gases.

In terms of the two phase bubbling bed model (Section 11.8), the unsteady state equations in the absence of chemical reaction are:

Bubble phase: $\quad \epsilon_b \dfrac{\partial C_b}{\partial t} + U \dfrac{\partial C_b}{\partial z} = k_q a_b \epsilon_b \, (C_d - C_b) \qquad (11.27)$

Dense phase: $\quad (1-\epsilon_b)[\epsilon_{mf} + \epsilon_p(1-\epsilon_{mf})] \dfrac{\partial C_d}{\partial t} = k_q a_b \epsilon_b (C_b - C_d)$

$$(11.28)$$

*Note that our criticism of the dispersed plug flow model above applied to the case where it is used for the bed as a whole, not to cases where it is used to describe an individual phase. There is some justification for expecting small random mixing steps in the dense phase.

This model has been used in conjunction with a step input change by Yoshida and Kunii (1968). Solution requires numerical integration.

In principle, either of these models can be used for process control of fluidized bed chemical reactors when the reaction terms are reinserted in the above equations from Sections 11.7 and 11.8. In practice, simpler control models are usually adopted for this purpose.

Although the two-phase models provide a good basis for treating axial dispersion in gas fluidized beds, there are no proven methods for dealing with radial dispersion. Experimental measurements are scarce. When radial dispersion coefficients have been reported, they tend to be about an order of magnitude smaller than values which have been ascribed to axial dispersion coefficients. The principal cause of radial mixing in bubbling systems is lateral motion of bubbles assocated with coalescence of non-aligned bubbles. Solids circulation patterns may also play a significant role.

Both axial and radial mixing can be altered by adding baffles to fluidized beds. In general, any baffle which impedes bubble and solids motion in a given direction or which tends to make bubbles more uniform will reduce the mixing in that direction. For example, vertical tubes tend to reduce radial dispersion, while horizontal screens or perforated plates reduce axial mixing and cause gas residence times to be more uniform. For a multi-stage unit (stages in series), the overall gas flow pattern can be made to approach plug flow conditions.

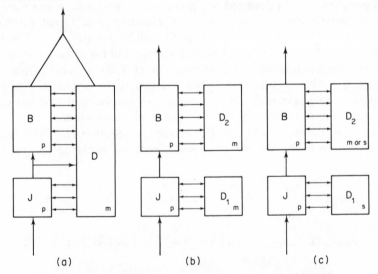

Figure 11.6 Schematic diagrams showing grid models: (a) Behie and Kehoe (1973); (b) Wen (1979); (c) Grace and deLasa (1978). J = jet region; B = bubble phase; D = dense phase; m = well mixed; p = plug flow; s = stagnant region.

11.10   REACTION IN THE GRID REGION

The models considered in the previous sections give reasonable descriptions of the bubbling region of fluidized beds operated in the bubbling bed regime. However, they do not give a good description of the hydrodynamics and contacting in the entry region just above the gas distributor or grid.

The first separate model for the grid region was proposed by Behie and Kehoe (1973). The model is shown schematically in Fig. 11.6(a). Gas entering the column through upward facing horizontal orifices is assumed to pass initially through a solids-free jet region (designated J). At the top of the jets, the gas is distributed partly to the bubble phase (denoted by B) and partly to the dense phase (designated D), according to the two-phase theory of fluidization (see Chapter 4). Interphase mass transfer is assumed to occur both between the jets and the dense phase and between the bubbles and dense phase, with transfer being significantly more rapid in the former case. Perfect mixing of gas was assumed to apply to the entire dense phase. In an alternative formulation (Wen, 1979) (see Fig. 11.6b) the dense phase was divided into two compartments, both well mixed, one extending up to the top of the jets and the other corresponding to the bubbling zone.

The above model was modified by Grace and deLasa (1978) to allow for solids entrained in the jets and to consider the case where gas in the corresponding dense phase region is stagnant, rather than perfectly mixed. The model is represented schematically in Fig. 11.6(c). Either of the models considered in Section 11.7 or 11.8 may be used for the bubbling region higher in the bed. Considering only the lower part of the bed corresponding to the jet region, we may write separate mass balances for the two regions or phases. For a first-order gas phase reaction under steady state isothermal conditions, with one-dimensional flow assumed in each phase:

Jets: $$U\frac{dC_{Aj}}{dz} + k_{qj}a_j(C_{Aj}-C_{Ad}) + k_1\epsilon_j\gamma_j C_{Aj} = 0 \qquad (11.29)$$

Dense phase: $k_{qj}a_j(C_{Ad} - C_{Aj}) + k_1(1-\epsilon_j)\,(1-\epsilon_{mf})C_{Ad} = 0 \qquad (11.30)$

where a_j is the jet interfacial area per unit bed volume, ϵ_j is the fraction of the cross-sectional bed area occupied by jets, and γ_j is the volume fraction of the jets occupied by catalyst particles. The final terms must be modified for non-first-order reactions. Integration of Eqs. (11.29) and (11.30) with the boundary conditions $C_{Aj} = C_{Ain}$ at $z = 0$ yields the following concentration at the level $z = H_j$, corresponding to the top of the jet or grid region:

$$C_{AHj} = C_{Ain}\exp\left\{\frac{-X_j k_1^j}{X_j + k_1^j} - \frac{k_1^j\,\epsilon_j\gamma_j}{(1-\epsilon_j)(1-\epsilon_{mf})}\right\} \qquad (11.31)$$

where:

$$X_j = \frac{k_{qj}a_jH_j}{U} \qquad (11.32)$$

and

$$k_1^j = \frac{k_1(1-\epsilon_j)(1-\epsilon_{mf})H_j}{U} \qquad (11.33)$$

are analogous to X and k_1' or k_1^* in the bubbling bed models above.

Equation (11.31) can be used to estimate the concentration of the reacting species A at the top of the grid region; the bubbling bed models outlined earlier in this chapter can then be used to predict reaction rates in the remainder of the bed. There are, however, several difficulties with these models. Rowe, MacGillivray, and Cheesman (1979) studied the entry region above single orifices for beds of fine and intermediate size particles with the aid of X-ray photography and showed that the region directly above the orifice is characterized by periodically forming elongated voids rather than steady jets. This implies that models like that represented by Eqs (11.29) to (11.31) can only represent the grid region behaviour in a time mean sense. However, this difficulty is also present, and probably to a greater degree, for the bubbling bed models.

A more serious problem is the estimation of parameters to use in the model. There is a lack of data upon which to base most of these variables. Of the various correlations for jet penetration, H_j (see Chapter 4), that of Merry (1975) appears to have the soundest basis. To a first approximation, the average jet diameter can be taken as $\sqrt{2}$ to $2 \times d_{or}$ where d_{or} is the orifice diameter. Hence ϵ_j may be estimated to lie in the range 2 to $4f_{or}$ and a_j in the range 4 $\sqrt{2}f_{or}/d_{or}$ to $8f_{or}/d_{or}$, where f_{or} is the fractional free (orifice) area in the grid plate. The volume fraction of solids entrained in the jets, γ_j, is expected to lie in the range 0.001 to 0.006 (Merry, 1976); except for fast reactions, these values are generally too low to have an appreciable effect on conversion. Of greater important is the jet to dense phase interphase mass transfer coefficient k_{qj}. Behie and Kehoe (1973) based some sample calculations on mass transfer data obtained in a 0.61 diameter column of 60 μm cracking catalyst (Behie, Bergougnou, and Baker, 1976). The values of k_{qj} were of order 1 m/s, such a high value that the consequences for reactor modelling were considerable at least for reasonably rapid kinetic rate constants (Behie and Kehoe, 1973; Grace and deLasa, 1978). However, values of k_{qj} are very difficult to measure, and other workers (deMichaele, Elia, and Massimilla 1976; Sit, 1981) have inferred or measured transfer coefficients which are lower by an order of magnitude. The following illustrative problems demonstrates the importance of the interphase transfer step.

Example 11.4: Chemical reaction in the grid region

Estimate the conversion in the grid region of a fluidized bed reactor assuming:
(a) $k_{qj} = 1.0$ m/s and (b) $k_{qj} = 0.10$ m/s for a bed of 100 μm catalyst particles of
density 2,000 kg/m³. The superficial gas velocity is 1.0 m/s and the gas density is
1.0 kg/m³. The grid is a multi-orifice plate containing 12.7 mm diameter holes
on a square lattice with a hole-to-hole spacing of 0.102 m. The kinetic rate
constant is 12.8 s⁻¹. Assume that $\epsilon_{mf} = 0.48$.

Solution

The fractional free area in the grid is given by:

$$f_{or} = \frac{\pi \times (12.7)^2}{4 \times 102^2} = 0.01218$$

The gas velocity through the orifices is therefore:

$$u_{or} = \frac{U}{f_{or}} = \frac{1.0}{0.01218} \text{ m/s} = 82.1 \text{ m/s}$$

The height of the grid region can be estimated from the jet penetration
correlation of Merry (1975); i.e.:

$$H_j = 5.2 d_{or} \left(\frac{\rho_g d_{or}}{\rho_p d_p} \right)^{0.3} \left\{ 1.3 \left(\frac{u_{or}^2}{g d_{or}} \right)^{0.2} - 1 \right\}$$

$$= 5.2 \times 0.0127 \times \left(\frac{1.0 \times 0.0127}{2000 \times 0.0001} \right)^{0.3} \times \left\{ 1.3 \left(\frac{82.1^2}{9.8 \times 0.0127} \right)^{0.2} - 1 \right\} \text{m}$$

$$= 0.303 \text{ m}$$

$\epsilon_j \simeq (2 \text{ to } 4) f_{or}$; take $\epsilon_j = 3 f_{or} = 0.0365$

$a_j \simeq (4\sqrt{2} \text{ to } 8) f_{or}/d_{or}$; take $a_j = 4\sqrt{3} f_{or}/d_{or} = 6.64 \text{ m}^{-1}$

$\gamma_j = 0.001 \text{ to } 0.006$; take $\gamma_j = 0.0035$

From Eq. (11.33):

$$k_1^j = \frac{12.8 \times (1-0.0365) \times (1-0.48) \times 0.303}{1.0} = 1.943$$

(a) For $k_{qj} = 1.0$ m/s:

From Eq. (11.32):

$$X_j = \frac{1.0 \times 6.64 \times 0.303}{1.0} = 2.01$$

Therefore, from Eq. (11.31):

$$\frac{C_{AHj}}{C_{Ain}} = \exp\left\{\frac{-2.01 \times 1.943}{(2.01 + 1.943)} - \frac{1.943 \times 0.0365 \times 0.0035}{(1-0.0365) \times (1-0.48)}\right\} = 0.37$$

(b) Similarly, for $k_{qj} = 0.10$ m/s:

$$X_j = 0.201 \quad \text{and} \quad \frac{C_{AHj}}{C_{Ain}} = 0.83$$

from Eqs (11.32) and (11.31). The final result is insensitive to the choice of γ_j and not strongly affected by the choice of a_j and ϵ_j for the conditions under consideration. However, there is a very strong influence of the interphase mass transfer coefficient (the estimated conversion in the grid region varying from 63 to 17 per cent. for a tenfold variation in k_{qj}). Note that these conversions can be appreciable with respect to those in a much deeper bubbling bed (cf. Example 11.3).

The type of mass transfer envisaged in the model outlined above is probably dominant only for relatively fine particles (for example $d_p < 200$ μm). For coarse particles (e.g. > 1 mm), the grid region, at least for upward facing horizontal orifices, may resemble a series of parallel spouted beds (Mathur and Epstein, 1974). The primary mass transfer is then via the bulk outflow of gas, as predicted by Mamuro and Hattori (1968). Models and experimental results for a catalytic first-order reaction in isothermal spouted beds are given by Piccinini, Grace, and Mathur (1979) and Rovero et al. (1983).

Considerable work remains to be performed before reaction in the grid region can be predicted with confidence. Even for the simplest grid geometry of a multi-orifice plate, there is a need for definitive interphase mass transfer results and reliable parameter estimation methods, and there is little to go on for intermediate sizeparticles. For more complex grid geometries, that reaction can be rapid in the grid region and that minor changes in grid character can have significant effects (Cooke et al., 1968; Walker 1970; Hovmand, Freedman, and Davidson, 1971; Chavarie and Grace, 1975; Bauer and Werther, 1981).

1.11 REACTION KINETICS: GAS–SOLID REACTIONS

In previous sections we have treated only gas–phase reactions catalysed by solid particles. While many applications of fluidised beds are for reactions of this type, there are also many applications in which both the gas and solid take part in the reaction (see Table 11.1). Gas–solid reactions of this type include coal conversion processes (combustion, gasification, and pyrolysis), roasting of ores, calcination reactions, and hydrogen reduction of ores. The equation for such a reaction may be written:

$$A(gas) + bB(solid) \rightarrow \text{products (gas and/or fluid)} \tag{11.34}$$

where b is a stoichiometric constant.

The physical properties of a solid particle undergoing reaction tend to change with time. In some cases the particle may shrink during the course of reaction, e.g a coal particle undergoing combustion evolves volatile matter and then is converted into gases (CO, CO_2, and H_2O) and ash. In other cases one of the products of reaction is itself a porous solid, or the reaction leaves behind a porous structure, e.g. in the calcination of limestone or in the burning of coke deposits from porous cracking catalyst particles. Alternatively, the reaction may produce a solid product of larger volume than the original, thereby tending to block the pores, as in the sulphation of calcined limestone sorbent particles during coal combustion. It is clear that the rate expressions in these different cases may differ significantly.

Kinetics of gas–solid reactions are considered in detail in a number of texts and references (e.g. Smith, 1970; Levenspiel, 1972; Carberry, 1976). In the general case, the following steps occur in series during reaction of a solid particle:

(a) Mass transfer of gaseous reactant component from the bulk gas to the outside of the solid particle

(b) Diffusion of gaseous reactant into the pores of the particle

(c) Reaction of the gaseous reactant with the solid surface at some distance inside the particle

(d) Diffusion of gaseous products back to the outer surface of the solid particle

(e) Mass transfer of gaseous products from the exterior surface into the bulk gas

Not all of these steps are present in all cases. Steps (a) and (e) tend to encounter less resistance than steps (b) and (d), respectively. However, the relative resistance to steps (b), (c), and (d) depends on many factors including temperature, particle size, pore structure, and the nature of the reaction, reactants, and products. The most common reaction model for single particles and one that is the best simple representation for most gas–solid reactions (Wen, 1968) is the shrinking core model (also called the unreacted core model or shell progressive model). Reaction is visualized as occurring at a sharp front, separating unreacted material from product; this front begins at the outside of the particle and advances inwards as the reaction proceeds. This model adopts a pseudo steady state assumption which relies on the reaction front migrating inwards at a rate much lower than the speed at which gas can diffuse to the front. In most cases one of steps (b), (c), or (d) may be assumed to be rate controlling, i.e. to provide the bottleneck which determines the overall rate of reaction. When the chemical reaction step controls, it can be shown (e.g. Levenspiel, 1972) that the time required for complete conversion of a spherical particle under isothermal conditions exposed at time 0 to a gas concentration C_A^* is:

$$t_{cr} = \frac{\rho_B d_p}{2b k_{1s} C_A^* M_B}$$ (11.35)

Here the reaction is assumed to be first order, k_{1s} being the reaction rate constant based on the unit surface area of unreacted core; i.e.:

$$r_A = -\frac{1}{4\pi r_{uc}^2} \frac{dN_A}{dt} = k_{1s} C_A^*$$ (11.36)

where r_{uc} is the radius of the unreacted core.

When the reaction is fast enough that the resistance to chemical reaction and diffusion are of similar magnitude, it is possible (Kunii and Levenspiel, 1969) to retain Eq. (11.35), but with k_{1s} replaced by an effective rate constant k_{ef}. If the diffusion is through a layer of porous solid or ash material, then:

$$k_{ef} = \frac{1}{1/k_{1s} + d_p/12 \mathscr{D}_s}$$ (11.37)

where $\mathscr{D}_s$ is the effective diffusivity of the gas through the porous solid. For particles which shrink as the reaction proceeds so that the reaction front is always at the exterior surface, then:

$$k_{ef} = \frac{1}{1/k_{1s} + 1/k_d}$$ (11.38)

where k_d is the mass transfer coefficient between the particle outer surface and the bulk gas surrounding the particle.

11.12   FLUID BED REACTOR MODELS FOR GAS–SOLID REACTIONS

Heterogeneous gas–solid reactions tend to be more complex than catalytic gas–phase reactions in the following ways:

(a) The particles usually undergo significant physical changes (e.g. shrinkage growth, density changes, shape changes) during the course of the reaction.

(b) The kinetics of the reaction may be complex, as outlined in the previous section. The controlling mechanism may change as the reaction proceeds or be different for large and small particles.

(c) Energy balances are often required in addition to material balances, especially where the reactor finds its own temperature level (as in catalyst regenerators), where heat transfer is the rate-limiting step (as in calcination of limestone), or where reacting particles are at a different temperature from inactive particles (as in coal combustion).

(d) Since the solids (in addition to the gas) are involved in the reaction, residence time distributions and mass balances are required for the particles as well as for the gas phase components. The mass balance for the solids must be related to that of the gas through the stoichometry of the reaction or reactions. To account for the behaviour of different size fractions (feed and exit streams, growth or shrinkage, elutriation, attrition, and mixing), solids population balances are often written.

Many of the reactor models which have been applied for gas–solid reactions have been devised for a specific reaction. Table 11.7 lists a number of these models from the literature together with the principal assumptions. Note that a number of areas where assumptions must be made are common with the models for gas phase solid catalysed reactions (cf. Table 11.2). Because of the additional areas where assumptions must be made, the number of possible models is even larger than before. Many of the models proposed have never been subjected to direct experimental testing.

In order to illustrate the procedure to be followed in predicting the conversion for a continuous gas–solid reaction, we will assume that the reaction follows the shrinking core model and that the reactor is isothermal. For the time being, we further adopt a hydrodynamic model which parallels that given in Section 11.7 for a gas phase reaction, i.e. we assume no particles in the bubble phase, plug flow of gas in the bubble phase, and perfect mixing of gas in the dense phase. In addition, the solids are assumed to be perfectly mixed in a single-stage system. The procedure then involves a sequence of steps:

1. Given the bed diameter, particle and gas properties, bed depth and gas properties, work out the required hydrodynamic features. In particular, evaluate the bubble diameter, bubble velocity, and fraction of bed volume occupied by bubbles (d_{eq}, u_b, and ϵ_b) at a representative height, generally $z = 0.4H$. Hence find $a_b = 6/\bar{d}_{eq}$, the bubble area per unit volume, and H_{mf}. Note that the particles will generally be nearly completely reacted so that the particle properties should be based on the solid product, not on the feed material.

2. Calculate the interphase mass transfer coefficient k_q from Eq. (11.6) or Eq. (11.11).

3. Calculate the dimensionless groups β and C from Eqs (11.10) and (11.17),

Table 11.7 Key features of gas–solid fluid bed reactor models

Reference	Kinetics[a]	Phases[b]	Interphase transfer	Dense phase gas mixing[c]	Solids mixing[c]	Heat balance	Application and comments
Kunii and Levenspiel (1969)	SC,CR	3:B,C, and E	As in their catalytic model	St.	PM	No	General model; consider various limiting cases
Yoshida and Wen (1970)	SC	2:B and D	$11/d_{eq}$	C in S	C in S	No	General model; test for ZnS roasting: bubbles grow with height; allow for solids in bubbles; single, consecutive, and parallel reactions
Avedesian and Davidson (1973)	Diff.	2:B and D	As Hovmand, Freedman, and Davidson (1971)	PM	Batch	No	Coal combustion; few active particles among inerts
Becker, Beér, and Gibbs (1975)	Diff.	2:B and D	Penetration + leakage of circulating gas	PF	PM	Yes	Coal devolatization and combustion; few active particles among inerts
Gibbs (1975)	Diff.	2:B and D	As Davidson and Harrison (1963)	PM	A,E	No	Coal combustion, mono-disperse and wide d_p distribution; some allowance for volatiles
Basu, Broughton, and Elliott (1975)	SC	2:B and D	As Davidson and Harrison (1963)	PM	PM,PB,E	No	Coal combustion
Fukunada et al. (1976)	SC	2:B and D	$11/d_{eq}$	PF	PM	No	Roasting of ZnS; allow for particles in bubbles; compare with experiments
Gordon and Amundson (1976)	3 rxs	3:B,I and P	B–I: as Kunii I–P: take Sh=2	PM	PC	Yes	Carbon combustion; consider multiplicity of steady states
Mori and Wen (1976)	1st O	2:B and D	$11/d_{eq}$	C in S	NS	No	General model; bubbles grow with height
Horio and Wen (1976)	2 rxs	2:B and D	$11/d_{eq}$ followed by Hovmand, Freedman, and Davidson (1971)	C in S	PM,E	No	Coal combustion and SO_2 – limestone rx.; bubbles grow with height
Chen and Saxena (1977)	2 rxs	3:B,C and E	As Kunii and Levenspiel (1969)	PF	PB,PM	No	Coal combustion and sulphation of additives; bubbles grow with height; compare with pilot plant data
Horio and Wen (1978)	1st O	2:B and D	$11/d_{eq}$	C in S	PB,E	Yes	Coal combustion; bubbles grow with height; compare with data
Rajan, Krushnan, and Wen (1978)	4 rxs	2:B and D	$11/d_{eq}$	C in S	C in S	Yes	Coal combustion including devolatilization and S removal; bubbles grow with height; compare experimental data
Baron, and Hodges (1978)	SC	2:B and D	As Davidson and Harrison (1963)	PM	PM,E	No	Coal burning; general comparison with literature data
Errazu, deLaza and Sarti (1979)	2nd O	2:J or B and D	J–D: Behie B–D: Kunii or Davidson	PM	PM	Yes	Catalyst regenerator; separate allowance for grid region; some experimental data
Chang, Rong and Fan (1982)	1st or Oth order with respect to solid	2:B and D	As Kunii and Levenspiel (1969)	PM	Dispersion model (lateral)	No	Shallow bed; general model applied to zinc roasting; steady and unsteady solutions

[a] CR = continuous reaction model; Diff. = control by diffusion through ash layer; SC = shrinking core model; O = order, rxs = reactions.

[b] B = bubble phase; C = cloud phase; D = dense phase; E = emulsion phase; I = interstitial gas; J = jet phase; P = particles.

[c] C in S = well-mixed compartments in series; PF = plug flow; PM = perfect mixing; PB = population balance; A = attrition allowance; E = entrainment allowance; St = stagnant; NS = not specified.

respectively. (The other dimensionless group, k'_n, is kept as an unknown and will be found later by trial and error.)

4. Find the gas conversion, $\chi_A = 1 - C_{Aout}/C_{Ain}$, as a function of k'_n from the appropriate expression in Table 11.5.

5. Find the concentration of gas to which the solids are exposed. For the Orcutt model with all the solids assigned to the dense phase and perfect mixing in the dense phase, the relevant concentration is simply C_{Ad}. For a first-order reaction, the resulting concentration is:

$$C_A^* = C_{Ad} = \frac{C_{Ain} (1 - \beta e^{-x})}{1 - \beta e^{-X} + k'_1} \qquad (11.39)$$

6. The above concentration can now be used with Eq. (11.35) to evaluate the time t_{cr} for complete reaction of the particle, i.e.:

$$t_{cr} = \frac{\rho_B d_p}{2 b k_{ef} C_A^* M_B} \qquad (11.40)$$

where ρ_B and M_B are the density and molecular weight of the solid reactant and k_{ef} is the effective kinetic rate constant given by Eq. (11.37) or Eq. (11.38). For a single-stage bed with a mean solids residence time τ much greater than the turnover time (see Chapter 5), the solids conversion is given (Kunii and Levenspiel, 1969) by:

$$\chi_B = 3 \left\{ \frac{\tau}{t_{cr}} - 2 \left(\frac{\tau}{t_{cr}} \right)^2 + 2 \left(\frac{\tau}{t_{cr}} \right)^3 [1 - e^{-t_{cr}/\tau}] \right\} \qquad (11.41)$$

$$\approx 1 - \frac{1}{4} \frac{t_{cr}}{\tau} \qquad \text{for } \tau > 5 t_{cr} \qquad (11.41a)$$

7. Obtain a new expression for the gas conversion from a mole balance based on the stoichiometry of the reaction, i.e.:

$b \times$ number of moles of A consumed = number of moles of B consumed

or

$$b U A C_{Ain} \chi_A = \frac{F_{in} \chi_B}{M_B} \qquad (11.42)$$

where F_{in} is the solids mass feed rate.

8. Now guess a value of k'_1 and go through steps 4 to 7; repeat until the estimates of χ_A from steps 4 and 7 agree with each other.

The sample problem below illustrates the above procedures for a reaction of industrial importance.

Example 11.5: Zinc Roaster

Operating conditions for a zinc roaster operated by Canadian Electrolytic Zinc Limited at Valleyfield, Québec, have been reported by Avedesian (1974) to be as follows:

Bed diameter: 6.38 m
Operating pressure: 101 kN/m^2
Operating temperature: 1,233 K
Air superficial velocity: 0.78 m/s
Minimum fluidizing velocity: 4.8 mm/s
Bed weight: ~ 30,000 kg (estimated from the bed pressure drop quoted by Avedesian as 10 kN/m^2)
Solids feed rate: 2.48 kg/s (dry basis)

Certain hydrodynamic features of this roaster have already been considered in Examples 4.3 and 4.4. For the illustrative purposes of this worked example, we assume uniform solids of diameter 60 μm and a mean bubble size of 0.1 m. The reaction is known to proceed according to the shrinking core model. According to Fukunada et al. (1976), the surface chemical rate constant is given by an Arrhenius equation:

$$k_{1s} = 2.96 \times 10^{13} \exp \left\{ \frac{-3.14 \times 10^5}{RT} \right\} \quad \text{m/s}$$

The effective diffusivity of O_2 and SO_2 through the solid ash may be taken as 9.0×10^{-6} m^2/s. The density of the concentrate is 4,100 kg/m^3. Estimate the conversion of both gas and solids assuming that the bed operates entirely in the bubble regime with a mean bubble diameter of 0.10 m.

Solution

We assume $\epsilon_{mf} = 0.50$ and estimate a molecular diffusivity of oxygen in air at 1,233 K of 2.5×10^{-4} m^2/s. The reaction may be written in the form of Eq. (11.34) as:

$$O_2(g) + \tfrac{2}{3} ZnS(s) \rightarrow \tfrac{2}{3} ZnO(s) + \tfrac{2}{3}SO_2(g)$$

so that $b = \tfrac{2}{3}$. The molecular weights of ZnS and ZnO are 0.0975 and 0.0814 kg/mol, respectively.

At 1,233 K:

$$k_{1s} = 2.96 \times 10^{13} \exp \left(\frac{-3.14 \times 10^5}{8.314 \times 1,233} \right) \quad \text{m/s}$$

$$= 1.47 \text{ m/s}$$

$$C_{Ain} = \frac{0.21P}{RT} = \frac{0.21 \times 1.013 \times 10^5}{8.314 \times 1,233} = 2.075 \text{ mol/m}^3$$

Now proceed through the steps given above:

1. $u_A = 0.71 \sqrt{(9.8 \times 0.10)} + 0.78 - 0.0048 \text{ m/s} = 1.479 \text{ m/s}$

$$a_b = \frac{6}{0.10} = 60 \text{ m}^{-1}$$

$$\epsilon_b \simeq \frac{0.78 - 0.0048}{1.479} = 0.524 \text{ (from Eq. 4.42); in practice we expect } \epsilon_b$$

never to exceed about 0.40. Therefore, assume $\epsilon_b = 0.40$.

We expect the solids in the bed to be highly reacted. Hence their density will be:

$$\rho_s \simeq \frac{4,100 \times 0.0814}{0.0975} \text{ kg/m}^3 \cdot$$

$$= 3,420 \text{ kg/m}^3$$

Hence:

$$H_{mf} = \frac{30,000}{3,420 \times (\pi/4) \times 6.38^2 \times 0.5} \text{ m} = 0.548 \text{ m}$$

and

$$H = \frac{0.548}{1 - 0.40} \text{ m} = 0.915 \text{ m}$$

Product solids flowrate from the bed $F_1 = \frac{2.48 \times 0.0814}{0.0975} \text{ kg/s}$

$$= 2.07 \text{ kg/s}$$

Mean solids residence time $= \frac{30,000}{2.07} = 14,500 \text{ s}$

2. From Eq. (11.6):

$$k_q = 0.3 \times 0.0048 + \left(\frac{4 \times 2.5 \times 10^{-4} \times 0.50 \times 1.479}{\pi \times 0.10} \right)^{1/2} \text{ m/s} = 0.050 \text{ m/s}$$

3. From Eq. (11.10):

$$\beta = \frac{1 - 0.0048}{0.78} = 0.994$$

From Eq. (11.17)

$$X = \frac{0.050 \times 60 \times 0.40 \times 0.915}{0.994 \times 0.78} 0.78 = 1.415$$

Therefore:

$$1 - \beta e^{-X} = 0.758$$

4. From Table 11.5:

$$\chi_A = \frac{0.758 k_1'}{0.758 + k_1'}$$

5. From Eq. (11.39):

$$C_{Ad} = \frac{2.075 \times 0.758}{0.758 + k_1'} \text{ mol/m}^3$$

6. $$k_{ef} = \frac{1}{1/1.47 + 60 \times 10^{-6}/(12 \times 9 \times 10^{-6})} = 0.809 \text{ m/s}$$

$$t_{cr} = \frac{4,100 \times 0.000060}{2 \times \frac{2}{3} \times 0.809 \times 0.0975 C_{Ad}} = \frac{2.34}{C_{Ad}} \text{ s}$$

$$\chi_B = 1 - \frac{t_{cr}}{4 \times 14,500} = 1 - \frac{t_{cr}}{58,000}$$

7. From Eq. (11.42):

$$\chi_A = \frac{2.48}{0.0975} \frac{\chi_B}{\frac{2}{3} \times 0.78 \times \pi/4 \times 6.38^2 \times 2.075}$$

$$= 0.737 \chi_B$$

After trial and error, $k_1' = 26$; $C_{Ad} = 0.0588$ mol/m^3; $t_{cr} = 39.8$s; $\chi_A = 0.737$; $\chi_B = 0.9993$.

Hence the solids are more than 99.9 per cent. converted while the oxygen is nearly 74 per cent. converted according to the model.

To replace the Orcutt model by the two-phase bubbling bed reactor model (outlined in Section 11.8 for a gas phase reaction), the steps are essentially the same. However, in step 3, β is set equal to unity; X is calculated from Eq.

(11.30); ϕ_b and ϕ_d are estimated as outlined in Section 11.8. For step 4, the gas conversion χ_A is now found as a function of k'_n from Table 11.6. The mean concentration of gas to which the solids are exposed C'_A, to be calculated in step 5, is found by integration. For a first-order irreversible reaction, it can be shown that:

$$C^*_A = \frac{\phi_b \bar{C}_{Ab} + \phi_d \bar{C}_{Ad}}{\phi_b + \phi_d} = \frac{C_{Ain} \chi_A}{(\phi_b + \phi_d) k^*_1} \qquad (11.43)$$

where:

$$\chi_A = 1 - \exp \left\{ -\frac{k^*_1 \{X (\phi_b + \phi_d) + k^*_1 \phi_b \phi_d\}}{X + k^*_1 \phi_d} \right\} \qquad (11.44)$$

All other steps are identical to those outlined above.

The model developed above and the example are for the case of uniform solids of unchanging particle size. In most fluidized bed processes, particles fed to the bed have a wide distribution of sizes, and fines are elutriated from the bed. To take these factors into account for particles of unchanging size, we make the following changes in the procedures given above for calculating the solids conversion:

1. The feed particle size distribution is represented by N size fractions of representative size d_{pi} and mass fraction x_i (generally determined by sieve analysis).
2. Calculate the time for complete reaction of each size fraction $(t_{cr})_i$ from Eq. (11.35) or another appropriate kinetic model with d_p replaced by d_{pi}.
3. Determine the mean residence time of particles in this size range from the equation:

$$\tau_i = \frac{M}{m_{under} + (1 - \eta_i) K^*_i A} \qquad (11.45)$$

where M is the mass of solids in the bed, m_{under} is the underflow or overflow mass withdrawal rate of product, K^*_i is the elutriation rate constant, and η_i is the efficiency of any cyclones used to recycle solids to the bed. Both K^*_i and η_i can be estimated from standard correlations.

4. The mean conversion of particle of size d_{pi} can be calculated from:

$$\chi_{Bi} = 1 - \int_0^{(t_{cr})_i} (1 - \chi_{Bi}) E(d_{pi}, t) dt \qquad (11.46)$$

where χ_{Bi}, the conversion of particles of size d_{pi} for a residence time t, is obtained from standard kinetic models and the exit age distribution is generally assumed to correspond to perfect mixing of each particle size fraction so that:

$$E(d_{pi},t) = \frac{\exp\{-(t/\tau_i)\}}{\tau_i} \tag{11.47}$$

Hence, for example, the shrinking core model with chemical reaction control χ_{Bi} is given by Eq. (11.41) with τ and t_{cr} replaced by τ_i and $(t_{cr})_i$, respectively.

5. Finally, to calculate the overall conversion, we must sum over the different size fractions fed to the bed:

$$(\chi_B)_{overall} = \Sigma \; x_i \; \chi_{Bi} \tag{11.48}$$

It is possible to extend the above procedure to cover more complex cases, e.g. multi-stage beds, beds in series, or other cases where the solids residence time distribution does not correspond to perfect mixing. For particles which shrink or grow as a result of chemical reaction, attrition, or agglomeration, population balances are required to account for the different size fractions. A number of these cases have been considered by Kunii and Levenspiel (1969) and in the models listed in Table 11.7.

As demonstrated by the models outlined above, modelling of the disappearence of gaseous reactants follows the same general lines as two-phase modelling of catalytic gas phase reactions. However, it is common for the gaseous reagents (e.g. oxygen in the case of coal combustion) to be present well in excess of the stoichiometric requirements. In this case, the residence time distribution of the solids, including losses by elutriation, plays a critical role in determining the overall conversion of solids. Where high conversions of solids are required, it is often desirable to place two or more units in series and to provide better solids recovery than is possible with a single cyclone.

11.13 REACTION IN THE FREEBOARD REGION

The freeboard region can play a significant role in determining the overall performance of fluidized bed reactors. For example, nitric oxide levels in coal combustors change rapidly in the freeboard region (Gibbs, Perlira, and Beér, 1975), while the temperature level, coke conversion, and afterburning in cracking catalyst regenerators depend strongly on the freeboard (deLasa and Grace, 1979).

The contribution of the freeboard arises from the following considerations:

(a) Additional contacting of gas and solids can take place in the freeboard region. Although the hydrodynamics of the region are not well understood, contacting appears to be favourable, with little or no tendency of the gas to separate into two phases (as in the bubbling bed) and near plug flow for the gas. This effective contacting and lack of backmixing, coupled with the fact that entrained particles tend to be finer than those in the bed, make up to some extent for the low hold-up of solids in the region above the dense bed surface.

(b) Whereas temperatures tend to be highly uniform in the dense bed, significant temperature gradients are possible in the freeboard region. This may lead to increased reaction rates, unwanted side reactions, or recycling of additional heat to the bed itself as solids return from cyclones or by gravity.

(c) Where heat transfer surfaces extend into the freeboard region, additional heat transfer occurs. Heat transfer coefficients are nearly as favourable near the bed surface as for immersed tubes, but the coefficients fall off rapidly with increasing height. Radiation losses from particles in the freeboard can also be substantial.

The contribution of the freeboard region will depend on a number of factors:

(a) The extent of reaction which has already occurred in the dense bed. (It is evident that if a reactant has already been used up before reaching the bed surface, no further reaction involving consumption of that reactant can take place in the freebaord.)

(b) The height and geometry of the freeboard region, including expanded areas and fixed baffles or tubes, if any.

(c) The fines content of the powder.

(d) The efficiency of the cyclone or other solid recovery system in returning entrained fines to the bed where they can be reentrained.

(e) The operating conditions, principally the superficial gas velocity, pressure, and temperature.

To a lesser extent, reaction in the freeboard may also depend on the depth of the dense bed and the presence of immersed tubes through their influence on the bursting pattern of voids at the bed surface and hence on the ejection of solids into the freeboard.

The freeboard itself may be divided into different regions, as shown in Fig. 11.7:

(a) Splash zone: this is immediately above the dense bed surface.

(b) Disengagement zone: This extends up to transport disengaging height level; it may be taken to include the splash zone since there is no distinct boundary between the splash zone and the disengagement zone.

(c) Region beyond TDH. In regions (a) and (b), particles travel both upwards and downwards, amid periodic bursts (or splashes) of particles. In zone (a), on the other hand, particle motion is almost solely upwards

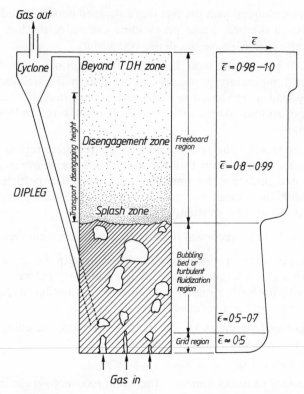

Figure 11.7   Freeboard region of fluidized bed and its zones.
Other regions are also indicated. Typical time-mean voidages
are shown for each region.

and the distribution of particles is much more uniform, although
streamers or clusters can often be identified. For relatively large particles
without fines, essentially all of the ejected particles return to the dense
bed surface, and region (c) may be neglected. Region (c) is also absent if
the column is successfully designed to have the freeboard height equal to
the TDH.

It is clear from the discussion above that the behaviour in the freeboard
region is complex and depends on many different factors. The general
inadequacy of correlations for entrainment and TDH (see Chapter 6)
complicates the treatment of the freeboard region still further.

Yates and Rowe (1977) adopted a simple model in which gas and solids
were assumed to remain isothermal. Radial gradients were ignored. The slip
velocity between solids and gas was assumed to be $U-v_t$ throughout the
entire freeboard. Two extreme cases, plug flow and perfect mixing, were
assumed for the gas mixing. For the more realistic case of plug flow and for a

first order gas phase solid catalysed reaction, their model leads to the following expression:

$$\frac{C_{Az}}{C_{AH}} = \exp\left\{\frac{-\xi\beta_W\,(Q_B/A)\,(1-\epsilon_{mf})\,(z-H)}{U(U-v_t)\,(1/k_1 + d_p/6k_d)}\right\} \qquad (11.49)$$

where β_w is the wake volume/bubble volume for bubbles reaching the bed surface. The volumetric flux of bubbles, Q_B/A, can be estimated as detailed in Chapter 4, while ξ, the fraction of wake particles ejected into the freeboard as bubbles burst at the bed surface, varies with d_{eq}, but is typically of order 0.4 (George and Grace, 1978). The mass transfer coefficient k_d can be obtained for steady flow of gas past the particle at a relative velocity of v_t as detailed by Clift, Grace, and Weber (1978). Sample calculations performed by Yates and Rowe (1977) show that the freeboard region can be more effective than an equal height of dense bed.

Example 11.6

Narrow size fraction 60 μm spherical catalyst particles are being used in a fluidized bed reactor. The particle density is 2,650 kg/m^3 and the catalyst activity is 10.0 s^{-1}, the reaction being first order. The fluidizing gas has the properties of air at 20°C and 101 kPa. The molecular diffusivity of the reacting gas in air is 2.0 × 10^{-5} m^2/s. Estimate the fraction of the gas leaving the expanded bed surface which is converted in the freeboard region if the total height of this region is 4.0 m. The superficial gas velocity is 1.0 m/s. Assume isothermal behaviour.

Solution

From Clift, Grace, and Weber (1978), $v_t = 0.26$ m/s and $k_d = 0.73$ m/s. Estimates for U_{mf} and ϵ_{mf} are 0.004 m/s and 0.50, respectively. From George and Grace (1978), we estimate $\xi = 0.4$. A reasonable estimate for β_W is 0.33. The visible bubble flowrate may, to a first approximation, be estimated from the two-phase theory so that:

$$\frac{Q_B}{A} = 1.0 - 0.004 = 0.996 \text{ m/s}$$

Hence, Eq. (11.49) yields:

$$\frac{C_{Az}}{C_{AH}} = \exp\left\{\frac{-0.4 \times 0.33 \times 0.996 \times 0.5 \times 4.0}{1.0 \times (1.0 - 0.26) \times \{1/10.0 + 0.000060/(6 \times 0.73)\}}\right\} = 0.029$$

Hence, the model indicates that only about 3 per cent. of the gas which is unconverted leaving the surface of the dense bed is still unconverted at the top of the freeboard region. This illustrates the prominent influence which the freeboard can exert.

Generally speaking, it is necessary to allow for temperature gradients in the freeboard region by writing a differential energy balance. For plug flow of gas, this may be written:

$$C_g\rho_g \, \frac{dT}{dz} + \frac{hP(T-T_w)}{UA} + \Delta H_r C_{Ain} \, \frac{d\chi_A}{dz} = 0 \qquad (11.50)$$

where the particles and gas in the freeboard are assumed to have the same temperature T locally (usually a reasonable approximation), while h is the freeboard suspension-to-wall heat transfer coefficient, P the column perimeter, T_w the column wall temperature, and ΔH_r the heat of reaction. The middle term can be neglected for well-insulated systems. Allowance may be made for radiation losses or cooling tubes in the freeboard by incorporating extra terms.

Miyauchi and Furusaki (1974) used an equation like Eq. (11.50) to show the effect of the freeboard on selectivity. The freeboard could be beneficial or harmful to the yield of a desired product, depending on the mechanism of reaction and whether the reaction is exothermic or endothermic. For example, for reactions of the type $A \rightarrow B \rightarrow C$ where B is the desired product, the freeboard region is helpful for endothermic reactions, but not for exothermic reactions.

DeLasa and Grace (1979) treated the freeboard region of a gas–solid heterogeneous reaction, the burning of coke deposits from deactivated cracking catalyst particles in a regenerator. Separate allowance was made for upward and downward moving solids. Material balances were written for both the solid reactant (coke) and the gaseous reactant (oxygen); energy balances were used in addition. The hydrodynamics of particle ejection and trajectories in the freeboard were taken from George and Grace (1978). Numerical integration was required to solve the equations for different monodisperse particle sizes and conditions matching those in an actual industrial regenerator. The freeboard was shown to have a profound effect for shallow beds, especially for particles whose terminal velocities are close to the superficial gas velocity. The model has been extended to describe non-steady state conditions by deLasa and Errazu (1980).

Chen and Wen (1982) modified the above approach by assuming axial dispersion of gas in the freeboard region and by assuming an exponential decay of solids concentration above the bed surface. The model was shown to be in favourable agreement with experimental results obtained in a 1.8 m square fluidized combustor. Fee et al. (1982) considered the part of the freeboard in which particles are ejected and fall back to the bed surface as a continuously stirred tank reactor. Significant sulphur capture was shown to occur in the freeboard of a fluidized bed combustor.

11.14   REACTORS OPERATING IN THE SLUG FLOW REGIME

Two phase models similar to those described earlier for the bubbling regime have also been written for fluidized beds operating in the slug flow regime. The best known slug flow reactor models are those of Hovmand et al (1971) and Raghuraman and Potter (1978). Yates and Gregoire (1980) found that the latter gave better agreement with experimental conversions obtained in a 0.10 m column. It is essential to allow for favourable gas–solid contacting in the region below the level where slug flow is reached (typically $z=2D$) in applying slug flow reactor models.

11.15   REACTORS OPERATING IN THE TURBULENT FLUIDIZATION REGIME

In industrial practice, many fluidized bed reactors operate in the turbulent fluidization regime. This is especially the case for solid catalysed gas phase reactions since most catalysts fall within the group A category where superficial gas velocities of the order of 0.3 m/s are reportedly (Avidan, 1982) sufficient to initiate this regime in reactors of commercial size. In smaller scale reactors, higher gas velocities are needed for this regime. Perhaps for this reason, there appear to be almost no data available in the open literature showing reaction rates for fluidized beds operated in the turbulent regime.

The appearance of turbulent fluidized beds (see Chapter 7), with tongues of gas and solids darting backwards and forwards, suggest that this regime should give rise to effective gas–solid contacting. Van Swaaij (1978) and Wen (1979) suggested that beds operated in this regime might be modelled according to a single-phase plug flow model (Section 11.5). On the other hand, Avidan (1982) suggests that appreciable axial gas mixing occurs and that an axially dispersed single phase model should be appropriate. Since many of the experimental results correlated by Van Deemter (1980) were for reactors, large and small, operated within this regime, his summary of results can provide estimates for the effective axial dispersion coefficients needed for this simple model. Experimental data are required to verify that there is negligible interphase mass transfer resistance.

11.16   REACTORS OPERATING IN THE FAST FLUIDIZATION REGIME

There has been considerable interest in recent years in circulating beds and in riser reactors for both gas–solid reactions and solid catalysed reactions. Operation within the fast fluidization regime (see Chapter 7) confers a number of advantages including effective gas–solid contacting, increased gas throughput, ability to treat cohesive and low density materials (e.g. peat,wood waste, caking coals), good radial mixing, improved turndown, temperature uniformity throughout the entire reactor volume, and (for combustion reactions) reduced NO_x emissions because of secondary air additions. Erosion, attrition and particulate emissions can all be controlled within reasonable limits by good design.

Wen (1979) suggested tentatively that a single-phase plug flow model (section 11.5) might be appropriate for this regime in view of the effective gas–solid contacting. On the other hand, there is strong evidence of phase segregation into either a core of dilute phase moving upwards at high velocity surrounded by a thin annular region of predominantly downward dense particle motion (Bierl, Gajdos, and McIver 1980) or of the occurrence of particle clusters (Yerushalmi, Cankurt, Geldart, and Liss, 1978). In either case, there may well be a need for a two-phase model. The two-region model of Van Deemter (1961), which allows for a dilute upflow and a dense region, may provide a good basis for such models. However, additional work on the hydrodynamics of the fast fluidization regime and reaction studies on beds operated within this regime are required to allow the parameters needed for the model to be specified. Energy balances can be incorporated when required, as for other regimes. For modelling of gas–solid reactions in circulating beds, it would appear to be appropriate to assume that the solids are perfectly mixed.

11.16 NOMENCLATURE

a_b	interfacial bubble area per unit bubble volume
a_j	interfacial jet area per unit bed volume
A	cross-sectional area of column
b	stoichometric constant defined in Eq. (11.34)
C_A, C_B	concentration of A and B
C_A^*	gas phase concentration to which a reacting particle is exposed
C_{Ab}, C_b	concentration of A in the bubble phase
C_{Ad}, C_d	concentration of A in the dense phase
C_{AH}	average concentration of A at $z=H$
C_{Ain}, C_{in}	inlet concentration of A
C_{Aj}	concentration of A in the jet phase
C_{Aout}	outlet concentration of A
C_g	specific heat of gas
d_{eq}	diameter of equivalent volume sphere = bubble diameter
d_{or}	diameter of grid orifice
d_p	particle diameter

D	column diameter
$\mathscr{D}$	molecular diffusivity of component in gas stream
$\mathscr{D}_s$	effective diffusivity of gaseous component through porous solid
$E(d_{pi}, t)$	exit age distribution function
f_{or}	fractional free area in the grid
F_A, F_B	functions defined in Table 11.6
F_{in}	solids mass feedrate
g	acceleration of gravity
G	function defined in Table 11.6
h	bed-to-wall or suspension-to-wall heat transfer coefficient
H	bed height
H_j	height of jet penetration
H_{mf}	bed height at minimum fluidization
k_1^j	dimensionless first-order rate constant defined in Eq. (11.33)
k_{1s}	first-order reaction rate constant based on unit surface area as defined by Eq. (11.36)
$k_A . k_B$	first-order rate constants for disappearance of A and B, defined in Table 11.5
k_A', k_B'	dimensionless rate constants, $k_A H_{mf} (1-\epsilon_{mf})/U$ and $k_B H_{mf} (1-\epsilon_{mf})/U$
k_A^*, k_B^*	dimensionless rate constants, $k_A H/U$ and $k_B H/U$
k_d	gas-to-particle mass transfer coefficient
k_{ef}	effective kinetic rate constant given by Eq. (11.37) or Eq. (11.38)
k_n	nth-order kinetic rate constant
k_n', k_n^*	dimensionless kinetic rate constants (Eqs. 11.16 and 11.22)
k_q	bubble to dense phase mass transfer coefficient
k_{qj}	jet to dense phase mass transfer coefficient
K_e	equilibrium constant defined in Table 11.5
K^*	elutriation rate constant
M_B	molecular weight of B
n	reaction order
N_A	number of moles of reactant A
P	column perimeter
Q_B	volumetric flowrate of void units
r_A, r_B	time rate of disappearance of reactant A or B
r_{uc}	radius of unreacted core of reacting particle

R	gas constant = 8.314 J/mol K
t	time
t_{cr}	time for complete reaction of a solid particle
T	absolute temperature
T_w	wall temperature
u_A	true velocity of bubbles relative to wall in a freely bubbling bed
u_b	rise velocity of single isolated bubble
u_{or}	gas velocity through grid orifice
U	superficial gas velocity
U_{mf}	minimum fluidization velocity
v_t	terminal settling velocity of a particle
V_b	bubble volume
V_c	cloud volume
V_p	particle volume
x	mass fraction of particles
X	number of mass transfer units (Eq. 11.7 or Eq. 11.20)
X_j	number of mass transfer units in grid region (Eq. 11.32)
z	vertical position coordinate measured upwards from the grid
α	ratio of bubble velocity to interstitial gas velocity (Eq. 11.2)
β	ratio of gas flow via bubble phase to total gas flow
β_w	wake volume per unit bubble volume
γ_b	volume fraction of bubbles occupied by dispersed particles ($= \phi_b/\epsilon_b$)
γ_j	volume fraction of jets occupied by dispersed particles
ΔH_r	heat of reaction
ϵ_b	fraction of bed volume occupied by bubbles
ϵ_j	fraction of grid entry region occupied by jets
ϵ_{mf}	bed voidage at minimum fluidization
ϵ_p	internal particle porosity
η	cyclone separation efficiency
ρ_B	density of solid reactant B
τ	solids mean residence time
ϕ_b	fraction of bed volume occupied by solids associated with the bubble phase
ϕ_d	fraction of bed volume occupied by solids associated with the dense phase

χ_A gas phase conversion ($= 1 - C_{Aout}/C_{Ain}$)

χ_B solid phase conversion

Subscript

i   pertaining to particles in size interval i

11.17 REFERENCES

Anwer, J., and, Pyle, D.L. (1974). 'Gas motion around bubbles in fluidized beds, *La Fluidization et ses Applications,* Soc. Chim. Ind., Paris, pp. 240–253.

Avedesian, M.M. (1974). 'Roasting zinc concentrate in fluidized beds — practice and principles', 24th CSChE Meeting, Ottawa.

Avedesian, M.M., and Davidson, J.F. (1973). 'Combustion of carbon particles in a fluidized bed', *Trans. Instn Chem. Engrs,* **51**, 121–131.

Avidan, A. (1982). 'Turbulent fluid bed reactors using fine powder catalyst, AIChE–CIESC Mtd., Beijing, Sept.

Barnstone, L.A., and Harriott, P., (1967), Frequency response of gas mixing in a fluidized bed reactor, AIChE J. **13**, 465–475.

Baron, R.E., Hodges, J.L., and Sarofim, A.F. (1978). Mathematical model for predicting efficiency of fluidized bed steam generators', A.I.Ch.E. Symp. Ser., **74** (176), 120–125.

Basu, P., Broughton, J., and Elliott, D.E. (1975). 'Combustion of single coal particles in fluidized beds', *Inst. Fuel Symp. Ser.,* **A3** (1), 1–10.

Bauer, E., and Werther, J. (1981). 'Scale-up of fluid bed reactors with respect to size and gas distributor design — measurements and model calculations', *Proc. 2nd World Congress of Chem. Engng.* 3, 69–72.

Becker, H.A., Beér, J.M., and Gibbs, B.M. (1975). 'A model for fluidized bed combustion of coal', *Inst. Fuel Symp. Ser.,* **A1**(1).

Behie, L.A., Bergougnou, M.A., and Baker, C.G.J. (1976). 'Mass transfer from a grid jet in a large gas-fluidized bed', in *Fluidization Technology* (Ed. D.L. Keairns), Vol. 1, Hemisphere, Washington, pp. 261–278.

Behie, L.A., and Kehoe, P. (1973). 'The grid region in a fluidized bed reactor', A.I.Ch.E. J., **19**, 1070–1072.

Bierl, T.W. Gajdos, L.J., McIver, A.E., and McGovern, J.J. (1980). DOE Rep. FE-2449-11.

Bukur, D., Caram, H.S., and Amundson, N.R. (1977). 'Some model studies of fluidized bed reactors', in *Chemical Reactor Theory: A Review*, (Ed. L. Lapidus and N.R. Admundson), Prentice-Hall, Englewood Cliffs, NJ, pp. 686–757.

Bywater. R.J. (1978). 'Fluidized bed catalytic reactor according to a statistical fluid mechanic model', *A.I.Ch.E. Symp. Ser.,* **74** (176), 126–133.

Carberry, J.J. (1964). 'Designing laboratory catalytic reactors', *Ind. Engng. Chem.*, **56** (11), 39–46.

Carberry, J.J. (1976). *Chemical and Catalytic Reaction Engineering,* McGraw-Hill, New York.

Chang, C.C., Rong, S.X., and Fan, L.T. (1982). 'Modelling of shallow fluidized bed reactors', *Can. J. Chem. Eng.,* **60**, 781–790.

Chavarie, C., and Grace, J.R. (1975). 'Performance analysis of a fluidized bed reactor', *Ind. Eng. Chem. Fund.*, **14**, 75–91.

Chavarie, C., and Grace, J.R. (1976). 'Interphase mass transfer in a gas-fluidized bed', *Chem. Eng. Sci.*, **31**, 741–749.

Chen, L.H., and Wen, C.Y. (1982). 'Model of solid gas reaction phenomena in the fluidized bed freeboard'. *A.I.Ch.E.J.*, **28**, 1019–1027.

Chen, T.P., and Saxena, S.C. (1977). 'Mathematical modelling of coal combustion in fluidized beds with sulphur emission control by limestone and dolomite', *Fuel*, **56**, 401–413.

Clift, R., Grace, J.R., and Weber, M.E. (1978). *Bubbles, Drops and Particles*, Academic Press, New York.

Cooke, M.J., Harris, W., Highley, J., and Williams, D.F. (1968). 'Kinetics of oxygen consumption in fluidized bed carbonizers', *Inst, Chem. Engrs Symp. Ser.*, **30**, 21–27.

Darton, R.C. (1979). 'A bubble growth theory of fluidized bed reactors', *Trans. Instn Chem. Engrs*, **57**, 134–138.

Davidson, J.F., and Harrison, D. (1963). *Fluidized Particles,* Cambridge University Press, Cambridge.

deLasa, H.I., and Errazu, A. (1980). In *Fluidization*, Eds. J.R. Grace and J.M. Matsen, Plenum, New York, pp. 563–570.

deLasa, H.I., and Grace, J.R. (1979). 'The influence of the freeboard region in a fluidized bed catalytic cracking regenerator', *A.I.Ch.E. J.*, **25**, 984–991.

deMichaele, G., Elia, A., and Massimilla, L. (1976). 'The interaction between jets and fluidized beds', *Ing. Chim. Ital.*, **12**, 155–162.

Errazu, A.F., deLasa, H.I., and Sarti, F. (1979). 'A fluidized bed catalytic cracking regenerator model: grid effects', *Can. J. Chem. Eng.*, **57**, 191–197.

Fan, L.T., Fan, L.S., and Miyanami, K. (1977). 'Reactant dynamics of catalytic fluidized bed reactors characterized by a transient axial dispersion model with varying physical quantities', *Proc. Pachec Conference*, pp. 1379–1388.

Fee, D.C., Myles, K.M., Marroquin, G., and Fan L–S., (1982). 'An analytical model for freeboard and in-bed limestone sulfation in fluidized-bed coal combustors', *Proc. 7th Int. Fluidized Bed Combustion Conference*, 1121–1126.

Fontaine, R.W., and Harriott, P. (1972). 'The effect of molecular diffusivity on mixing in fluidized beds', *Chem. Eng. Sci.*, **27**, 2189–2197.

Fryer, C., and Potter, O.E. (1972a). 'Countercurrent backmixing model for fluidized bed catalytic reactors; applicability of simplified solutions', *Ind. Eng. Chem. Fund.*, **11**, 338–344.

Fryer, C., and Potter, O.E. (1972b). 'Bubble size variation in two-phase models of fluidized bed reactors', *Powder Technol.*, **6**, 317–322.

Fryer, C., and Potter, O.E. (1976), 'Experimental investigation of models for fluidized bed catalytic reactors', *A.I.Ch.E. J.* **22**, 38–47.

Fukunda, Y., Monta, T., Asaki, Z., and Kondo, Y. (1976). 'Oxidation of zinc sulphide in a fluidized bed', *Metal. Trans.*, **7B**, 307–314.

George, S.E., and Grace J.R. (1978). 'Entrainment of particles from aggregative fluidized beds, *A.I.Ch.E. Symp. Ser.*, **74** (176), 67–74.

Gibbs, B.M. (1975). A mechanistic model for predicting the performance of a fluidized bed coal combustor, *Inst. Fuel Symp. Ser.*, (1) A5.

Gibbs, B.M., Pereira, F.J., and Beér, J.M. (1975). 'Coal combustion and NO formation in an experimental fluidized bed', *Inst. Fuel Symp. Ser.*, (1) D6.

Gordon, A.L., and Amundson, N.R. (1976). 'Modeling of fluidized bed reactors: combustion of carbon particles', *Chem. Eng. Sci.*, **31**, 1163–1178.

Grace, J.R. (1971). 'An evaluation of models for fluidized bed reactors', *A.I.Ch.E. Symp. Ser.*, **67** (116), 159–167.

Grace, J.R. (1984). 'Generalized models for isothermal fluidized bed reactors', in *Recent Advances in the Engineering Analysis of Chemically Reacting Systems* (Ed. L.K. Doraiswamy), Wiley, New Delhi.

Grace, J.R., and deLasa, H.I. (1978). Reaction near the grid in fluidized beds *A.I.Ch.E. J.*, **24**, 364–366.

Heimlich, B.N., and Gruet, I.C. (1966). 'Transfer function analysis of fluidized bed residence time distribution data', *Chem. Eng. Prog. Symp. Ser.*, **62** (67), 28–34.

Horio, M., and Wen, C.Y. (1976). 'Analysis of fluidized-bed combustion of coal with limestone injection', in *Fluidization Technology* (Ed. D.L. Keairns), Vol. 2, Hemisphere, Washington, pp. 289–320.

Horio, M., and Wen, C.Y. (1978). 'Simulation of fluidized bed combustors: Combustion efficiency and temperature profile', *A.I.Ch.E. Symp. Ser.*, **74** (176), 101–111.

Hovmand, S., and Davidson, J.F. (1971). 'Pilot plant and laboratory scale fluidized reactors at high gas velocities; the relevance of slug flow', Chapter 5 in *Fluidization* (Eds J.F. Davidson and D. Harrison), Academic Press, London, pp. 193–259.

Hovmand, S., Freedman, W., and Davidson, J.F. (1971). 'Chemical reaction in a pilot-scale fluidized bed', *Trans. Instn Chem. Engrs,* **49**, 149–162.

Jackson, R. (1971). 'Fluid mechanical theory', Chapter 3 in *Fluidization* (Eds. J.F. Davidson and D. Harrison, Academic Press, London.

Kato, K., and Wen, C.Y. (1969). 'Bubble assemblage model for fluidized bed catalytic reactors', *Chem. Eng. Sci.*, **24**, 1351–1369.

Kunii, D., and Levenspiel, O. (1969). *Fluidization Engineering,* Wiley, New York.

Latham, R., Hamilton, C., and Potter, O.E. (1968). 'Backmixing and chemical reaction in fluidized beds', *Brit. Chem. Eng.*, **13**, 666–671.

Levenspiel, O. (1972). *Chemical Reaction Engineering,* 2nd ed., Wiley, New York.

Lewis, W.K., Gilliland, E.R., and Glass, W. (1959). 'Solid catalysed reaction in a fluidized bed', *A.I.Ch.E. J.,* **5**, 419–426.

Lignola. P.G., Donsi, G., and Massimilla, L. (1983). 'Mass spectrometric measurements of gas composition profiles associated with bubbles in a two-dimensional bed', *A.I.Ch.E. Symp. Ser.*, **79** (222), 19–26.

Mamuro, T., and Hattori, H. (1968). Flow pattern of fluid in spouted beds, *J. Chem. Engng Japan,* **1**, 1–5.

Mathur, K.B., and Epstein, N. (1974). *Spouted Beds*, Academic Press, New York.

May, W.G. (1959). 'Fluidized-bed reactor studies', *Chem. Engng. Prog.*, **55** (12), 49–56.

Merry, J.M.D. (1975). 'Penetration of vertical gas jets into fluidized beds', *A.I.Ch.E. J.*, **21**, 507–510.

Merry, J.M.D. (1976). 'Fluid and particle entrainment into vertical jets in fluidized beds', *A.I.Ch.E. J.*, **22**, 315–323.

Mireur, J.P. and Bischoff, K.B. (1967). 'Mixing and contacting models for fluidized beds', *A.I.Ch.E.J.*, **13**, 839–845.

Miyauchi, T., and Furusaki, S. (1974). 'Relative contribution of variables affecting the reaction in fluid bed contactors', *A.I.Ch.E. J.*, **20**, 1087–1096.

Mori, S., and Wen, C.Y. (1976). 'Simulation of fluidized bed reactor performance by modified bubble assemblage model', in *Fluidization Technology* (Ed. D.L. Keairns), Vol. 1, Hemisphere, Washington, pp. 179–203.

Murray, J.D. (1965). 'On the mathematics of fluidization', *J. Fluid Mech.,* **21**, 465–493, and **22**, 57–80.

Orcutt, J.C., Davidson, J.F., and Pigford, R.L. (1962). 'Reaction time distributions in fluidized catalytic reactors', *Chem. Eng. Prog, Symp. Ser.*, **58** (38), 1–15.

Partridge, B.A., and Rowe, P.N. (1966). 'Chemical reaction in a bubbling gas fluidized bed', *Trans. Instn Chem. Engrs*, **44**, 335–348.

Pereira, J.A.F. (1977). Ph.D. Thesis, University of Edinburgh.

Piccinini, N., Grace, J.R., and Mathur, K.B. (1979). 'Vapour-phase chemical reaction in spouted beds: verification of theory'. *Chem. Eng. Sci.*, **34**, 1257–1263.

Pyle, D.L., and Rose, P.L. (1965). 'Chemical reaction in bubbling fluidized beds', *Chem. Eng. Sci.*, **20**, 25–31.

Raghuraman, J., and Potter, O.E. (1978), Countercurrent backmixing model for slugging fluidized bed reactors, *A.I.Ch.E. J.*, **24**, 698–704.

Rajan, R., Krishnan, R., and Wen, C.Y. (1978). 'Simultation of fluidized bed combustors: coal devolatilization and sulphur oxides retention', *A.I.Ch.E. Symp. Ser.*, **74** (176), 112–119.

Rovero, G., Piccinini, N., Grace, J.R., Epstein, N., and Brereton, C.M.H. (1983). 'Gas phase solid-catalysed chemical reaction in spouted beds', *Chem. Eng. Sci.*, **38**, 557–566.

Rowe, P.N. (1964). 'Gas–solid reaction in a fluidized bed', *Chem. Eng. Prog.*, **60** (3), 75–80.

Rowe, P.N., MacGillivray, H.J., and Cheesman, D.J. (1979). 'Gas discharge from an orifice into a gas fluidized bed', *Trans. Instn Chem. Engrs*, **57**, 194–199.

Rowe, P.N., Partridge, B.A., and Lyall, E. (1964). 'Cloud formation around bubbles in fluidized beds', *Chem. Eng. Sci.*, **19**, 973–985.

Shaw, I., Hoffman, T.W., and Reilly, P.M. (1974). 'An experimental evaluation of two-phase models describing catalytic fluidized bed reactors', *A.I.Ch.E. Symp. Ser.*, **70** (141), 41–52.

Shen, C.Y., and Johnstone, H.F. (1955). 'Gas–solid contact in fluidized beds', *A.I.Ch.E. J.*, **1**, 349–354.

Sit, S.P. (1981). Grid region and coalescence zone gas exchange in fluidized beds', Ph.D. Dissertation, McGill University.

Sit, S.P., and Grace, J.R. (1978). 'Interphase mass transfer in an aggregative fluidized bed', *Chem. Eng. Sci.*, **33**, 1115–1122.

Sit, S.P., and Grace, J.R. 1981. 'Effect of bubble interaction on interphase mass transfer in gas fluidized beds', *Chem. Eng. Sci.*, **36**, 327–335.

Smith, J.M. (1970. *Chemical Engineering Kinetics*, 2nd ed., McGraw-Hill, New York.

Tajbl, D.G., Simons, J.B., and Carberry, J.J. (1966). 'Heterogeneous catalysis in a continuous stirred tank reactor', *Ind. Eng. Chem. Fund.*, **5**, 171–175.

Toei, R., Matsuno, R., Nishitani, K., Hayashi, H., and Imamoto, T. (1969). 'Gas interchange between bubble phase and continuous phase in gas–solid fluidized beds at coalescence', *Kagaku Kogaku*, **33**, 668–674.

Van Deemter, J.J. (1961). 'Mixing and contacting in gas-solid fluidized beds', *Chem. Eng. Sci.*, **13**, 143–154.

Van Deemter, J.J. (1980). 'Mixing patterns in large-scale fluidized bed, in *Fluidization* (Eds J.R. Grace and J.M. Matsen), Plenum Press, New York, pp. 69–89.

Van Swaaij, W.P.M. (1978). 'The design of gas–solids fluid bed and related reactors', *A.C.S. Symp. Ser* (Eds. D. Luss and V.W. Weekman), **72**, 193–222.

Walker, B.V. (1970). '*Gas–solid contacting in bubbling fluidized bed'*, Ph.D. Dissertation, Cambridge University.

Walker, B.V. (1975). 'The effective rate of gas exchange in a bubbling fluidized bed', *Trans. Instn Chem. Engrs*, **53**, 225–266.

Wen, C.Y. (1968). 'Noncatalytic heterogeneous solid fluid reaction models', *Ind Eng. Chem.*, **60**, 34–54.

Wen, C.Y. (1979). 'Chemical reaction in fluidized beds', in *Proc. N.S.F. Workshop on Fluidization and Fluid-Particle Systems* (Ed. H. Littman), Rensselaer Polytechnic Inst., pp. 317–387.

Wen, C.Y., and Fan, L.T. (1975). *Models for Flow Systems and Chemical Reactors*, Marcel Dekker, New York.

Werther, J. (1980), 'Mathematical modeling of fluidized bed reactors', *Int. Chem. Eng.*, **20**, 529–541.

Yates, J.G., and Grégoire, J.Y. (1980), 'An experimental test of slugging-bed reactor models', *In Fluidization*, eds. J.R. Grace and J.M. Matsen, Plenum, New York, pp. 581–588.

Yates, J.G., and Rowe, P.N. (1977). 'A model for chemical reaction in the freeboard region above a fluidized bed', *Trans. Instn Chem. Engrs*, **55**, 137–142.

Yerushalmi, J., Cankurt, N.T., Geldart, D., and Liss, R., (1978). 'Flow regimes in vertical gas–solid contact systems', *A.I.Ch.E. Symp. Ser.*, **74** (176), 1–13.

Yoshida, K., and Kunii, D. (1968). 'Stimulus and response of gas concentration in bubbling fluidized beds', *J. Chem. Eng. Japan*, **1**, 11–16.

Yoshida, K., and Wen, C.Y. (1970). 'Noncatalytic solid–gas reaction in a fluidized bed reactor', *Chem. Eng. Sci.*, **25**, 1395–1404.

Gas Fluidization Technology
Edited by D. Geldart
Copyright © 1986 John Wiley & Sons Ltd.

CHAPTER 12

Solids Transfer in Fluidized Systems

T.M. KNOWLTON

12.1 INTRODUCTION

In chemical processes employing the continuous circulation or transfer of particulate solids, the success or failure of the venture as a whole is often dependent upon the successful operation of the solids transfer system. The large fluidized catalytic cracking (FCC) units used in the petroleum industry and the coal conversion plants to produce synthetic fuels from coal are just two examples of processes where this is true.

Because of the importance of the solids transfer system, its successful design is critical to a process. Unfortunately, the design of such systems can be extremely complex. Gas–solids mixtures (whether pneumatically conveyed or moved in gravity flow in standpipes) can be transported in several different regimes — each with its own inherent quirks and peculiarities. The problem facing the designer of such a system is to couple these various regimes to achieve a smooth, stable transfer of solids over a wide range of operating conditions. This is a difficult task at best but it is made even more difficult because the understanding of many aspects of two-phase gas–solids flow is sadly inadequate. Thus, lacking a sound scientific basis for design, in practice many transport systems are often still designed on the basis of operating experience or rules of thumb.

Pneumatic transport of solids can be classified into four different regimes: horizontal dilute phase flow, vertical dilute phase flow, horizontal dense phase flow, and vertical dense phase flow. The boundary between dilute and dense phase conveying is not clear-cut. The parameter of solids/gas loading (kilograms of solid per kilogram of gas) in the conveying line has often been used to distinguish between dilute and dense phase flow. Solids/gas loadings of 0.01 to 15 kg of solid per kilogram of gas were used to denote dilute phase

341

flow, while dense phase flow was characterized by solids/gas loadings of 15 to over 200 kg of solids per kilogram of gas. However, these are just very rough, rule-of-thumb guidelines, and there can be much overlap. An analysis of each type of conveying is given in the following section.

It is the objective of this chapter to present information which will increase the probability of a successful design of solids transfer systems. In only one chapter there is not space enough for exhaustive theory. Therefore, the emphasis will be placed primarily on practical design.

12.2 VERTICAL DILUTE PHASE CONVEYING

In the design of industrial vertical dilute phase pneumatic conveying systems, the main consideration is generally that of choosing the correct velocity at which to transport the solids. Too low a velocity will result in unacceptable, unstable slugging flow; too high a velocity will result in excessive gas requirements and high pressure drops. The general relationship between velocity and pressure drop per unit length, $\Delta P/L$, in a dilute phase vertical riser is shown schematically in Fig. 12.1.

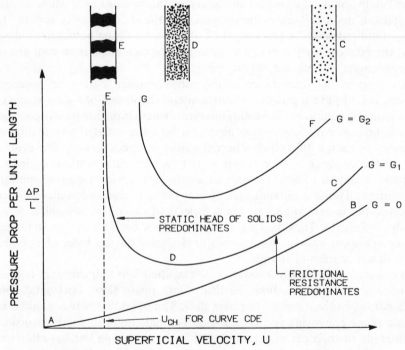

Figure 12.1 Phase diagram for dilute phase vertical pneumatic conveying.

Line AB is the pressure drop–velocity relationship for gas alone passing through a dilute phase pneumatic conveying line. Curve CDE is for a solids mass flux G_1, while curve FG is for a solids mass flux G_2 which is greater than G_1. At C, the gas velocity is very high and the conveying line is very dilute. As the conveying velocity is decreased from C to D, the gas and solids both rise more slowly. The solids inventory in the lift line rises, thus increasing the static head. However, the frictional resistance portion of the total pressure drop predominates in this region; thus, as the velocity decreases, so does the pressure drop.

In the region D to E, the decreasing velocity results in a rapid increase in solids inventory in the line. The static head of solids now predominates over the frictional resistance and the pressure drop rises. Near E, the bulk density of the mixture becomes too great for the gas velocity to support and the mixture begins to slug, or choke. The superficial gas velocity at E is termed the choking velocity U_{ch}, which is the minimum velocity which can be used to smoothly transport solids in vertical dilute phase conveying lines, and is obviously an important design parameter. The curve FG defines the pressure drop–velocity relationship for a higher mass flux, G_2. Feeding the conveying line at the solids rate G_2 will cause the conveying line to choke at a higher gas velocity. Thus, choking can be reached by decreasing the gas velocity at a constant solids flowrate or increasing the solids flowrate at a constant gas velocity.

Although operation near choking will result in the minimum gas requirement for pneumatic solids transfer, the choking region is a very unstable one. A small decrease in the gas velocity near choking causes the average pressure drop in the lift line to rise rapidly. This is accompanied by large fluctuations in the pressure drop as the line starts to choke. In large diameter conveying lines the slugging of the solids can also cause excessive vibration which can be structurally damaging if allowed to persist for any length of time.

A good operating point for vertical dilute phase conveying lines is at, or slightly to the right of, D. At this point, the $\Delta P/L$ versus velocity curve is relatively flat, and small perturbations in the system will not cause large changes in the conveying line pressure drop. Lift gas requirements at D are still low and yet far enough away from choking to be 'safe'. Perhaps most importantly, this is the point of minimum lift-line pressure drop.

If centrifugal fans are used to supply the gas for the riser, there is another reason why operation in the region D to E is unsatisfactory. Centrifugal fans are characterized by reduced capacity as outlet pressure is increased. Leung, Wiles, and Nicklin (1971a) have noted that if a slight system perturbation caused an increase in solids flowrate, the pressure drop in the lift line would increase. This would raise the pressure at the outlet of the fan and would result in a decrease in the output of the fan, producing a lower velocity in the lift line. This would shift the system operating point on the $\Delta P/L$ versus velocity curve towards E, producing a further increase in the system pressure drop. This cycle would continue until choking conditions were reached, resulting in unstable flow in the pipe. This destructive cycle does not occur if the fan operates at velocities to the right of D.

Zenz and Othmer (1960) are generally considered to be the first to describe choking. They considered it to be the degeneration of dilute gas–solids flow into dense, slugging flow. Matsen (1981) and Yang (1982) have recently suggested that choking occurs because of the formation of clusters or particle aggregates. Indeed, Grace and Tuot (1979), through instability analysis, showed that a uniform gas–solid suspension is unstable and that there is a tendency for clusters to form (see Chapter 7).

However, not every gas–solids mixture will choke. It is possible for solids to undergo a transition from a dilute phase directly to a denser non-choking, fluidized bed type of transport. Yousfi and Gau (1974) were the first to observe such a transition, and developed a criterion to precit when choking would occur; i.e.:

$$\frac{v_t^2}{gd_v} > 140 \quad \text{for choking to occur} \tag{12.1}$$

Yang (1976a) extended this concept further and developed the following criterion for choking:

$$\text{Fr} = \frac{v_t^2}{gD} \begin{array}{l} < 0.35 \quad \text{no choking} \\ > 0.35 \quad \text{choking occurs} \end{array} \tag{12.2}$$

This criterion (which is based on the maximum stable bubble size theory of Davidson and Harrison, 1963) predicts that light and small materials are less likely to choke, and takes into account the diameter of the line being employed.

A third criterion for choking was proposed by Smith (1978):

$$\frac{v_t \, \epsilon^{n^{-1}} \, n \, (1 - \epsilon)}{(gD)^{0.5}} > 0.41 \quad \text{for choking to occur} \tag{12.3}$$

Leung (1980) considered these three choking criteria and concluded that Yang's criterion was most useful since it contained the important parameter of pipe size and was most consistent with the experimental data.

The concept that not all materials will choke is also supported by the work of Canning and Thompson (1980). They found that large diameter particles invariably formed slugs in vertical (and horizontal) pneumatic conveying, while finer solids did not tend to slug.

The 'fast fluidized bed' (see Chapter 7) researched extensively by workers at the City College of New York (Yerushalmi, Turner, and Squires, 1976) appears to be a pneumatic conveying line operating in the relatively dense phase region near choking. This has been proposed by both Yang (1982) and Matsen (1981). Indeed, Gajdos and Bierl (1978), using probe measurements and X-ray exposures, reported that fast fluidization was phenomenologically no different than classical dilute phase conveying. Both regimes were found to consist of high velocity dilute core, surround by a low velocity dense annulus. This annular core flow in a riser has also been reported by Batholomew and Casagrande (1957), Van Zoonen (1961), and Saxton and Worley (1970).

12.2.1 Particle Velocity

The particle velocity in vertical dilute phase conveying lines is an important parameter because it determines the residence time of the solids in the line. The solids velocity in the conveying line differs from the gas velocity in the line by the slip velocity:

$$v_s = |v_g - v_{slip}| \tag{12.4}$$

The slip velocity is often approximated by the terminal velocity of the particles, and if this assumption is made, Eq. (12.4) becomes:

$$v_s = |v_g - v_t| \tag{12.5}$$

Measurements by Hinkle (1953) using high speed photography and coarse particles, showed that this relationship applied within ±20 per cent. Capes and Nakamura (1973) trapped a wide variety of solids in a 7.6 cm diameter riser to determine particle velocities and concluded: (a) that slip velocity was often greater than the terminal velocity and (b) that the deviation between v_{slip} and v_t was greater for particles with high terminal velocities. This was attributed to particle-to-wall friction and particle recirculation in the lift line. However, their data also indicate that for particles with terminal velocities below 7.6 m/s, Eq. (12.5) can be used as a reasonable approximation.

Matsen (1981) has proposed that the slip velocity is basically only a function of the voidage of the dilute, flowing suspension. He argues that as the voidage decreases (due to increased solids loading, for example) the slip velocity increases. He attributes this to cluster formation, but a solids annular flow model would also explain this voidage-slip velocity relationship.

12.2.2 Choking Velocity Correlations

In spite of the extensive literature on all aspects of pneumatic conveying, the most reliable way to determine design parameters for a particular pneumatic conveying system is still by experiment. This generally requires the construction and operation of a test rig, but this is not always possible because of insufficient time and/or funds. There are, however, numerous correlations for predicting conveying design parameters — so many, in fact, that a designer could easily feel like the proverbial mosquito in a nudist colony — he just would not know where to begin! Comparisons of many of these correlations have been published, and although somewhat limited, the comparisons provide a reference point for discussion.

The choking velocity defines the lower limit of the gas velocity for a dilute phase vertical pneumatic conveying system. A comparative study of several choking velocity correlations was carried out by the Institute of Gas Technology for the US Department of Energy (Institute of Gas Technology, 1978). This report has an extensive review of the available choking velocity correlations. The correlations of Zenz (1964), Rose and Duckworth (1969), Leung, Wiles, and Nicklin (1971a), Yousfi and Gau (1974), Knowlton and Bachovchin (1976), Yang (1975), and Punwani, Modi and Tarman (1976) were evaluated using both low pressure and high pressure data. The correlation of Punwani, Modi, and Tarman (1976) was recommended for use in predicting choking velocities. This correlation is basically the Yang (1975) correlation modified to take into account the considerable effect of gas density on choking velocity shown by Knowlton and Bachovchin (1976). The substantial dependence of U_{ch} on gas density is illustrated in Fig. 12.2. The Punwani correlation is shown below:

$$\frac{2gD\ (\epsilon_{ch}^{-4.7} - 1)}{(v_{ch} - v_t)^2} = 8.72 \times 10^{-3}\ \rho_g^{0.77} \tag{12.6}$$

$$\frac{G_s}{\rho_p} = (v_{ch} - v_t)\ (1 - \epsilon_{ch}) \tag{12.7}$$

where:

$$v_{ch} = \frac{U_{ch}}{\epsilon_{ch}} \tag{12.8}$$

To calculate the actual gas velocity at choking, Eqs (12.6) and (12.7) must be solved simultaneously for ϵ_{ch} and v_{ch}. This correlation predicted the data tested within 25 per cent.

Yang (1982) has recently modified his earlier choking velocity correlation to include the effect of gas density. The modified Yang choking velocity correlation is also shown below:

$$\frac{2gD\ (\epsilon_{ch}^{-4.7} - 1)}{(v_{ch} - v_t)^2} = 6.81 \times 10^5 \left(\frac{\rho_g}{\rho_p}\right)^{2.2} \tag{12.9}$$

$$\frac{G_s}{\rho_p} = (v_{ch} - v_t)\ (1 - \epsilon_{ch}) \tag{12.10}$$

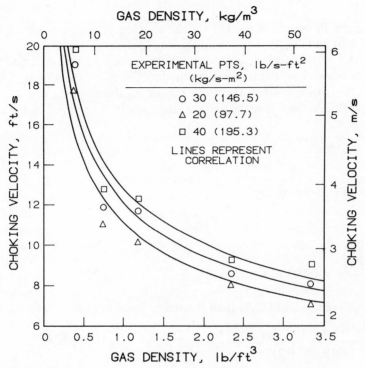

Figure 12.2 The effect of gas density on choking velocity (Knowlton and Bachovchin, 1976).

A worked example of how to calculate the choking velocity using the Punwani correlation is shown below.

Example 12.1: Choking velocity calculation — Punwani correlation

Calculate the choking velocity for coal particles with an average particle size of 300 μm being conveyed at a rate of 1,816 kg/h using nitrogen in a 75 mm diameter pipe at a system temperature of 20°C and a system pressure of 17 bar gauge.

ρ_g = 21.6 kg/m^3

ρ_p = 1,200 kg/m^3

μ = 1.84 $\times$ 10^{-5} N s/m^2

D = 75 mm

G = 114 kg/m^2 s

Solution

Solving the Punwani correlation for the choking velocity consists of solving Eqs. (12.6) and (12.7) simultaneously for v_{ch} and ϵ_{ch}.

In order to calculate the choking velocity a value must be obtained for the terminal velocity v_t, the method outlined in Chapter 6, Section 6.3, may be used; alternatively, since the flow regime is transitional, Eq. (12.11) is appropriate. Assume that the particle size given in the example is the volume diameter d_v.

$$v_t = \frac{0.153 g^{0.71} d_v^{1.14} (\rho_p - \rho_g)^{0.71}}{\rho_g^{0.29} \mu^{0.43}} \qquad (12.11)$$

For the conditions given:

$$v_t = 0.5 \text{ m/s}$$

Also:

$$v_s = \frac{G_s}{\rho_p} = 0.095 \text{ m/s}$$

Equations (12.6) and (12.7) then become:

$$\frac{2 \times 9.81 \times 0.076 \, (\epsilon_{ch}^{-4.7} - 1)}{(v_{ch} - 0.5)^2} = 8.72 \times 10^{-3} \times (21.6)^{0.77}$$

$$(v_{ch} - 0.5)(1 - \epsilon_{ch}) = 0.095$$

Combining the equations we have:

$$(\epsilon_{ch}^{-4.7} - 1)(1 - \epsilon_{ch})^2 = 5.69 \times 10^{-4}$$

Using the trial and error method:

$$\epsilon_{ch} = 0.9528$$

and

$$v_{ch} = \frac{0.095}{1 - \epsilon_{ch}} + 0.5$$

$$= 2.51 \text{ m/s}$$

$$U_{ch} = v_{ch} \, \epsilon_{ch} = 2.51 \times 0.9528 = 2.39 \text{ m/s}$$

12.2.3   *Vertical Dilute Phase Pressure Drop Correlation*

Many correlations are also available for use in determining the pressure drop in vertical pneumatic conveying lines. Twenty such correlations were found and presented in the *Coal Conversion Systems Technical Data Book* (Institute of Gas Technology, 1978). Several of these correlations (Hinkle, 1953; Currin and Gorin, 1968; Konno and Saito, 1969; Rose and Duckworth, 1969; Richards and Wiersma, 1973; Knowlton and Bachovchin, 1976; Leung, 1976; and Yang, 1976) were compared in the *Data Book* using the low pressure data of Curran and Gorin (1968) and the high pressure data of Knowlton and Bachovchin (1976). The correlation of Yang was found to predict the low pressure data better than the other correlations. However, overall, the modified Konno and Saito correlation was found to predict the data somewhat better than the Yang correlation and, although much less sophisticated than the Yang correlation, is much simpler to use. The modified Konno and Saito correlation is presented below:

$$\Delta P_T = \underset{(1)}{\frac{U^2 \rho_g}{2}} + \underset{(2)}{G_s v_s} + \underset{(3)}{\frac{2 f_g \rho_g U^2 L}{D}} + \underset{(4)}{\frac{0.057 U \rho_g \theta L g}{\sqrt{(gD)}}} + \underset{(5)}{\frac{G_s L g}{v_s}} + \underset{(6)}{\rho_g L g}$$

where: (12.12)

$$v_s = v_g - v_t \quad \text{and } \theta = \frac{G_s}{U \rho_g}$$

and where the numbered terms have the following significance:

(1) pressure drop due to gas acceleration,
(2) presure drop due to particle acceleration,
(3) pressure drop due to gas-to-pipe friction,
(4) pressure drop related to solid-to-pipe friction,
(5) pressure drop due to the static head of the solids,
(6) pressure drop due to the static head of the gas.

If the gas and the solids are already accelerated in the lift line, then the first two terms should be omitted from the calculation of the pressure drop.

A worked example of how to calculate the pressure drop in vertical dilute phase pneumatic conveying using the Konno and Saito correlation is given below.

Example 12.2: Pressure drop in vertical dilute phase conveying — Konno and Saito method

Determine the pressure drop in a dilute phase pneumatic conveying line in which coal of an average particle size of 300 μm is being conveyed at a velocity of 12 m/s with nitrogen in a 75 mm diameter conveying line at a rate of 1,861 kg/h. The temperature of the line is 21°C and is at a pressure of 17 bar guage. The line is 15 m long.

Solution

Assume the gas is already accelerated when the particles enter the line. Since the gas is already accelerated then, from Eq. (12.12):

$$\Delta P_T = G_s v_s + \frac{2f_g \rho_g U^2 L}{D} + \frac{0.057 U \rho_g \theta L g}{\sqrt{(gD)}} + \frac{G_s L g}{v_s} + \rho_g L g \qquad (12.13)$$

$$v_s = v_g - v_t \qquad (v_t \text{ is the same as calculated in Example 12.1})$$

$$= (12 - 0.5) \text{ m/s}$$

$$= 11.5 \text{ m/s}$$

$$\theta = \frac{G_s}{U \rho_g} = \frac{114}{12 \times 21.6} = 0.44 \text{ kg solid/kg gas}$$

$$\Delta P_T = 114 \times 11.5 + \frac{2f_g \times 21.6 \times (12)^2 \times 15}{0.075} + \frac{0.057 \times 12 \times 21.6 \times 0.44 \times 15 \times 9}{\sqrt{(9.81 \times 0.075)}}$$

$$+ \frac{114 \times 15 \times 9.81}{11.5} + 21.6 \times 15 \times 9.81$$

To calculate the Reynolds number to determine the friction factor we have:

$$\text{Re} = \frac{DU\rho_g}{\mu} = \frac{0.075 \times 12 \times 21.6}{1.84 \times 10^{-5}} = 1.06 \times 10^6$$

and

$$f_g = 0.004$$

Therefore:

$$\Delta P_T = 1,311 + 4,976 + 1,115 + 1,459 + 3,178 = 12,039 \text{ N/m}^2$$

Since $10,000 \text{ N/m}^2 = 1.45 \text{ lb/in}^2$:

$$\Delta P_T = 1.74 \text{ lb/in}^2$$

12.3 HORIZONTAL DILUTE PHASE CONVEYING

As in vertical dilute phase pneumatic conveying, the problem confronting the designer of a horizontal dilute phase transfer line is selecting a suitable gas velocity. The general relationship between superficial velocity and $\Delta P/L$ for a horizontal dilute phase transfer line is shown in Fig. 12.3. In many ways this relationship is similar to that for a vertical dilute phase conveying line. Line AB represents the curve obtained for gas alone travelling through the pipe,

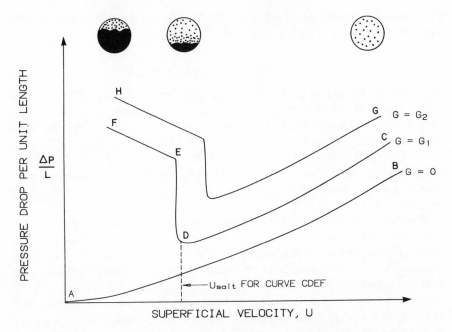

Figure 12.3 Phase diagram for dilute phase horizontal pneumatic conveying.

curve CDEF for a mass flux G_1, and curve GH for a mass flux G_2 greater than G_1. At C, the gas velocity in the horizontal line is sufficiently high to carry solids at a rate G_1 through the pipe in a very dilute suspension. If the gas velocity is reduced while continuing to feed at the constant mass flux G_1, the frictional resistance, and thus the pressure drop in the line, will decrease. The solids will move more slowly and their concentration in the pipe will increase. At D, the gas velocity is insufficient to keep the solids in suspension, and they begin to settle to the bottom of the pipe. The gas velocity at D is termed the saltation velocity, U_{salt}, and is a strong function of solids loading.

If the gas velocity is reduced further to E, the pressure drop will rise rapidly as the solids continue to deposit on the bottom of the pipe constricting the space available for gas flow. As the gas velocity is reduced between E and F, the depth of the particle layer and the pressure gradient both increase. In this regime, some particles may move slowly in dense phase flow along the bottom of the pipe, while, simultaneously, others travel in suspension in the gas in the upper part of the pipe.

The saltation velocity sets a lower limit on velocity for any horizontal dilute phase pneumatic transfer system. It is desirable, however, to operate at as low a velocity and pressure drop as possible and still remain far enough from saltation to be 'safe' if a system upset should occur. Therefore, the saltation velocity needs to be determined to set a lower bound on this operating point.

12.3.1   Saltation Velocity Correlation

There have been many efforts to correlate the saltation velocity. Two comparisons of many of the published saltation velocity correlations have been made (Jones and Leung, 1978; and Arastoopour et al., 1979). The Jones and Leung comparison was made using eight different correlations (Thomas, 1962; Doig and Roper, 1963; Zenz, 1964; Rose and Duckworth, 1969; Rizk, 1973; Matsumoto et al., 1974a, 1975b; Mewing, 1976). The comparisons were made using air-solids data only, and at low pressure. Jones and Leung recommended that the Thomas correlation be used for design.

The Zenz correlation predicts that as particle size is decreased, the saltation velocity at first decreases, then passes through a minimum, and then increases again. This latter effect is explained as being due to small particles becoming 'trapped' in the laminar sublayer and needing a higher velocity to dislodge the particle. However, for small particles (less than about 80 μm in size), the Zenz correlation at times predicts unbelievably high values of the saltation velocity. It is best applied to particles larger than this size.

A worked example of how to calculate the saltation velocity using the Zenz correlation is shown below.

Example 12.3:   Calculation of saltation velocity—method of Zenz

Calculate the saltation velocity of −10+100 US mesh sand being conveyed at a rate of 1,362 kg/h through a 150 mm diameter tube with nitrogen at 21°C and 0.34 bar gauge.

ρ_p = 2644 kg/m^3
ρ_g = 1.61 kg/m^3
μ = 1.84 × 10^{-5} N s/m^2
D = 150 mm
G_s = 21.4 kg/m^2 s
d_{pi} (10 mesh particle) = 2,000 μm
d_{pi} (100 mesh particle) = 150 μm

Solution

The Zenz correlation utilizes a graphical procedure in conjunction with various calculations to determine the saltation velocity U_{salt}. First a calculation of the single-particle saltation velocity, U_{so}, and a factor called S_Δ must be made. In Fig. 12.4, U_{so}/ω is plotted versus d_p/Δ where:

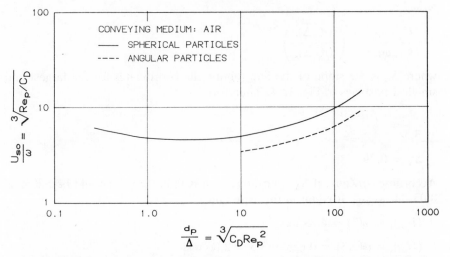

Figure 12.4 Single-particle saltation velocities as a function of particle size in a 63.5 mm diameter lucite tube (Zenz, 1964).

$$\omega = \left\{ \frac{4g\mu(\rho_p - \rho_g)}{3\rho_g^2} \right\}^{1/3} \quad \text{(the units of } \omega \text{ are in metres per second)}$$

$$(12.14)$$

and

$$\Delta = 12 \left\{ \frac{3\mu^2}{4g\rho_g(\rho_p-\rho_g)} \right\}^{1/3} \quad \text{(the units of } \Delta \text{ are in metres)} \quad (12.15)$$

Substituting values:

$\omega = 0.626$ m/s
$\Delta = 1.82 \times 10^{-5}$ m $= 18.2$ μm

For the largest particle:

$$\left(\frac{d_p}{\Delta} \right)_1 = \frac{2,000}{18.2} = 109.9$$

For the smallest particle:

$$\left(\frac{d_p}{\Delta} \right)_s = \frac{150}{18.2} = 8.24$$

Then from Fig. 12.4 for angular particles:

$$\left(\frac{U_{so}}{\omega} \right)_1 = 7 \quad \text{and} \quad \left(\frac{U_{so}}{\omega} \right)_s = 3.3$$

Now:

$$\frac{(U_{so}/\omega)_1}{(U_{so}/\omega)_s} = \left(\frac{(d_p/\Delta)_1}{(d_p/\Delta)_s}\right)^{S_\Delta}$$

where S_Δ is the slope of the line joining the coordinates for the largest and smallest particles in Fig. 12.4: Therefore:

$$\frac{7}{3.3} = \left(\frac{109.9}{8.24}\right)^{S_\Delta}$$

$$S_\Delta \simeq 0.29$$

According to Zenz, if S_Δ is negative or less than 0.05, it should be taken as 0.05. However, it is not in this case and;

$$(U_{so})_1 = \omega(7) = 4.38 \text{ m/s}$$

$$(U_{so})_s = \omega(3.3) = 0.626 \times 3.3 = 2.06 \text{ m/s}$$

Notice that $(U_{so})_1$ is larger than $(U_{so})_s$. However, $(U_{so})_s$ can sometimes be larger than $(U_{so})_1$. This can occur because the small particles can be trapped in the laminar boundary layer. Always use the largest U_{so} to calculate U_{salt}, however.

Figure 12.4 was determined from experiments in a 63.5 mm diameter tube. The single-particle saltation velocities will now have to be corrected for the larger sized 150$_{mm}$ diameter tube. U_{so} has been shown to be proportional to $D^{0.4}$. Therefore:

$$\frac{(U_{so})_{150}}{(U_{so})_{63.5}} = \left(\frac{150}{63.5}\right)^{0.4}$$

$$[(U_{so})_s]_{150} = 4.38 \times (2.36)^{0.4} = 6.17 \text{ m/s}$$

The saltation velocity at loaded conditions is related to the single-particle saltation velocity by:

$$\frac{G_s}{\rho_p} = 0.213 \, (S_\Delta)^{1.5} \, \frac{(U_{salt} - U_{so})}{U_{so}} \tag{12.16}$$

$$\frac{21.4}{2,644} = 0.213 \times (0.29)^{1.5} \frac{(U_{salt} - 6.17)}{6.17}$$

$$U_{salt} = 7.67 \text{ m/s}$$

12.3.2 Horizontal Dilute Phase Pressure Drop Correlations

Several pressure drop correlations developed for horizontal dilute phase pneumatic conveying were also compared in the *Coal Conversion Systems*

Technical Data Book (Institute of Gas Technology, 1978) (Culgan, 1952; Hinkle, 1953; Engineering Eqipment User's Association, 1963; Curran and Gorin, 1968; Rose and Duckworth, 1969; and Yang, 1976a). The correlations of Yang and Hinkle were found to predict the data far better than the other correlations. The accuracy of both correlations was about the same for the data tested. The Hinkle correlation, however, is much simpler than the Yang correlation and is presented below:

$$\Delta P_T = \frac{U^2 \rho_g}{2} + G_s v_s + \frac{2 f_g \rho_g U^2 L}{D} \left[1 + \frac{f_s v_s}{f_g U} \frac{G_s}{U \rho_g} \right] \tag{12.17}$$

where:

$$v_s = U (1 - 0.0638 d^{0.3} \rho_p^{0.5}) \tag{12.18}$$

$$f_s = \frac{3}{8} \frac{\rho_g}{\rho_p} C_D \frac{D}{d_p} \left(\frac{U - v_s}{v_s} \right)^2 \tag{12.19}$$

An illustration of how to calculate horizontal dilute phase pressure drop using the Hinkle correlation is given in Example 12.4

Example 12.4: Pressure drop in horizontal dilute phase conveying — Hinkle correlation

Calculate the pressure drop in a horizontal dilute phase pneumatic conveying line in which coal with an average particle size of 300 μm and density 1,200 kg-m^3 is being conveyed at a velocity of 18 m/s with nitrogen in a 75 mm diameter conveying line at a rate of 1,816 kg/h. The temperature of the line is 21°C and the line is at a pressure of 1.36 bar gauge. The line is 24 m long. Assume the gas is already accelerated when the solids enter the line.

Since the gas is already accelerated when the solids are added to the line:

$$\Delta P_T = \frac{U^2 \rho_g^0}{2} + G_s v_s + \frac{2 f_g \rho_g U^2 L}{D} \left[1 + \frac{f_s}{f_g} \frac{v_s}{U} \frac{G_s}{U \rho_g} \right] \tag{12.20}$$

$\rho_g = 2.83$ kg/m^3

$\mu = 1.84 \times 10^{-5}$ N s/m^2

$v_s = U (1 - 0.0638\ d_p^{0.3}\ \rho_p^{0.5})$

$\quad = 18 \{1 - 0.0638\ (300 \times 10^{-6})^{0.3}\ (1,200)^{0.5}\}$

$\quad = 14.5$ m/s

$G_s = \dfrac{1.816}{3,600} \dfrac{4}{\pi (0.075)^2} = 114$ kg/m^2s

Solution

At high Reynolds numbers, C_D in Eq. (12.19) is a constant at 0.43 and:

$$f_s = \frac{3 \times 2.83 \times 0.43 \times 0.075}{8 \times 1{,}200 \times 300 \times 10^{-6}} \left(\frac{18 - 14.5}{14.5}\right)^2$$

$$f_s = 0.0056$$

From Eq. (12.20):

$$\Delta P = 114 \times 14.5 \; + \frac{2f_g(2.83 \times 18^2 \times 24)}{0.075} \left(1 + \frac{0.0056}{f_g} \frac{14.5}{18} \frac{114}{18 \times 2.83}\right)$$

$$\mathrm{Re} \; = \frac{DU\rho_g}{\mu} = \frac{0.075 \times 18 \times 2.83}{1.84 \times 10^{-5}} = 2.08 \times 10^5$$

and

$$f_g \quad = 0.004$$

$$\frac{f_s v_s}{f_g U} = \frac{0.0056 \times 14.5}{0.004 \times 18} = 1.13$$

Zenz (1964) recommends that if $f_s v_s/f_g U > 1$ then assume $f_s v_s/U f_g = 1$ in the calculation of the pressure drop; i.e.:

$$\Delta P = 1{,}653 + 4{,}694 = 6{,}347 \text{ N/m}^2 = 0.92 \text{ lb/in}^2$$

The various correlations described above were found to be the 'best' correlations for predicting U_{salt}, U_{ch}, the horizontal dilute phase pressure drop, and the vertical dilute phase pressure drop, based on the data used. However, for the following reasons, care must be taken in applying these correlations and automatically assuming that they will accurately predict the design parameters for a particular system:

(a) The correlations were evaluated for a very limited amount of data.

(b) There is disagreement between studies as to which correlation is 'best'. Note that the Jones and Leung (1978) analysis resulted in a different 'best' correlation for U_{salt} than did the Arastoopour *et al.* (1979) study.

(c) The 'best' correlations can generally be expected to predict parameters no more accurately than 30 to 40 per cent.

(d) If the correlations are to be applied to high temperature and/or high pressure systems, their accuracy may be much worse than 30 to 40 per cent. since nearly all of the correlations were developed from data obtained at ambient conditions.

Therefore, when applying these correlations, bear in mind that the state of the art (as applied to pneumatic conveying correlations) still leaves much to be desired.

12.4 BENDS

Bends complicate the design of pneumatic dilute phase transfer systems, and when designing a conveying system it is best to use as few bends as possible. Bends increase the pressure drop in a line, and also are the points of most serious erosion and particle attrition.

Solids normally in suspension in straight, horizontal, or vertical pipes tend to salt out at bends due to the centrifugal force encountered while travelling around the bend. Because of this separation, the particles slow down and are then reentrained and reaccelerated after they pass through the bend, resulting in the higher pressure drops associated with bends.

There is a greater tendency for particles to salt out in a horizontal pipe which is preceded by a downflowing vertical-to horizontal bend than in any other configuration (Patterson, 1959). If this type of bend is present in a system, it is possible for solids to remain on the bottom of the pipe for very long distances following the bend, before they redisperse (Fig. 12.5b). The particles in a horizontal pipe following an upflowing vertical-to-horizontal bend are not as likely to salt out since they can easily redisperse into the gas stream due to gravity (Fig. 12.5a). Therefore, it is recommended that

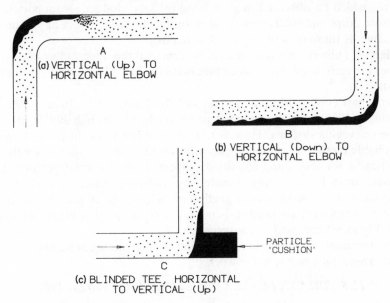

Figure 12.5 Solids flow profiles for three bend configurations.

downflowing vertical-to-horizontal bends be avoided if at all possible in dilute phase pneumatic conveying systems.

As noted above, bends increase the pressure drop in a conveying system. In addition to the pressure drop due to the bend itself, because the bend slows down the solids there is also a pressure drop due to subsequent particle reacceleration by the gas. The length of straight pipe needed before the solids reattain their steady state velocity can be considerable (Scott, Richard, and Mooij, 1976).

In the past, designers of dilute phase pneumatic conveying systems intuitively thought that gradually sloped, long radius elbows would reduce the erosion and increase bend service life relative to 90° elbows. Zehz (1960), however, recommended that blinded tees (Fig. 12.5c) be used in place of elbows in pneumatic conveying systems. The theory behind the use of the blinded tee is that a 'cushion' of stagnant particles collects in the blinded or unused branch of the tee, and the conveyed particles then impinge upon the stagnant particles in the tee rather than on the metal surface, as in a long radius or short radius elbow.

Bodner (1982) determined the service life and pressure drop of various bend configurations. He found that the service life of the blinded tee configuration was far better than any other configuration tested and that it gave a service life 15 times greater than that of radius bends or elbows. This was due to the 'cushioning accumulation of particles' in the blinded branch of the tee which he observed in glass bend models. Bodner also reported that pressure drops and solid attrition rates for the blinded tee were approximately the same as those observed for radius bends.

Marcus, Hilbert, and Klinzing (1984) reported that long radius bends have higher pressure losses than short radius bends due to the extra length of the bend.

Schuchart (1968), Morikawa *et al.* (1978), Tsuji (1980), Jung (1958), and Weidner (1955) also investigated pressure drops through bends in dilute phase conveying systems. However, in spite of this work, at present there is no reliable method of predicting accurate bend pressure drops other than by experiment for the actual conditions expected. In industrial practice, bend pressure drop is often approximated by assuming that it is equivalent to approximately 7.5 m of vertical section pressure drop. In the absence of any reliable correlation to predict bend pressure drop, this crude method is probably as reliable and as conservative as any.

The formulation of an accurate method of determining bend pressure drop is one area which needs *much* work.

12.5 DILUTE PHASE CONVEYING IN INCLINED PIPES

The only data available on the dilute phase flow of solids through inclined pipes are due to Zenz (1960). At very high velocities where the solids were well dispersed, the inclined pipe pressure drop was found to be less than the vertical dilute phase pressure drop. At lower gas velocities the solids formed nodes on the bottom of the pipe. A further reduction in the conveying velocity resulted in slugging or choking in the pipe. This choking condition occurred at a higher velocity than observed in vertical pipes. No useful correlations exist to predict choking velocities and pressure drops in inclined pipes.

12.6   COMBINED SYSTEMS

Many times a dilute phase pneumatic conveying system will consist of a combination of horizontal and vertical transfer lines with various bends in the line. This system is more complex than the simple vertical or horizontal lines considered earlier. If no slanted lines or downward flowing vertical-to-horizontal bends are used in the system, the lower velocity limit for the system will be the saltation velocity, since this velocity is always greater than or equal to the choking velocity for a particular gas–solids mixture (Zenz, 1960). The determination of the final design of the transfer system depends upon the pressure drop balance in the total circulation loop.

12.7   EFFECT OF PIPE DIAMETER

It is important to know the effect of pipe diameter on dilute phase conveying design parameters (that is U_{ch}, U_{salt}, and ΔP) when scaling up in size. Surprisingly, there is very little information available in the literature on the effect of pipe diameter on these parameters, even though most test rigs operate in the 5 to 7.5 cm diameter range and commercial conveying lines are in the 30 to 122 cm diameter range.

Analysis of the data of Patterson (1959) on his work with pulverized coal shows that the pressure drop in his dilute phase conveying line was greater for 30.5 cm diameter pipe than for 20.3 cm diameter pipe.

Woebcke and Cofield (1978) found that the dilute phase pressure drop in horizontal conveying increased with increasing pipe diameter. They also reported that the 'minimum velocity' (saltation velocity) in their tests increased with increasing pipe diameter over the range of 3.2 to 15.2 cm diameter pipes.

The Zenz (12964) correlation for saltation velocity also predicts an increase in U_{salt} with pipe diameter to the 0.4 power, i.e.:

$$U_{salt} \propto D^{0.4}$$

although this dependency was based on work with rather small tubes, 3.2 to 6.4 cm in diameter.

The Yang (1976a), Punwani, Modi, and Tarman (1976), and modified Yang (1982) correlations for the choking velocity also predict an increase in U_{ch} with pipe diameter. This dependency is approximately pipe diameter to the 0.25 power, i.e.:

$$U_{ch} \sim D^{0.25}$$

In summary, theory and sketchy literature results both predict an increase in U_{ch} and U_{salt} with pipe diameter. This increase is generally proportional to the 0.25 to 0.4 power of the pipe diameter. Even sketchier literature results indicate that dilute phase pneumatic conveying pressure drop increases with pipe diameter. More systematic studies are definitely needed in this area.

12.8   DENSE PHASE PNEUMATIC CONVEYING

Dilute phase conveying is by far the most common type of conveying. It can also be the most economical. Sandy, Daubert, and Jones (1970) calculated that the minimum power requirements for dilute phase vertical pneumatic conveying were approximately an order of magnitude less than the power requirements for dense phase vertical pneumatic conveying of an air-fused alumina system. The large power requirements needed for vertical dense phase transport were a result of the high pressure drops needed to move solids upwards in dense phase flow.

However, it is not always the case that dilute phase pneumatic conveying will be less costly than dense phase pneumatic conveying as it depends strongly on the particular gas–solids system being used. In general, the trade-off which must be made is between the higher pressure drop per unit length ($\Delta P/L$) associated with dense phase pneumatic conveying versus the higher gas flows needed for dilute phase pneumatic transport.

In spite of the high pressure drop per unit length needed, dense phase conveying offers certain advantages over dilute phase conveying which may tip the scales in its favour for a particular process. Since the solids velocities are much lower in dense phase conveying than in dilute phase conveying, solids attrition and pipe erosion are considerably reduced. This feature is particularly desirable when expensive or fragile solids are to be transported, and it is desired to minimize particle breakup and elutriation losses. Dense phase conveying also decreases the demand on a particle collection system.

Gas requirements for dense phase pneumatic conveying (as noted above) are much lower than for dilute phase pneumatic conveying. Sandy, Daubert, and Jones (1970) found that vertical dilute phase gas requirements for their air/fused alumina system were over 20 times those for vertical dense phase conveying. In instances where the conveying gas is extremely expensive, or where it is desired to minimize the amount of diluting transfer gas fed to a reactor, dense phase transport is an alternative choice.

In this chapter, dense phase conveying will denote all solids conveying in which the solids are not completely suspended in the gas stream. Solids can be conveyed in dense phase both vertically and horizontally. In vertical dense phase conveying the entire line is filled with solids in either a packed bed or fluidized bed slugging mode. In horizontal dense phase conveying the line may or may not be completely filled with solids.

12.8.1 Packed Bed Vertical Dense Phase Conveying

In vertical dense phase packed bed conveying, the solids are transported through a tube by forcing transporting gas upwards through the interstices of the particles. When the frictional forces between the gas and the particles are sufficient to overcome the weight of the particles plus the frictional forces of the solids sliding on the tube walls, the solids will start to move upwards. This occurs at a pressure drop per unit length, $\Delta P/L$, approximately equal to that at minimum fluidization for that particular material. This $\Delta P/L$ value is also approximately equal to the bulk density of the solids.

In order to prevent fluidization of the particles in the pipe or tube the outlet of the conduit must be restricted in some manner. This can be accomplished by placing a valve, orifice, or an expanded section of a non-fluidized bed of solids at the outlet of the pipe. This effectively prevents the solids from expanding and becoming fluidized.

Because of the restriction at the outlet of the pipe, the relative velocity between gas and solids may now be greater than the minimum fluidization velocity, and the solids will move upwards as a packed bed. This operating mode is shown as lines GH and IJ in the phase diagram of Fig. 12.6, where the mass flux G_1 is less than G_2. The familiar fluidization and dilute phase vertical conveying $\Delta P/L$ versus velocity curves (ODE and ABC, respectively) are also shown in this figure for reference. Curve OD is the packed bed $\Delta P/L$ versus velocity curve. At D the $\Delta P/L$ is sufficient to cause the bed to expand and normally become fluidized. However, if the bed is restricted (e.g. by a screen small enough to let only gas pass through it) the $\Delta P/L$ continues to increase with velocity at zero solids flow. Thus, ODF is the $\Delta P/L$ versus velocity curve for $G = 0$ in vertical dense phase transport.

Berg (1954a, 1954b, 1954c, 1954d) received several patents relating to the vertical transport of solids in a dense phase packed bed regime in order to reduce particle attrition and loss of expensive catalyst material. Because of bubble formation at high solids flowrates and excessive pressure drops in the transport line, both due to gas expansion, Berg had to taper or increase the diameter of the transport line every few feet. Alternatively, he also found that removing some of the gas would also prevent excessive pressure drops in the

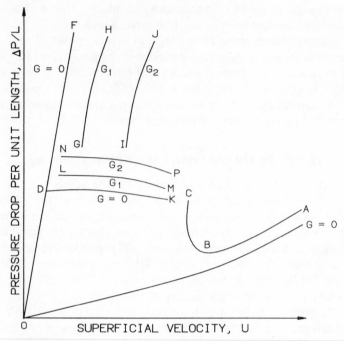

Figure 12.6 .Phase diagram for dense phase vertical pneumatic conveying.

line. Sandy, Daubert, and Jones (1970) were also forced to use increasingly large flow areas to keep $\Delta P/L$ from becoming excessively high in their dense phase vertical transport line.

Extreme care must be exercised in placing an expanded section of pipe at the exit of the line in dense phase vertical transport (Knowlton and Hirsan, 1980) so that its area is not too large. A large diameter section requires a gas velocity large enough to produce a $\Delta P/L$ equal to the bulk density of the material before solids will flow. This amount of gas also has to pass through the smaller diameter section of pipe preceding the expanded section. If the ratio of the large diameter pipe to the small diameter pipe is too large, the $\Delta P/L$ in the smaller diameter section can become exorbitantly high.

Several correlations have been developed to predict the pressure drop in dense phase vertical packed bed transport (Berg, 1954a, 1954b, 1954c, 1954d; Sandy, Daubert, and Jones, 1970; Leung, Wiles and Nicklin, 1971b; and Wen and Galli, 1971). The Berg correlation is relatively complex. The Leung correlation is the simplest and is merely the Ergun equation applied to a moving packed bed. The Sandy and the Wen and Galli correlations also employ the Ergun equation to predict the pressure drop due to gas–solids friction, but also include other terms, primarily the pressure drop needed for the gas to support the solids. The Leung, Sandy *et al.* and Wen and Galli correlations are extremely sensitive to voidage and particle size. Their accuracy is strongly dependent upon the correct prediction of these properties. The Wen and Galli correlation is shown below:

$$\frac{\Delta P_T}{L} = \underbrace{\frac{150(1 - \epsilon)^2 \, \mu(U/\epsilon - v_s)}{(\psi d_v)^2 \epsilon^2}}_{(1)} + \underbrace{\frac{1.75(1-\epsilon)\rho_g(U/\epsilon - v_s)^2}{\psi d_v \epsilon}}_{(2)}$$

$$+ \underbrace{(1-\epsilon)(\rho_p - \rho_g)}_{(3)} + \underbrace{\frac{2\rho_g U^2 f_g}{D}}_{(4)} \qquad (12.21)$$

where:

(1) and (2) are the pressure drop due to gas–solids friction,
(3) is the pressure drop for the gas to support the solids,
(4) is the gas–wall friction loss.

12.8.2 Fluidized Bed Dense Phase Vertical Conveying

If the solids in a dense phase vertical transport line are not restricted as in dense phase packed bed flow, they can be transported in dense phase in fluidized bed flow. However, because of gas expansion, this flow generally reverts to bubbling or slug flow unless the transport line is tapered.

This type of flow is characterized by lines LM and NP on the phase diagram of Fig. 12.6, where NP represents a solids mass flux, G_2, greater than G_1. Although these curves are informative, they are not as useful as they are in dilute phase conveying. In dilute phase conveying, the gas velocity can be reduced while maintaining a constant mass flux. The solids are simply transferred at the same flow rate, but in a denser state. This is usually not the case in dense phase vertical transport with fluidized (or packed bed) flow. This is because the transport gas rate also generally determines the solids flow rate. When the superficial transport gas velocity is changed, the solids flowrate also changes. Thus, a knowledge of how the solids flowrate varies with gas velocity is also needed for the design of dense phase pneumatic conveying systems. A typical solids flowrate versus velocity curve for an unrestricted vertical dense phase transfer line operating in slug flow is shown in Fig. 12.7, along with a curve showing the variation in pressure drop in the transport line as a function of solids flowrate (Fig. 12.8). The curves were obtained in conjunction with a non-mechanical J-valve study by Knowlton and Hirsan (1980).

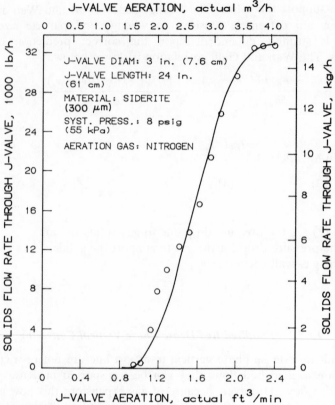

Figure 12.7   Dense phase vertical fluidized solids flowrates as a function of motive gas (Knowlton and Hirsan, 1980).

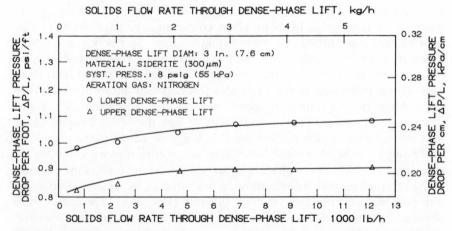

Figure 12.8   Pressure drop in dense phase vertical fluidized flow versus solids flowrate (Knowlton and Hirsan, 1980).

12.8.3 *Dense Phase Horizontal Conveying*

In horizontal dense phase conveying, it is not necessary that the solids completely fill the pipe cross-section. In fact, it is usually the case that they will not. At any velocity less than that necessary to keep the solids in suspension (i.e. less than the saltation velocity) the solids are considered to be in dense phase flow.

The saltation velocity is the velocity at D in Fig. 12.3 for a mass flux, G_1. The portion of curve CDEF described by DEF in Fig. 12.3 is considered to be in dense phase flow. As the velocity is reduced from D to E, a particle layer appears on the bottom of the pipe. From E to F the depth of the particle layer increases and the pressure drop in the pipe builds up.

Wen and Simons (1959) studied horizontal dense phase gas solids flow in 0.5, 0.75. and 1 in diameter glass pipe over a wide range of solids/gas loadings (80 to 800 kg of solids per kilogram of gas). They visually observed that the solids (glass beads and coal ranging from approximately 70 to 750 μm) went through four separate flow regimes, depending on the solids/gas loading or the gas velocity in the pipe.

At high gas velocities the solids were observed to be in dilute phase suspension (Fig. 12.9a). At lower gas velocities, the solids settled in the pipe

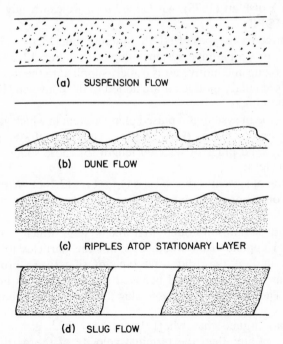

(a) SUSPENSION FLOW

(b) DUNE FLOW

(c) RIPPLES ATOP STATIONARY LAYER

(d) SLUG FLOW

Figure 12.9 Solids flow regimes in dense phase horizontal pneumatic conveying.

and dune formation occurred (Fig. 12.9b). Decreasing the gas velocity even more (or, equivalently, increasing the solids/gas loading) caused more solids to settle to the bottom of the pipe and the solids were observed to travel in small dunes or ripples across the top of a thick, relatively stationary, salted layer (Fig. 12.9c). A further increase in solids loading eventually caused a complete blockage of the pipe.

In some cases, Wen and Simons observed that intermittent slug flow of solids (Fig. 12.9d) occurred instead of dune formation (Fig. 12.9b). This was accompanied by unstable, wildly fluctuating pressure drops across the line.

Because of the uncertainties associated with horizontal dense phase conveying (HDPC) and the high probability of plugging the pipe, most designers have shied away from horizontal dense phase conveying. They have opted instead to use the dilute phase conveying for its known reliability. In fact, HDPC has long been considered somewhat of a 'black art' by many people. This attitude has grown because there has been no systematic procedure with which to predict what system to use with a particular material. Some materials could be conveyed relatively easily in the HDPC mode, while others would plug the line and the system would fail.

A paper by Canning and Thompson (1980) appears to offer the first systematic procedure which can be used to determine what horizontal dense phase system to use with a particular type of material. This work, along with the earlier work of Flatt (1976), has helped to significantly advance the state of the art of HDPC.

There are three basic types of HDPC systems (Fig. 12.10). The simplest is the blow tank, or simple-pipe system., depicted in Fig. 12.10(a). With this system, pressurizing and motive air is usually supplied to the top of a tank and the solids and secondary motive air are added to the conveying line through a valve below the tank.

The second type of system is a pulsed phase system in which gas is added to the conveying line in pulses (Fig. 12.10b). This type of system is used to artificially form solid plugs of selected length for those solids which do not naturally form slugs.

The third type of system is the by-pass type (Fig. 12.10c). With this system, some of the conveying air is by passed around the solids so that the plug length can be limited and plugging will not occur.

Canning and Thompson found that the system which was best for a particular solid depended on whether or not the particles to be conveyed formed slugs, and, if they did what type of slugs were produced. They also found that a particular solid behaved similarly in both horizontal and vertical flow. They therefore analysed slug formation using choking–velocity theory.

Yang (1976a) argued that when the velocity of a gas slug in vertical conveying was greater than the terminal velocity of the particles, then no

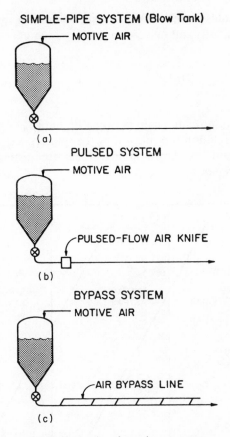

Figure 12.10 A schematic representation of three primary dense phase feed systems.

stable slugs were possible. In this case the particles could be fast fluidized without slugging or choking. Therefore, for unstable or asymmetric slugs, the terminal velocity must be less than the slug rise velocity, or:

$$v_t < 0.35(gD)^{0.5} \qquad (12.22)$$

Similarly, for stable slugs or choking to exist:

$$v_t > 0.35(gD)^{0.5} \qquad (12.23)$$

For fine particles, the terminal velocity can be written:

$$v_t = \frac{(\rho_p - \rho_g)d_v^2 g}{18\mu} \qquad \text{for } Re_t < 0.2 \qquad (12.24)$$

For higher Reynolds numbers other expressions for v_t should be used. Combining Eqs (12.22) and (12.24) we have:

$$\frac{(\rho_p - \rho_g)\, g d_v^2}{18\mu} < 0.35(gD)^{0.5} \tag{12.25}$$

Thus, we now have a relationship between d_v and $(\rho_p - \rho_g)$ for various pipe diameters. This relationship may now be plotted on a Geldart (1973) fluidization diagram, as shown in Fig. 12.11, to distinguish between the various slugging regions.

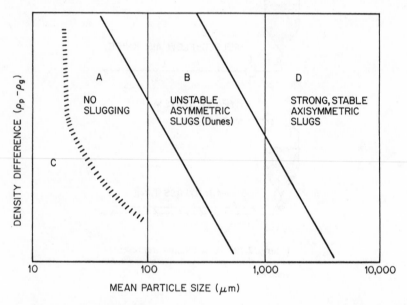

Figure 12.11 The Geldart fluidization diagram applied to dense phase pneumatic conveying (Canning and Thompson, 1980).

The overriding concern in HDPC is to avoid blocking the pipe. Blockages are most likely to occur with materials that form weak, asymmetric, unstable slugs. These materials are found in the group B region of the Geldart fluidization diagram. With these materials, the gas slugs move faster than the interstitial gas and the material conveys in dune flow. When the dunes grow to fill the pipe, severe vibration and eventual blockage of the pipe may result. In the group D region of the diagram, the gas slugs move more slowly than the interstitial gas and stable slugs can form which completely fill the pipe diameter.

Canning and Thompson reported that with the aid of the fluidization diagram it was possible to choose a suitable HDPC system for a particular powder based only on the fluidization classifiaction of the powder. The system best suited for a particular powder type is described below.

Group C Solids

These are materials which are generally very difficult to convey in HDPC because of their cohesiveness.

Group A Solids

These materials are the best candidates for HDPC. They retain air and remain fluidized, and can be generally conveyed at low velocities and at high loadings with little risk of pipe blockages. All three types of systems shown in Fig. 12.10 can be used with these materials. For long (greater than 200 m distances) it may be advisable to use a by-pass system.

Group B Solids

These materals are the worst type to use in HDPC. At high loadings they tend to cause pipe vibrations and blockage. The by-pass system can be used to convey these materials at higher loadings and at lower velocities than a dilute phase system.

Group D Solids

These materials can be conveyed easily if they have a narrow particle size distribution. All three system types may be used to convey these particles.

12.8.4   Extrusion Flow

Zenz and Rowe (1976) studied a group of materials which can be made to flow through pipes in extrusion-type flow. In this type of flow, the solids occupy the entire pipe cross-section and flow much like plastic being extruded through a round die. Zenz and Rowe found that only certain solids would flow in this manner. These solids were generally of low bulk density and very small particle size. The solids also had to exhibit a property called 'bulk deformability', characterized in this study by a test wherein a steel rod was placed on top of the solids. If the rod fell through the solids without aerating them, the solids could be conveyed in extrusion flow.

12.8.5   Air-Assisted Gravity Conveyor

Another type of system which can transform solids in HDPC is the air-assisted gravity conveyor. With air-assisted gravity conveying, the solids are fluidized with air distributed along the length of the conveying pipe or channel (Woodcock and Mason, 1976). The solids then flow by gravity along the bottom of the channel. This type of conveying uses a very small amount of energy. Although solids can be made to move against gravity with this type of conveying, it is generally used only with 'downhill' conveying.

12.9   STANDPIPES

The standpipe has been in use for over 40 years. It was 'invented' by a research team from Jersey Standard in the 1940s (Campbell, Martin, and Tyson, 1948) working on feeding cracking catalyst into fluidized beds during the Second World War.

The purpose of a standpipe is to transfer solids from a region of low pressure to a region of higher pressure. This is indicated in Fig. 12.12 where solids are being transferred from a fluidized bed at pressure P_1 to another fluidized bed operating at pressure P_s, which is higher than P_1.

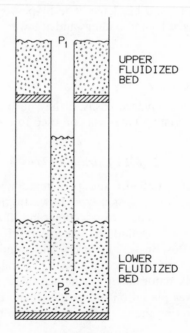

Figure 12.12   Standpipe.

Solids can be transferred against a pressure if gas flows upwards *relative* to the downward flowing solids, thus generating the required 'sealing' pressure drop. The direction of the actual gas flow relative to the standpipe wall, however, can be either up or down and still have the relative gas–solids velocity v_r directed upwards.

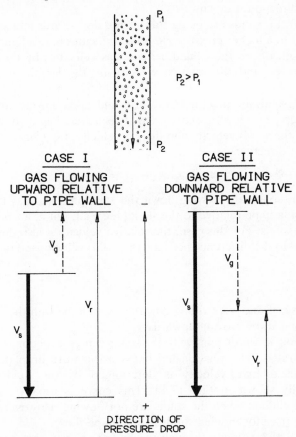

Figure 12.13 Schematic representation of gas flow down and gas flow up relative to the standpipe wall.

This can be shown with the aid of Fig. 12.13 and the definition of the relative velocity. Mathematically:

$$v_r = |v_s - v_g| \qquad (12.26)$$

where:

v_r = relative gas solids velocity (also called v_{slip})

v_s = solid velocity

v_g = interstitial gas velocity (U/ϵ)

In solids transfer systems, most people tend to visualize what is occurring in a particular pipe by mentally travelling along with the solids. Therefore, in this chapter the *positive* reference direction in determining v_r will always be the *direction the solids are flowing*. For vertical pneumatic conveying this positive reference direction is generally upwards. For standpipe flow, the positive reference direction is downwards.

In Fig. 12.13, solids are being transferred downwards in a standpipe from pressure P_1 to a higher pressure P_2. Solid velocities in this figure are denoted by the length of the thick-lined arrows, gas velocities by the length of the dashed arrows, and the relative velocity by the length of the thin-lined arrows.

In case I, solids are flowing downwards and gas is flowing upwards relative to the standpipe wall. The relative velocity is directed upwards and is equal to the sum of the solids velocity and the gas velocity, i.e. (for downwards being the positive direction):

$$v_r = v_s - (-v_g) = v_s + v_g \qquad (12.27)$$

For case II, solids are flowing down the standpipe and gas is also flowing down the standpipe relative to the standpipe wall, but at a velocity less than that of the solids. For this case, the relative velocity is also directed upward and is equal to the difference between the solids velocity and the gas velocity, i.e.:

$$v_r = v_s - v_g \qquad (12.28)$$

For this case, if one were riding down the standpipe with the solids, the gas would appear to be moving upwards.

Gas flowing upwards relative to the solids causes a frictional pressure drop to be generated. The relationship between pressure drop per unit length $(\Delta P/L)$ and relative velocity is determined by the fluidization curve, schematically shown in Fig. 12.14. This curve is usually generated in a fluidization column when the solids are not flowing. However, the relationship also applies for moving solids in standpipe flow.

As the relative velocity through the bed of solids is increased from zero, $\Delta P/L$ increases linearly with v_r. This is the packed bed region. At some v_r, the ΔP generated is equal to the weight of the solids per unit area and the particles become fluidized. The relative velocity at this point is called the interstitial minimum fluidized velocity, v_{mf}, or U_{mf}/ϵ_{mf}. The $\Delta P/L$ at v_{mf} is designated as $\Delta P/L)_{mf}$ and is often referred to as the fluidized bed 'density' since $\Delta P/L$ has the units of density.

Increases in v_r above v_{mf} do not lead to further increases in $\Delta P/L$. For Geldart group B materials, any increase in gas flow in excess of that required at v_{mf} goes into the formation of bubbles. Therefore, as v_r increase, $\Delta P/L$ stays relatively constant and then begins to drop as the bubble volume in the

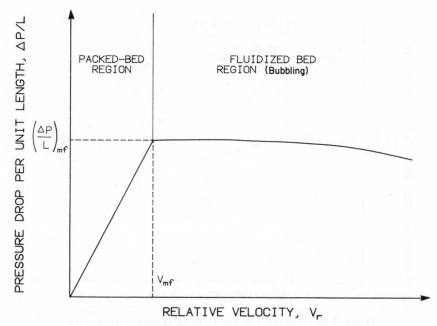

Figure 12.14   Typical fluidization curve for Geldart group B solids.

bed increases. The $\Delta P/L$ versus v_r curve shown in Fig. 12.14 is for group B solids.

The fluidization curve for Geldart froup A materials is slightly different from that for group B solids. As v_r is increased above v_{mf} for these materials, the solids expand without bubbles forming for a certain velocity range, and $\Delta P/L$ drops over this velocity range. The velocity where bubbles begin to form is called the minimum bubbling velocity, v_{mb}, to differentiate it from v_{mf}. A typical fluidization curve for group A solids is shown in Fig. 12.15.

Standpipes generally operate in one of three basic flow regimes — packed bed flow and fluidized bed flow, and streaming flow:

1. *Packed bed flow*. In packed bed flow, v_r is less than v_{mf} i.e. (U_{mf}/ϵ_{mf}) and the voidage in the standpipe is essentially constant. As v_r is increased, $\Delta P/L$ increases in packed bed flow. Often when a standpipe is operating in the moving packed bed flow regime a flow condition is reached which causes the standpipe to vibrate and a loud 'chattering' of the solids occurs. This flow mode is not present at low solid velocities and also decreases in severity as the relative gas/velocity approaches v_{mf}. This flow regime should be avoided but, unfortunately, no method exists to predict when it will occur. Little work has been conducted on this problem, and this is also an area which needs much work.

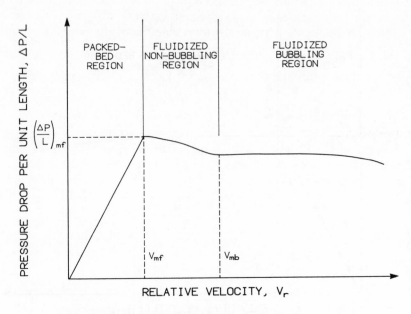

Figure 12.15 Typical fluidization curve for Geldart group A solids.

2. *Fluidized bed flow.* In fluidized bed solids flow, v_r is equal to, or greater than, v_{mf}. The voidage in the standpipe can change along the length of the standpipe and $\Delta P/L$ does not increase with v_r. There are two kinds of fluidized bed flow:

(a) bubbling fluidized bed flow and
(b) non-bubbling fluidized bed flow.

When a group B solid is operating in fluidized bed flow it always operates in the bubbling fluidized bed mode since bubbles are formed at all relative velocities above v_{mf}. However, for group A solids, there is an operating 'window' corresponding to a relative or slip velocity between v_{mf} and v_{mb}, where the solids are fluidized but no bubbles are formed in the standpipe. A standpipe operating with a relative velocity above v_{mb} and with group A solids operates in the bubbling fluidized bed flow.

   In the bubbling fluidized bed mode of standpipe operation there are four different types of regimes which can be distinguished:

Type 1.   Emulsion gas flow *up*, bubble flow *up*, net gas flow *up*
Type 2.   Emulsion gas flow *down*, bubble flow *up*, net gas flow *up*
Type 3.   Emulsion gas flow *down*, bubble flow *up*, net gas flow *down*
Type 4.   Emulsion gas flow *down*, bubble flow *down*, net gas flow *down*

These four types of bubbling fluidized bed flow are discussed individually below. In the discussion, it is assumed that the two-phase theory of fluidization holds (see Chapter 4). this theory holds that:

(a) All gas in excess of v_{mf} travels through the bed in the form of bubbles.

(b) Gas travels through the emulsion at an interstitial velocity equal to v_{mf}.

Type 1 flow. For this type of bubbling fluidized bed flow, both the gas flowing through the emulsion and the bubbles flow upwards relative to the standpipe wall, as indicated in Fig. 12.16(a). This type of bubbling fluidized flow occurs in a standpipe when v_s is less than v_{mf}. For fluidized bed bubbling type 1 flow:

$$v_s < v_{mf} \tag{12.29}$$

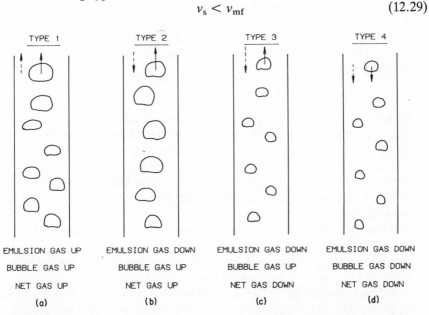

| TYPE 1 | TYPE 2 | TYPE 3 | TYPE 4 |

EMULSION GAS UP	EMULSION GAS DOWN	EMULSION GAS DOWN	EMULSION GAS DOWN
BUBBLE GAS UP	BUBBLE GAS UP	BUBBLE GAS UP	BUBBLE GAS DOWN
NET GAS UP	NET GAS UP	NET GAS DOWN	NET GAS DOWN
(a)	(b)	(c)	(d)

Figure 12.16 Schematic representation of four types of bubbling fluidized bed standpipe flow.

Type 2 flow. In this regime, schematically depicted in Fig. 12.16(b), v_s is greater than v_{mf} so that emulsion gas will flow down the standpipe relative to the wall. In order for the bubbles in the standpipe to rise, v_s must also be less than the bubble rise velocity, u_b. The volumetric rate of upward bubble flow (Q_b) is greater than the volumetric rate of downward-moving dense phase gas flow (Q_d) for this flow regime, making the net gas flow upwards. Therefore, for bubbling fluidized bed type 2 flow:

$$v_{mf} < v_s < u_b \tag{12.30}$$

and

$$Q_b > Q_d \qquad (12.31)$$

Type 3 flow. This type of bubbling fluidized bed flow is similar to type 2 flow in that v_s is greater than v_{mf} but less than u_b, as shown in Fig. 12.16(c). However, for this type of flow, Q_b is less than Q_d, resulting in a net gas flow downwards. Therefore, for type 3 bubbling fluidized bed flow:

$$v_{mf} < v_s < u_b \qquad (12.32)$$

and

$$Q_b < Q_d \qquad (12.33)$$

Type 4 flow. In this case (Fig. 12.16d), the solids velocity is so great that it exceeds u_b. The bubbles cannot rise in the standpipe and are carried down the standpipe relative to the standpipe wall at a velocity equal to $v_s - u_b$. The net gas flow is obviously downward and so for type 4 bubbling fluidized bed flow:

$$v_s > u_b \qquad (12.34)$$

Bubbles, especially large bubbles, are undesirable in a standpipe. If a standpipe is operating in the bubbling fluidized bed mode such that the solids velocity, v_s, is less than the bubble rise velocity u_b, then bubbles will rise and grow by coalescence. The bubbles rising against the downward flowing solids hinder or limit the solids flowrate (Knowlton and Hirsan, 1978; Eleftheriades and Judd, 1978). The larger the bubbles, the greater the hindrance.

When the solids velocity in the standpipe is greater than the bubble rise velocity, the bubbles will travel down the standpipe relative to the standpipe wall. It is also possible in this case (Zenz, 1984) for bubbles to coalesce and hinder solids flow through the standpipe.

Bubbles also reduce the $\Delta P/L$ or 'density' of the solids in the standpipe. Thus, a standpipe operating in a strongly bubbling regime will require a longer length to seal the same differential pressure than for a standpipe in which the same solids are at slightly above minimum fluidizing conditions.

In light of the above discussion, it is evident that for optimum fluidized bed standpipe operation:

(a) v_r in a standpipe operating with group B solids should be maintained just slightly greater than v_{mf};

(b) v_r in a standpipe operating with group A solids should be maintained between v_{mf} and v_{mb}.

3. *Streaming flow.* Standpipes can also operate in a streaming, dilute phase flow characterized by high voidages. In general, this type of flow

throughout the entire standpipe length is not desired because the full dense phase 'head' of the solids is not developed.

Other subclassifications of these basic flow regimes have been noted and mathematically defined by Leung and coworkers (Leung, Wiles, and Nicklin, 1971a, 1971b; Leung and Wilson, 1973; Leung, 1976; Leung and Wiles, 1976; Leung and Jones, 1978a, 1978b, 1978c; Leung, 1980), who have studied standpipe flow extensively over the past ten years. Their efforts have significantly advanced the understanding of solids flow in standpipes. Their detailed analyses of standpipe operation have demonstrated that multiple steady state standpipe solids flow regime combinations are possible. More recently, Jones, Teo, and Leung (1980) have analysed the stability of standpipe operation. Any serious, in-depth study of standpipe operation should include a review of their work.

Basically, the three flow regimes described above are the ones most frequently existing in a standpipe. Various combinations of these flow regimes can also exist. Fluidized bed flow can exist above packed bed flow and streaming flow is extremely common above fluidized bed flow and even packed bed flow.

12.9.1   Overflow and Underflow Standpipes

There are two types of standpipe configurations which can be employed — the overflow standpipe (Fig. 12.17a) and the underflow standpipe (Fig. 12.17b). The overflow standpipe is so named because solids overflow from the top of the fluidized bed into the standpipe. In the underflow standpipe, the solids are introduced into the standpipe from the underside, or bottom, of the bed or hopper.

In any gas–solids flow system a pressure drop 'loop' can be defined such that the sum of the pressure drops around the loop will be zero. In any such loop in which a standpipe is a component, the standpipe is usually the *dependent* part of the loop. This means that the pressure drop across the standpipe will automatically adjust to balance the pressure drop produced by the other pressure drop components in the loop. How the standpipe pressure drop adjustment is made is different for overflow compared to underflow standpipes.

Consider the overflow standpipe system shown in Fig. 12.18(a). Solids are being transferred from the upper fluidized bed to the lower fluidized bed against the pressure differential $P_2 - P_1$. This pressure differential is composed of the pressure drop in the lower fluidized bed from the standpipe exit to the top of the bed (ΔP_{LB}), the pressure drop across the grid of the upper fluidized bed (ΔP_{grid}), and the pressure drop across the upper fluidized bed (ΔP_{UB}), i.e.:

$$P_2 - P_1 = \Delta P_{LB} + \Delta P_{grid} + \Delta P_{UB} \qquad (12.35)$$

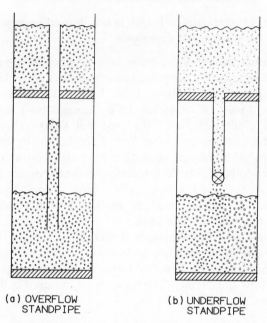

(a) OVERFLOW
    STANDPIPE

(b) UNDERFLOW
    STANDPIPE

Figure 12.17 Overflow and underflow standpipe configurations.

This pressure drop must be balanced by the pressure drop in the overflow standpipe. If the standpipe is operating in the fluidized bed mode at a $\Delta P/L$ equal to $(\Delta P/L)_{mf}$, the solids height in the standpipe H_{sp} will adjust so that the standpipe pressure drop ΔP_{sp} will equal the product of $(\Delta P/L)_{mf}$ and H_{sp}, i.e.:

$$\Delta P_{sp} = P_2 - P_1 = \left(\frac{\Delta P}{L}\right)_{mf} H_{sp} \tag{12.36}$$

The pressure diagram for this system is also shown in Fig. 12.18(a).

If the gas flowrate through the two beds is increased, ΔP_{grid} will increase while the pressure drops across the two beds will remain essentially constant, so that $P_2 - P_1$ increases to $P_2' - P_1$. The pressure drop across the standpipe will then also adjust to $P_2' - P_1$. This occurs because the height of solids in the standpipe will increase from H_{sp} to H_{sp}' so that:

$$\Delta P_{sp} = P_2' - P_1 = \left(\frac{\Delta P}{L}\right)_{mf} H_{sp}' \tag{12.37}$$

This is illustrated in Fig. 12.18(b) and is also reflected in the pressure diagram. If the increase in pressure drop is ever such that H_{sp} must increase to

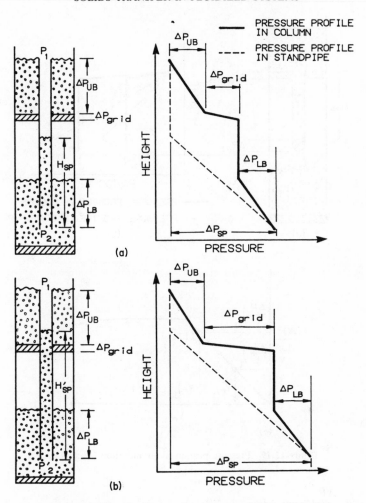

Figure 12.18 Pressure profiles in an overflow standpipe.

a value greater than the standpipe height available, the standpipe will not operate.

In Fig. 12.19(a), solids are being transferred through an underflow standpipe from the upper fluidized bed to the lower fluidized bed against the differential pressure, $P_2 - P_1$. The differential pressure, $P_2 - P_1$, consists of the pressure drop across the grid of the upper fluidized bed ΔP_{grid}. However, there is also a pressure drop across the solids flow control valve ΔP_v for this standpipe configuration. Therefore, the standpipe pressure drop ΔP_{sp} must equal the sum of ΔP_v and ΔP_{grid}, i.e.:

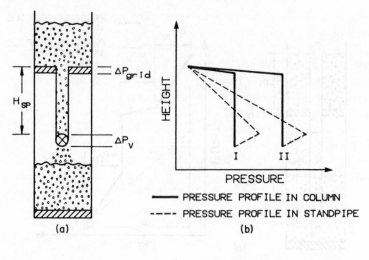

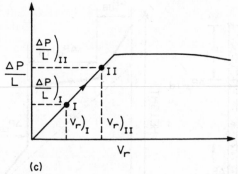

Figure 12.19 Pressure profiles in an underflow standpipe.

$$\Delta P_{sp} = \frac{\Delta P}{L} H_{sp} = \Delta P_{grid} + \Delta P_{v} \qquad (12.38)$$

$$= P_2 - P_1 + \Delta P_{v} \qquad (12.39)$$

Thus, for this underflow standpipe case, the standpipe must generate a pressure drop greater than $P_2 - P_1$. This is shown as case I in the pressure diagram of Fig. 12.19(b).

If the gas flowrate through the fluidized beds is increased, ΔP_{grid} will increase. If it is assumed that ΔP_{v} remains constant, then ΔP_{sp} must increase to balance the pressure drop loop. This is shown as case II in the pressure diagram of Fig. 12.19(b). Unlike the previous case discussed for the overflow standpipe, the solids level cannot rise to increase the pressure drop in the standpipe. Therefore, $\Delta P/L$ in the standpipe must increase to balance the pressure drop loop. This will occur (if the standpipe had been operating in the packed bed mode before the increase in ΔP_{grid}) by a simple increase in v_r in the standpipe. This can be visualized with the aid of Fig. 12.19(c).

For Case I, the pressure drop loop in the bed was satisfied by having the standpipe operate at point I on the $\Delta P/L$ versus v_r curve, as shown in Fig. 12.19(c). When the pressure drop in the system increased, the v_r in the standpipe automatically adjusted to generate a higher $\Delta P/L$, $(\Delta P/L)_{II}$, to balance the higher system pressure drop.

If the pressure drop across the grid would have increased to a value such that the product of $(\Delta P/L)_{mf}$ and the standpipe height, H_{sp} were less than the sum of ΔP_v and ΔP_{grid}, then the standpipe would not seal and, therefore, would not operate.

From the discussion above, it is evident that the pressire drop in an overflow standpipe adjusts because of a solids level change in the standpipe. the pressure drop in the packed bed underflow standpipe adjusts by a $\Delta P/L$ change due to a changing v_r.

In addition to the way the two standpipes adjust to balance pressure drops in the loop, there are other distinguishing characteristics of overflow and underflow standpipes.

Overflow Standpipes

(a) It is necessary to immerse the exit end of an overflow standpipe in a fluidized bed. If not immersed, the standpipe will empty of solids, a pressure seal will not form, and solids will not flow through the standpipe.

(b) The overflow standpipe operates 'automatically'. This means that the solids which enter the top of the overflow standpipe are discharged at the same rate into the fluidized bed without the need of a valve, i.e. 'automatically'.

(c) The overflow standpipe usually operates with solids in the fluidized regime. However, it is not necessary that it do so. The overflow standpipe can also operate in the packed bed mode.

(d) The overflow standpipe generally operates only partially full of solids. This is because an overflow standpipe is designed so that it will have a height 25 to 30 per cent. greater than needed to develop the design pressure drop.

(e) Since the solids enter the overflow standpipe from the top of the bed, large solids, or agglomerates, may collect at the bottom of the feed fluidized bed since they cannot be discharged.

Underflow Standpipes

(a) Solids enter the underflow standpipe from the bottom of a fluidized or packed bed. Therefore, it is necessary to have a valve at the exit of the underflow standpipe to maintain solids in the bed. This valve must control the solids flowrate and the standpipe usually does not operate 'automatically'.

(b) The underflow standpipe operates completely full of solids.

(c) The underflow standpipe usually operates in the packed bed mode with group B solids. With group A solids, either the fluidized bed or the packed bed mode can be used.

(d) The underflow standpipe does not need to be buried in a fluidized bed to operate. The valve at the bottom of the standpipe enables the overflow standpipe to also discharge the solids above the bed, if desired.

(e) The underflow standpipe will discharge agglomerates which collect near the bottom of the bed if they are not larger than the standpipe diameter.

One of the most common types of standpipes is the underflow standpipe operating with group A solids in the fluidized bed mode. This type of standpipe is used extensively in the petroleum industry to feed cracking catalyst from fluidized beds into dilute phase pneumatic conveying lines (Fig. 12.20).

In this type of underflow standpipe, the standpipe pressure drop will not adjust to balance changes in system pressure because the standpipe is operating in the fluidized bed mode. This standpipe is designed so that the standpipe length is great enough to generate more 'head' than needed. The excess pressure is then 'burned up' across the slide valve used to regulate the solids flow. The slide valve is also used to adjust the pressure drop in the system loop to balance pressure drop changes in the other pressure drop components.

The solids velocity v_s in the standpipe carrying cracking catalyst is generally of the order of 1 to 2 m/s, while v_{mf} is approximately 1 cm/s. This means that gas is also flowing down the standpipe at a velocity v_g of about 1 to 2 m/s.

12.9.2 Friction Loss

For a standpipe operating in a fluidized mode, the pressure buildup in the standpipe is nearly entirely due to the hydrostatic 'head' of solids. The pressure loss due to solids/wall friction is basically negligible in the fluidized bed mode. This is not the case for packed bed flow in standpipes.

When the relative velocity in a standpipe is less than v_{mf}, the solids-wall friction can be significant. Basically, the percentage of the solids weight not supported by the gas will be supported by the friction of the solids sliding on

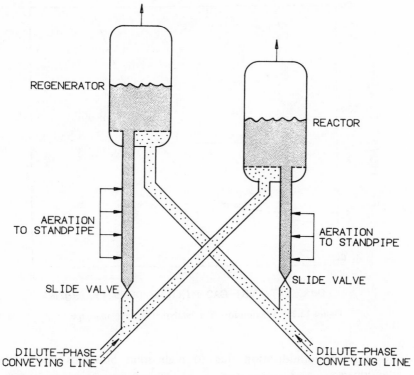

Figure 12.20 Typical solids flow loop in a catalytic cracker with underflow standpipe.

the standpipe wall. The cracking catalyst friction loss in an 8 in. (20 cm) diameter standpipe, as reported by Matsen (1976), is shown in Fig. 12.21. In this figure, the standpipe friction loss is presented as a function of the true density of the catalyst–gas mixture in the standpipe (usually measured with radiation in commercial standpipes). When the true density of the mixture in the standpipe increases above that at minimum fluidization, the friction loss due to solids sliding on the standpipe wall increases sharply.

12.9.3 Standpipe Aeration

The purpose of adding aeration gas to a standpipe is generally to maintain the solids in the standpipe in a fluidized state near the minimum fluidization condition. This results in the highest standpipe density, or $\Delta P/L$, possible, thus minimizing the standpipe length and also the solids/wall friction loss.

There are two primary instances where aeration is added to standpipes to maintain fluidized flow:

(a) To prevent defluidization due to gas compression in long standpipes,

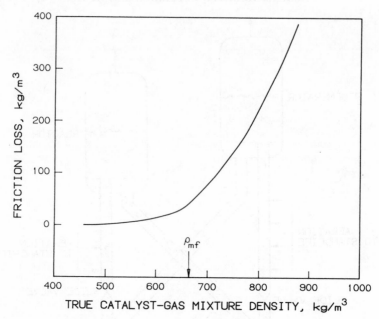

Figure 12.21　Friction loss in a catalytic cracking standpipe.

(b) To prevent defluidization due to high solid flowrates in overflow standpipes.

Defluidization Due to Gas Compression

This is by far the most common reason for adding aeration to standpipes. Consider the underflow, fluidized group A cracking catalyst standpipe system shown in Fig. 12.20, where gas and solids are both flowing downwards relative to the standpipe wall. The length of this type of standpipe can be considerable (30 to 45 m) in commercial units. In such a long standpipe at low pressures, the percentage change in gas density can be significant from the top of the standpipe to the bottom. Typically, the absolute pressure may vary from 1.36 to 3.4 bar, giving an increase in gas density of 250 per cent.

Kojabashian (1958) showed that the solids voidage in the standpipe decreases along the standpipe as the pressure increases. This is because the gas volume decreases or 'shrinks' and the solids move closer together, thus causing the voidage ϵ to decrease. The voidage can change so much because of the pressure change that the flow regime in the lower part of the standpipe can change from fluidized bed flow to packed bed flow. Kojabashian (1958) and Leung (1976) have shown that several different pressure profile variations can exist, as shown in Fig. 12.22.

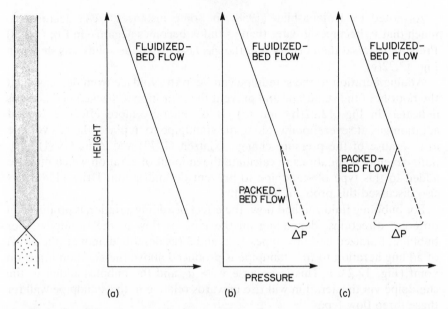

Figure 12.22 Standpipe pressure profiles as a function of height in an underflow standpipe in which voidage decreases with height.

If the standpipe is fluidized over its entire length and if the fluidized density is constant throughout the entire length of the standpipe, then the standpipe pressure profile will take the form shown in Fig. 12.22(a).

As noted above, since ϵ decreases as the gas density increases down the standpipe, the solids flow regime can be in fluidized bed flow at the top of the standpipe and then revert to packed bed flow at the bottom of the standpipe. The effect that decreasing ϵ has on v_r can be seen by expressing the relative velocity equation:

$$v_r = v_s - v_g \tag{12.28}$$

in the form:

$$v_r = \frac{G_s}{\rho_p \, (1-\epsilon)A} - \frac{G_g}{\rho_g \, \epsilon A} \tag{12.40}$$

If ϵ decreases percentage-wise faster than ρ_g increases, then v_s decreases and v_g increases — causing v_r to decrease also. If v_r decreases to a value less than v_{mf} at the bottom of the standpipe, but gas is still flowing upwards relative to the solids, then the pressure profile will take the form shown in Fig. 12.22(b). the pressure buildup in the standpipes for this case is less than that shown in Fig. 12.22(a) by the difference, ΔP, indicated in the figure.

As noted by Kojabashian (1958), in some instances ϵ can decrease so much that v_g becomes greater than v_s and v_r becomes negative in Eq. (12.28) Pressure reversal then occurs (i.e. the gas faster than the solids), as shown in Fig. 12.22(c).

Adding aeration to the standpipe can keep the voidage from decreasing at the bottom of the standpipe and prevent the transition to packed bed flow, as indicated in Fig. 12.22(b) and (c). It is generally more effective to add aeration gas at several points along the standpipe to 'replace' the gas volume lost because of the pressure change. Matsen (1973) and Zenz (1984) have both developed equations to calculate the amount of aeration which must be added to this type of standpipe to prevent defluidization. Dries (1980) has also discussed this problem in detail.

For bubbling fluidized bed flow, the effect of adding aeration is produced in different directions, depending on the type of flow in the standpipe. For bubbling fluidized bed flow types 1, 2, and 3 (as described above), the effect of adding aeration to the standpipe is produced above the aeration injection point (Fig. 12.23a). This is because $v_s > u_b$ and the bubbles added to the standpipe via the aeration will rise upwards relative to the standpipe wall for these three flow types.

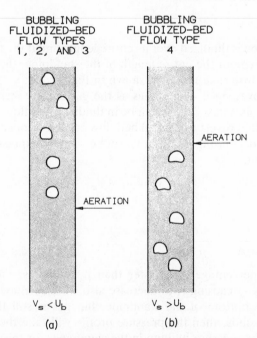

BUBBLING
FLUIDIZED-BED
FLOW TYPES
1, 2, AND 3

BUBBLING
FLUIDIZED-BED
FLOW TYPE
4

AERATION

AERATION

$V_s < U_b$

$V_s > U_b$

(a)

(b)

Figure 12.23 Direction of aeration effectiveness for $v_s < u_b$ and $v_s > u_b$.

For type 4 bubbling fluidized bed solids flow where $v_s < u_b$, bubbles added to the standpipe by aeration are carried down the standpipe with the solids, and the effect of the aeration is produced below the aeration injection point (fig. 12.23b).

It must be emphasized here that the *minimum* amount of aeration necessary to maintain fluidization in the standpipe is the optimum amount of aeration to use. Too much aeration gas added to the standpipe will cause large bubbles to form which will hinder and decrease the flow of solids down the standpipe.

Defluidization Due to Solid Flow Rate Increases

Sauer, Chan, and Knowlton (1984) have recently shown that overflow standpipes operating in a bubbling fluidized bed mode can defluidize if the solids flowrate is increased above a limiting value. Defluidization in this case is not caused by solids/wall friction but is thought to occur because of the force produced by the dilute phase streaming solids falling on the top of the dense phase solids in the standpipe. This force may 'compress' the fluidized column of solids, decrease ϵ, and cause defluidization. Sauer, Chan and Knowlton (1984) showed that adding aeration to the bottom of the standpipe at a single location could prevent defluidization. This is because the solids velocities in the standpipes they were operating were always less than the bubble rise velocity (bubbling fluidized bed flow types 1, 2, and 3). If v_s had exceeded u_b in their standpipes (bubbling fluidized bed flow type 4), aeration would have had to be added at the top of the standpipe.

12.9.4 Angled Standpipes

Angled (or inclined) standpipes are used in many cases in industry to transfer solids between two points which are separated horizontally, as well as vertically. Angled standpipes would seem to be a very convenient way to transfer solids on paper; operationally, angled standpipes are not recommended.

Sauer, Chan, and Knowlton (1984) studied angled standpipes and a hybrid combination of an angled standpipe and a straight standpipe operating in fluidized bed bubbling flow. They used transparent lines to observe the flow of solids in the angled sections in these standpipes. Invariably, they found that the gas and solids separated, the gas channelling via bubbles along the upper half of the pipe while the solids flowed along the bottom portion of the standpipe. This is illustrated in Fig. 12.24.

Sauer, Chan and Knowlton (1984) also found that the pressure buildup in angled standpipes was much lower than in straight standpipes due to gas–solids separation. Because of this separation, the gas does not completely

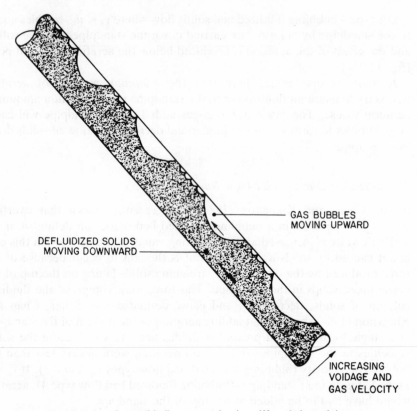

GAS BUBBLES
MOVING UPWARD

DEFLUIDIZED SOLIDS
MOVING DOWNWARD

INCREASING
VOIDAGE AND
GAS VELOCITY

Figure 12.24 Gas–solids flow separation in a 60° angled standpipe.

fluidize the solids and produce the full hydrostatic head of solids. This is illustrated in Fig. 12.25, which compares the $\Delta P/L$ produced by a straight overflow standpipe and an angled overflow standpipe as a function of the solids flowrate through them. As can be seen, the straight or vertical standpipe produces a much higher standpipe density.

If at all possible, underflow standpipes to be operated in the fluidized bubbling bed mode should be designed to be straight. The section of overflow standpipes operating in dense phase flow should also be designed to be straight. The section of an overflow standpipe operating in dilute phase streaming flow can be angled with no adverse effects, however. This is indicated in Fig. 12.26.

12.9.5 Stripping in Standpipes

During the transfer of solids from one fluidized bed to another via a standpipe, a substantial amount of gas can be transferred with the solids in the

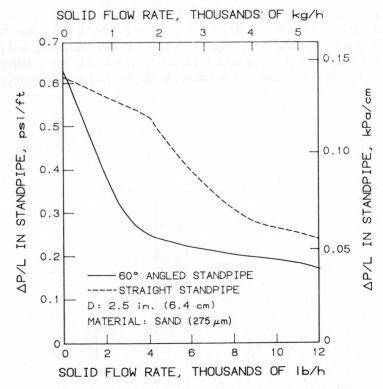

Figure 12.25 Comparison of $\Delta P/L$ in a 60° angled and vertical standpipe.

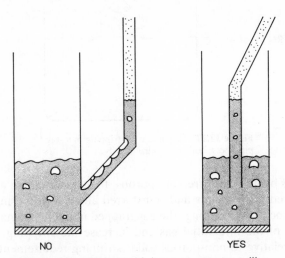

Figure 12.26 Correct and incorrect way to utilize angles in an overflow standpipe.

interstices. For safety reasons, or to prevent contamination or dilution of product gas, it is frequently desired to prevent gas being transferred via a standpipe from coming into contact with the gas in the receiving bed (Fig. 12.27). To accomplish this, a neutral gas is usually added to the standpipe to replace, or 'strip', the normal downcomer gas in the void spaces between the solids.

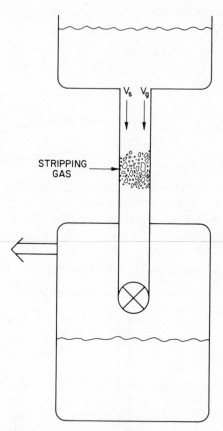

Figure 12.27 Schematic of stripping in a stand-pipe between two fluidized beds.

If the solids being transferred are porous, then a substantial amount of gas may be trapped in the pores and transferred along with the interstitial gas. During the process of stripping, the gas trapped in the pores may also diffuse into the bulk of the interstitial gas and increase the stripping requirements significantly relative to non-porous solids stripping requirements.

Stripping gas can be added to standpipes operating in both the moving packed bed and fluidized bed modes.

Moving Packed Bed Stripping

Knowlton *et al.* (1981) have published the only systematic study of stripping in moving packed bed standpipes. They found that a standpipe that operated in a moving packed bed regime with non-porous solids and that had stripping gas added to it could be analysed using a relative gas–solids velocity concept. They found that two cases must be considered: (a) gas travelling down with the solids and (b) gas travelling up the standpipe.

Gas Flow Down

For the standpipe shown in Fig. 12.28(a), the gas is initially travelling

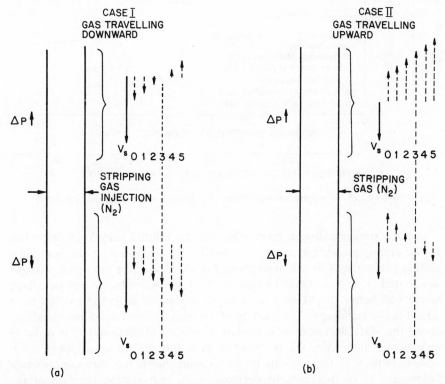

Figure 12.28 Relative gas–solid velocities in a stripping standpipe: (a) gas flowing downwards and (b) gas flowing upwards.

downwards relative to the standpipe wall. The dotted arrows designated '0' for case I in Fig. 12.28(a) indicate the magnitude and direction of the gas above and below the stripping gas injection point (SGIP) when no stripping gas is being added to the standpipe. When the stripping gas is added to the standpipe, the pressure drop in the standpipe above the SGIP rises, whereas the pressure drop in the standpipe below the SGIP falls, as shown in Fig. 12.29. However, the total standpipe pressure drop (Fig. 12.29) remains constant.

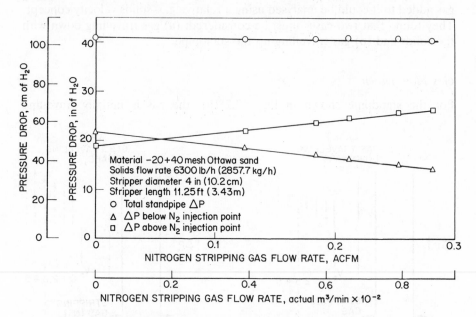

Figure 12.29 Standpipe pressure drops as a function of stripping gas flowrate.

Because pressure drop is proportional to the relative velocity, v_r increases in the standpipe section above the SGIP and decreases in the standpipe section below the SGIP when stripping gas is added to the standpipe. Arrows designated '1' in Fig. 12.28(b) represent the gas velocities in the standpipe above and below the SGIP after a small amount of stripping gas has been added to the standpipe. The amount of gas being carried down the standpipe above the SGIP decreases at condition '1' relative to condition '0' in order to increase v_r, and hence the pressure drop in this section as stripping gas is added. However, the amount of gas flowing down the standpipe section below the SGIP increases to decrease v_r in that section because of the decrease in pressure drop upon the addition of stripping gas.

As more stripping gas is added to the system:

(a) The standpipe pressure drop above the SGIP increases.
(b) v_r increases in the standpipe above the SGIP.
(c) The amount of gas being carried down the standpipe in the section above the SGIP decreases even more.

This trend continues until no gas is carried down the standpipe with the solids (point '3' in Fig. 12.28a). At this point, $v_g = 0$ above the SGIP. If more stripping gas is added, stripping gas will now flow up the standpipe above the SGIP, as well as down the standpipe section below the SGIP. At this point, the gas from the upper fluidized bed is completely stripped; i.e. it cannot travel down the standpipe.

Gas Flow Up

In Fig. 12.28(b), the gas in the standpipe is initially moving upwards. When stripping gas is added to this standpipe, the standpipe pressure drop above the SGIP rises and the pressure drop across the standpipe below the SGIP falls. Thus, the v_r in the standpipe above the SGIP rises and the v_r in the standpipe below the SGIP decreases. Therefore, all the stripping gas that is initially added goes up the standpipe above the SGIP, diluting the gas already moving upwards.

In the standpipe section below the SGIP the pressure drop falls; therefore, the amount of gas moving up the pipe in this section has to decrease. As more stripping gas is added, this trend continues until no gas moves up the pipe (point '3' in Fig. 12.28b) and the standpipe is effectively stripped.

Thus, in a packed bed of non-porous particles, the interstitial gas can be stripped with another gas regardless of whether the interstitial gas was initially moving upwards or downwards. The actual stripping section is below the SGIP in the former case and above it in the latter.

Stripping Gas Injection Point Location

Changing the location of the SGIP in a standpipe affects the amount of gas needed to strip. It does so because the pressure drop distribution in the standpipe changes.

Moving the SGIP can, therefore, reduce the amount of gas needed to strip in a standpipe. For gas travelling down the standpipe, moving the SGIP closer to the top of the standpipe reduces the amount of stripping gas needed.

For gas travelling up the standpipe, moving the SGIP towards the bottom of the standpipe reduces the stripping gas requirement.

In actual practice, stripping efficiency may fall off if the SGIP is located too close to the ends of the standpipe. This is because there may be a minimum standpipe length necessary for complete stripping because of mixing, diffusion, or other factors.

Stripping in Fluidized Standpipes

If v_s in a standpipe is greater than v_{mf} then the interstitial gas in the standpipe cannot be stripped before the solids fluidize. Knowlton *et al.* (1981) described the results of several tests in which the standpipe $\Delta P/L$ above the SGIP became high enough so that v_r exceeded v_{mf} and the solids began to fluidize in that section. Two cases must again be considered: gas travelling down with the solids and gas travelling up the standpipe.

Emulsion Gas Down

When stripping gas is added to a fluidized standpipe, the pressure drops above and below the SGIP remain essentially constant. Any excess stripping gas added to the standpipe will, therefore, travel up the standpipe as bubbles. In Types 2 and 3 bubbling fluidized flow, before the stripping gas is added, interstitial gas is still travelling down the standpipe in the emulsion phase with the solids.

For a very brief period standpipe gas is travelling downwards in the emulsion phase, while stripping gas is travelling upwards in the bubble phase. The stripping gas in the bubble phase and the standpipe gas in the emulsion phase then interchange over the height of the stripper. The result of this situation is that there is a gradient in the stripper wherein the standpipe gas concentration is lower at the bottom than at the top. This condition exists until the stripping gas travelling up the standpipe effectively 'absorbs' all of the standpipe gas and the standpipe is 'stripped'. For Type 4 bubbling bed flow, complete stripping cannot be achieved for the same reasons noted for Type 1 flow. No matter how much stripping gas is added to the standpipe, the interstitial standpipe gas flow downwards above the SGIP will continue to flow down the standpipe. Adding stripping gas results only in dilution of this gas.

Emulsion Gas Up

When the standpipe section above the SGIP becomes fluidized for gas flowing in Type 1 bubbling fluidized flow, complete stripping cannot be achieved. If more stripping gas is added to the fluidized standpipe for this case, the standpipe gas will be diluted by the bubbles of stripping gas. However, no matter how much stripping gas is added, the initial interstitial standpipe gas flowing upwards below the SGIP will continue to flow up the standpipe and complete stripping will not occur. Thus, for the case of gas flowing initially up the standpipe operating in the fluidized bed mode, the interstitial gas cannot be stripped, only diluted.

Stripping Section Design

The minimum amount of stripping gas needed to strip interstitial gas in a moving packed bed standpipe would be when $v_g = 0$ initially in the standpipe.

At this condition, solids would be falling through stagnant gas, and theoretically no stripping gas would be needed. Realistically, however, this will not occur because solids flowrates do not stay constant, and there is a certain degree of mixing and diffusion in a 'real' standpipe. However, for optimum stripping a standpipe should be operated as close to $v_g = 0$ as possible. Therefore, $v_g = 0$ should be the criterion for optimum stripping section design.

12.10 NON-MECHANICAL VALVES

A non-mechanical valve is a device which uses only aeration gas in conjunction with its geometrical shape to control the flowrate of particulate solids through it. Non-mechanical valves offer several advantages over traditional mechanical valves:

(a) They have no moving mechanical parts which are subject to wear and/or seizure.

(b) Constructed of ordinary pipe and fittings, they are extremely inexpensive.

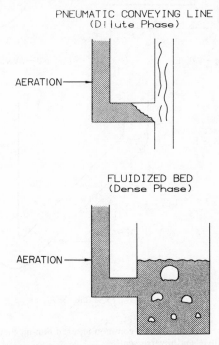

Figure 12.30 L-valve feeding into dilute phase and dense phase systems.

(c) They can be fabricated 'in-house', which avoids the long delivery times often associated with the purchase or replacement of mechanical valves.

(d) They can be used to control or feed solids into either a dense phase or dilute phase environment, such as a fluidized bed or a pneumatic conveying line (Fig. 12.30).

The three most commonly used types of non-mechanical valves are the L-valve, J-valve, and the reverse seal. These three devices are shown schematically in Fig. 12.31. The only difference between these devices are

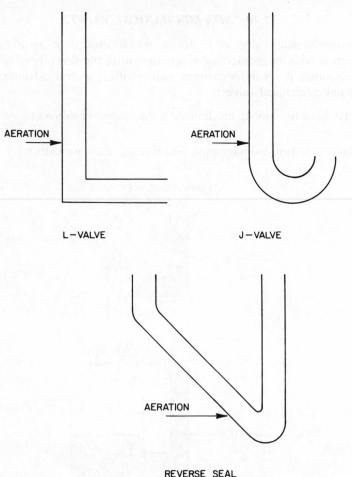

Figure 12.31 The three most common types of non-mechanical valves: L-valve, J-valve, and reverse seal.

their shape, the direction in which they discharge solids, and the distance between the control aeration injection point and the solids discharge point. All three devices operate on the same principle.

The most common of the non-mechanical valves is the L-valve (so named because it is shaped like the letter 'L') because its shape is easiest to construct and it is slightly more efficient than either the J-valve or the reverse seal. Since the principle of operation is the same for all non-mechanical valves, non-mechanical valve operation will be presented primarily through a discussion of the characteristics of the L-valve.

12.10.1 Principles of Operation and Design

When aeration gas is added to a non-mechanical valve, solids do not begin to flow immediately. There is a certain threshold amount of aeration which must be added before solids will begin to flow (Fig. 12.32). However, after the

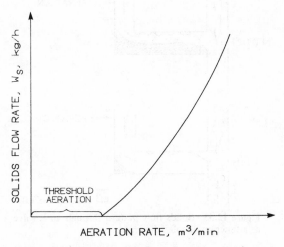

Figure 12.32 Typical solids flowrate versues aeration rate curve for a non-mechanical valve.

threshold aeration is reached, additional aeration gas added to the valve causes the solids flowrate to increase and decreasing the amount of aeration to the valve causes the solids flowrate to decrease. In general, there is no hysteresis in the aeration versus solids flowrate curve for a non-mechanical valve.

Solids flow through a non-mechanical valve because of drag forces on the particles produced by the aerating gas. When aeration gas is added to a non-mechanical valve, gas flows downwards through the particles and through the constricting bend. This relative gas–solids flow produces a frictional drag force on the particles in the direction of flow. When this drag force exceeds the force needed to overcome the resistance to solids flow in the constricting bend, the solids begin to flow through the valve.

The solids flow profile through a non-mechanical valve changes with the solids flowrate through the valve. This is illustrated in Fig. 12.33 for the case of the L-valve.

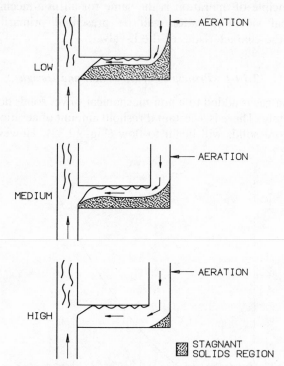

Figure 12.33   Solids flow pattern through an L-valve as a function of solids flowrate.

At low solids flowrates, the solids flow in a narrow zone near the inside of the constricting elbow. Most of the solids in the elbow are stagnant. Similarly, only a shallow upper layer of solids is moving in the horizontal section and the solids appear to be moving in 'ripples' near the top of the pipe.

At the medium solids flowrate shown in Fig. 12.33, the stagnant region of solids is much reduced. Also, a larger portion of the solids are now moving in the horizontal section of the L-valve.

At high solds flowrates the entire mass of solids in the horizontal section of the L-valve is in motion.

Non-mechanical valves work best with Geldart B materials. These are granular materials in the area shown on the Geldart fluidization diagram of Fig. 3.4. Group C materials (cohesive materials such as flour) do not work well in L-valves. In general, group C materials do not flow well in any type of pipe and are even difficult to fluidize.

Group A materials (materials which have a substantial deaeration time) also generally do not work well in L-valves. These materials retain air in their interstices and remain fluidized for a significant period of time and thus often flow uncontrollably through the L-valve like water.

Group D solids are very large solids with particle diameters generally $\frac{1}{4}$ in (6,350 μm) or greater. These materials can be made to flow through an L-valve, but the L-valve has to have a large diameter and much aeration has to be used. The large Group D solids also work best in an L-valve if they are mixed in with smaller material. The smaller particles fill up the void spaces between the larger particles and decrease the voidage. This increases the drag on the entire mass of solids when aeration is added and makes the solids easier to move through the constricting bend.

From the above discussion it is seen that L-valves are best utilized with granular material in the size range between 100 and about 8,000 μm.

Knowlton and Hirsan (1978) and Knowlton, Hirsan, and Leung (1978) have shown that the operation of a non-mechanical valve is dependent upon system pressure drop and geometry. In general, a non-mechanical valve will operate with an underflow standpipe operating in moving packed bed flow above it. The standpipe in turn is usually attached to a hopper.

Consider the L-valve solids feed system shown in Fig. 12.34 with its associated pressure drops. The aeration injection location is the high pressure point in the system; the low pressure common point is at the top of the feed hopper. The pressure drop balance is such that the sum of the L-valve, fluidized bed, and piping pressure drops must equal the standpipe and hopper pressure drops, i.e.:

$$\Delta P_{\text{standpipe}} + \Delta P_{\text{hopper}} = \Delta P_{\text{L-valve}} + \Delta P_{\text{bed}} + \Delta P_{\text{piping}} \quad (12.44)$$

The hopper pressure drop is usually negligible so that Eq. (12.41) may be written as:

$$\Delta P_{\text{standpipe}} \simeq \Delta P_{\text{L-valve}} + \Delta P_{\text{bed}} + \Delta P_{\text{piping}} \quad (12.42)$$

The standpipe is the dependent part of the pressure drop loop in that its pressure drop will adjust to exactly balance the pressure drop produced by the sum of the components on the independent side of the loop. However, there is a maximum pressure drop per unit length ($\Delta P/L$) that the moving packed bed standpipe can develop. This maximum value is the fluidized bed pressure drop per unit length, $(\Delta P/L)_{\text{mf}}$, for the material being used.

The independent pressure drop can be increased by increasing the bed pressure drop (increasing the bed height), the piping pressure drop

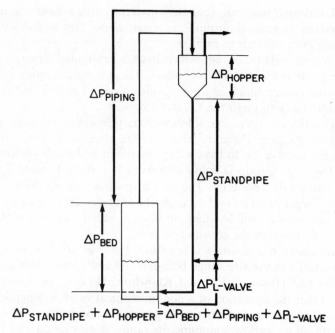

$$\Delta P_{STANDPIPE} + \Delta P_{HOPPER} = \Delta P_{BED} + \Delta P_{PIPING} + \Delta P_{L-VALVE}$$

Figure 12.34   L-valve solids feed system and pressure drop loop.

(increasing the gas flowrate), and the L-valve pressure drop (increasing the solids flowrate). For a constant gas velocity and bed height, as the solids flowrate into the bed is increased, the independent half of the pressure drop loop increases. The standpipe pressure drop adjusts to balance this increase. If the solids flowrate continues to increases, the standpipe pressure drop will also keep increasing to balance the pressure drop on the independent side of the loop. This will occur until the $\Delta P/L$ in the downcomer reaches $(\Delta P/L)_{mf}$.

Because of its lesser capacity to absorb pressure drop, a short standpipe will reach its maximum $\Delta P/L$ at a lower solids flowrate than a longer standpipe. Thus, the maximum solids flowrate through an L-valve depends on the length of the standpipe above it.

The pressure drop in a standpipe is generated by the relative velocity v_r between the gas and solids in the standpipe. When v_r reaches the value necessary for minimum fluidization of the solids, a transition from a packed bed flow to fluidized bed flow occurs. Any further increase in v_r results in the formation of bubbles in the standpipe. These bubbles hinder the flow of solids through the standpipe and thus decrease the solids flowrate.

To determine the minimum standpipe length necessary for an L-valve feeding into a fluidized bed, it is necessary to know the pressure drop on the independent side of the loop at the desired solids flowrate. The minimum length of standpipe necessary L_{min} is:

$$L_{min} = \frac{\Delta P_{independent}}{(\Delta P/L)_{mf}} \qquad (12.43)$$

The actual length of standpipe selected for the L-valve design should be greater than L_{min} to allow for the possibility of future increases in the solids flow requirements and to act as a safety factor. Typical standpipe lengths might be 1.5 to 2 times L_{min}.

Knowlton and Hirsan (1978, 1980) and Knowlton (1979) have also shown that the total aeration Q_T, which causes solids to flow through a non mechanical valve, is not just the external aeration added to the valve. It is the sum of the external aeration gas Q_A and the standpipe gas Q_{dc} which determines the solids flowrate through the valve. Thus, for gas travelling down the standpipe with the solids:

$$Q_T = Q_A + Q_{dc} \qquad (12.44)$$

For gas travelling up the standpipe:

$$Q_T = Q_A - Q_{dc} \qquad (12.45)$$

This difference is shown schematically in Fig. 12.35.

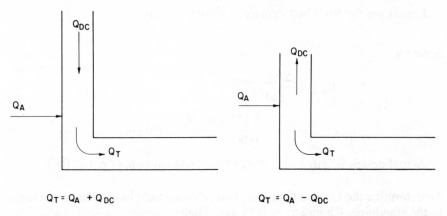

Figure 12.35 Effective aeration flows in an L-valve for gas flowing up and gas flowing down the standpipe.

To determine the diameter of an L-valve it is necessary to select the linear solids velocity desired in the standpipe. Nearly all L-valves usually operate over a linear particles velocity range of 0 to 0.3 m/s in the standpipe. Although an L-valve may be designed for any linear solids velocity, a typical value near 0.15 m/s is generally selected to allow for substantial increases or decreases in the solids flowrate, if necessary.

Unfortunately, at this time there are no published correlations to predict L-valve pressure drop and aeration requirements. The best way to determine these parameters is to estimate them from basic data from a small L-valve test unit or to extrapolate them from the data given by Knowlton and Hirsan (1978).

Example 12.5: L-valve design

The minimum standpipe length and the L-valve diameter for an L-valve feeding limestone into a fluidized bed from a hopper at a rate of 56,820 kg/h is to be determined. The configuration for this feed system is the same as that shown in Fig. 12.34 and:

$$\Delta P_{\text{standpipe}} = \Delta P_{\text{L-valve}} + \Delta P_{\text{bed}} + \Delta P_{\text{piping}} \tag{12.46}$$

Given:

Limestone bulk density = 1,520 kg/m^2

$\Delta P_{\text{L-valve}} \simeq 0.2$ bar

$\Delta P_{\text{bed}} \simeq 0.136$ bar

$\Delta P_{\text{piping}} \simeq 0.136$ bar

Limestone fluidized bed density = 0.0011 bar/cm

Solution

$$L_{\text{min}} = \frac{\Delta P_{\text{independent}}}{(\Delta P/L)_{\text{mf}}}$$

$$= \frac{0.2 + 0.136 + 0.136}{0.0011} = 429 \text{ cm}$$

Actual design length = 1.5 × 429 cm = 644 cm = 6.44 m (21 ft)

To determine the L-valve diameter, first select a linear limestone velocity, v_s, in the standpipe. Choose $v_s = 0.15$ m/s. Then:

$$W = \rho_b v_s A \tag{12.47}$$

$$\frac{56,820}{3,600} = 1,520 \times 0.15A$$

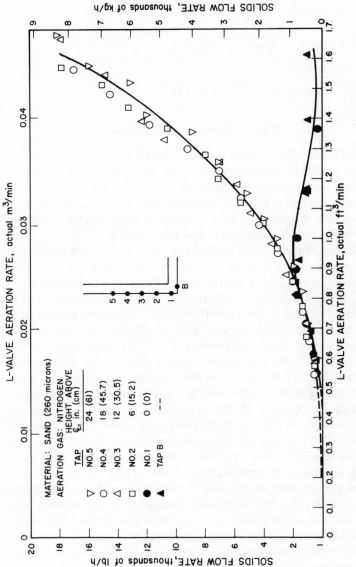

Figure 12.36 Solids flowrate versus L-valve aeration rate as function of aeration tap location.

$$A = \frac{\pi D^2}{4} = \frac{56,820}{3,600 \times 1520 \times 0.15} = 0.069 \text{ m}^2$$

$$D^2 = (0.;69) \frac{4}{\pi} = 0.088$$

$D \quad = 0.296 \text{ m} = 30 \text{ cm}$

$\quad\quad = 0.97 \text{ ft}$

Pick the pipe diameter to be 30 cm or 1 ft.

12.10.2 Aeration Tap Location

It is desirable to add aeration to a non mechanical valve as low in the standpipe as possible. This will result in the maximum standpipe length and minimum non mechanical valve pressure drop, both factors which result in increasing the maximum solids flowrate through the valve. However, if the aeration is added too low to a non mechanical valve, gas by-passing results and solids flow control is insensitive and not effective.

Knowlton and Hirsan (1978) studied the effect of aeration tap location on the operation of a 7.6 cm diameter L-valve. They found that aeration was most effective if it was added a distance of 15.2 cm (a length/diameter (L/D) ratio of 2) or more above the centerline of the horizontal section of the L-valve. Aeration added at the centerline of the horizontal section or at the bottom of the centerline of the standpipe was found to be ineffective. The aeration tap locations tested are shown in Fig. 12.36, together with the solids flow rate versus aeration rate curves for each aeration tap location tested.

As shown in Fig. 12.36, the solids flowrate was increased from zero to its maximum value by increasing the aeration rate to the L-valve. For taps B and 1 (located at the bottom of the standpipe and flush with the L-valve centerline, respectively), the solids flowrate increased to a maximum value of approximately 910 kg/h and then decreased when more aeration was added to the taps. For taps 2, 3, and 5, the solids flow rate increased with increasing aeration up to a value of approximately 8,180 kg/h.

The poor performance of the bottom two taps was due to gas by-passing from the aeration tap to the horizontal section of the L-valve and not being efficiently utilized to drag the solids through the constricting bend. The aeration added to the upper aeration taps was fully used to drag the solids through the bend. Therefore, all of these solids flowrate versus aeration curves for the upper taps were basically the same.

In summary, L-valve aeration should be added above the constricting bend. To assure good operation it should be added at an L/D of approximately 2 above the centerline of the horizontal section of the L-valve.

12.10.3 L-valve Horizontal Length

In order for an L-valve to operate properly, the L-valve horizontal section length must be kept between a minimum and maximum length. The minimum length must be greater than the horizontal length through which the solids flow due to their angle of repose, as shown in Fig. 12.37. Mathematically, the horizontal section must be greater than L where:

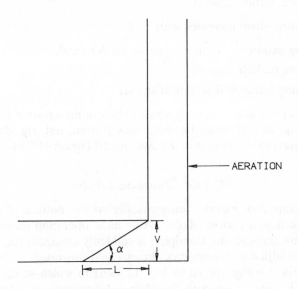

Figure 12.37   Minimum L-valve length.

$$L = \frac{V}{\tan \propto} \tag{12.48}$$

For design purposes the minimum horizontal length should actually be about 1.5L.

If the L-valve is designed properly, solids flow through it in micro pulses at a relatively high frequency, and for all practical purposes the solids flow is steady. For very long L-valve horizontal lengths, the solids flow becomes intermittent (surging, stopping, then surging again). In these 'long' L-valves, a dune builds up to such a size that it blocks the pipe and flow stops. Gas pressure builds up behind the dune and finally becomes so great that the dune collapses and the solids surge through the L-valve. Another dune then builds up after the gas pressure is released. This pattern then repeats itself. This cycle does not occur if the L-valve horizontal section length is kept below that equal to an L/D of about 12.

Knowlton and Hirsan (1978) have determined the effect of varying geometrical and particle parameters on the operation of the L-valve, which are summarized below.

L-valve aeration requirements increase with:

(a) Increasing vertical section diameter

(b) Increasing particle size

(c) Increasing particle density

L-valve pressure drop increases with

(a) Incresing aeration (i.e. increasing solids flowrate)

(b) Increasing particle density

(c) Decreasing horizontal section diameter

No equations presently exist with which to determine aeration requirements. Thus, L-valve design must be determined from test rig data or from extrapolation from the curves of Knowlton and Hirsan (1978).

12.10.4 Automatic L-Valve

An L-valve can also operate 'automatically' at the bottom of an overflow standpipe, such as a cyclone dipleg. Automatic operation means that if the solids flowrate through the standpipe is suddenly changed, the L-valve will automatically adjust to accomodate the changed flowrate.

Consider the L-valve shown in Fig. 12.38(a) in which solids are flowing through the L-valve at the rate W_1 while aeration is being added to the L-valve at the rate Q_A. The solids above the aeration point are flowing in the fluidized bed mode and are at an equilibrium height H_1. The solids below the aeration point are flowing in the moving packed bed mode. The fraction of aeration gas needed to move the solids at rate W_1 is flowing around the elbow with the solids, while the remaining portion of the aeration gas is flowing up the standpipe.

If the solids flowrate decreases from W_1 to W_3 (Fig. 12.38c), the solids level in the standpipe will initially rise because solids are being fed to the L-valve faster than they are being discharged. However, the increased height of solids in the standpipe offers an increased resistance to gas flow and so a larger fraction of the aeration gas Q_A will now flow around the elbow increasing the solids flowrate through the L-valve. If enough aeration gas is being added to the L-valve, the system will reach equilibrium at a point where a larger fraction of the aeration gas is flowing around the L-valve constriction causing solids to flow at rate W_2. The height in the standpipe above the L-valve will reach equilibrium at height H_2 greater than H_1, and the system is in balance.

If the solids flowrate decreases from W_1 to W_3 (Fig. 12.38c), the solids level

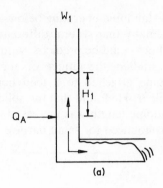

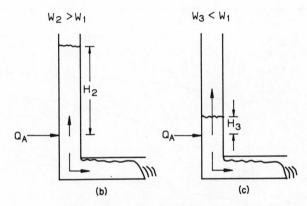

Figure 12.38 Automatic L-valve operation.

in the standpipe above the L-valve will fall, decreasing the resistance to gas flow, and a larger fraction of the constant aeration flow Q_A will pass up the standpipe. The reduced flow of gas through the L-valve will result in a lower solids flowrate. Balance is reached where solids are flowing through the L-valve at rate W_3 and the solids in the standpipe are at equilibrium height, H_3.

Very little work has been done with this type of L-valve system. Findlay (1981) has shown that this type of L-valve operation is possible, but much work will have to be done before the solids flowrate range and aeration limitation over which the L-valve will operate automatically can be estimated accurately.

12.11 SOLIDS FLOWRATE ENHANCEMENT FROM HOPPERS

Ginestra, Rangachari, and Jackson (1980) and Chen, Rangachari, and Jackson (1984) have developed one-dimensional equations to predict the relationship between pressure drop, soilds flowrate and control orifice diameter in a vertical standpipe operating below a conical mass flow hopper. They found experimentally that several different flow regimes are possible and that the flow regimes could be predicted with their mathematical model.

Knowlton (1980) studied the effect of increasing the length of an unconstrained standpipe attached to a feed hopper on the solids gravity flowrate from the hopper. He found that the solids flowrate from the hopper increased as the standpipe length was increased and that the solids flowrate increase was more pronounced for small particle sizes than for large particle sizes (Fig. 12.39).

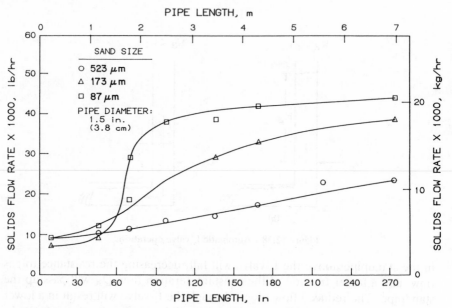

Figure 12.39 Gravity solids flowrate from a hopper as a function of standpipe length in an unconstrained standpipe.

For unrestricted standpipe flow, the solids flow through the pipe at a rate faster than their discharge from the hopper, and a dilute phase streaming flow results. The resulting profile in the hopper–standpipe system with an unconstrained standipe is shown in Fig. 12.40, for the case where the top of the hopper and the bottom of the standpipe are both at atmospheric pressure.

When the standpipe is operating in dilute phase flow, a constant pressure gradient over most of the standpipe length is developed. This produces a low pressure region at the top of the standpipe immediately below the hopper outlet. The longer the standpipe, the lower the pressure at the top of the standpipe.

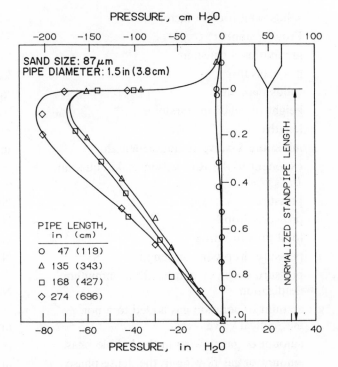

Figure 12.40 Pressure profiles in a hopper–unconstrained stand-pipe system.

The decrease in pressure at the top of the standpipe causes the pressure drop across the hopper outlet to increase. The gas flow through the particles in the hopper then has to increase, which increases the solids flowrate through the hopper outlet.

Chen, Rangachari, and Jackson (1984), Judd and Dixon (1978), and Judd and Rowe (1978) have also reported similar profiles in unrestricted standpipes. Chen, Rangachari, and Jackson (1984) have also predicted the flow profile and the increase in solids flowrate using their one-dimensional mathematical model.

12.12 NOMENCLATURE

A	area	m^2
d_p	particle size	m
d_v	volumetric particle diameter	m
D	tube or pipe diameter	m
f_g	gas–wall friction factor	—

f_s	solids–wall friction factor	—
F_r	Froude number, U^2/gD	—
g	gravitational constant	9.81 m/s^2
G_g	gas mass flux	$\text{kg/m}^2 \text{ s}$
G_s	solids mass flux	$\text{kg/m}^2\text{s}$
H_{sp}	height of solids in standpipe	m
L	length	m
L_{min}	minimum L-valve standpipe length	m
n	exponent in the Richardson–Zaki equation, $U/V_t = \epsilon^n$	—
P	pressure	N/m^2
ΔP	pressure drop	N/m^2
ΔP_T	total pressure drop	N/m^2
$\Delta P/L$	pressure drop per unit length	N/m^3
$(\Delta P/L)_{mf}$	pressure drop per unit length at minimum fluidization	N/m^3
Q_A	amount of aeration gas added to a non mechanical valve	$\text{m}^3\text{/s}$
Q_B	amount of gas flowing in the bubble phase	$\text{m}^3\text{/s}$
Q_d	amount of gas flowing in the dense phase	$\text{m}^3\text{/s}$
Q_{dc}	amount of gas flowing down or up a standpipe relative to the sandpipe wall	$\text{m}^3\text{/s}$
Q_T	amount of gas flowing around the L-valve bend	$\text{m}^3\text{/s}$
Re	pipe Reynolds number, $\rho_g DU/\mu$	—
Re_p	particle Reynolds number, $\rho_g d_p U/\mu$	—
S_Δ	exponent indicative of the 'powderiness' of a solid	—
u_b	bubble rise velocity	m/s
U	superficial gas velocity	m/s
U_{ch}	choking velocity, superficial	m/s
U_{salt}	superficial saltation velocity	m/s
U_{so}	single particle saltation velocity	m/s
v_{ch}	interstitial choking velocity, U_{ch}/ϵ	m/s
v_g	interstitial gas velocity, U/ϵ	m/s
v_{mb}	interstitial minimum bubbling velocity, U_{mb}/ϵ	m/s
v_{mf}	interstitial minimum fluidization velocity, U_{mf}/ϵ	m/s
v_r	relative gas solids velocity	m/s
v_s	solids velocity	m/s

v_{slip}	slip velocity between solids and gas	m/s
v_t	particle terminal velocity	m/s
W_s	solids flowrate	kg/s
$\propto$	angle of repose	—
Δ	pseudo-length used in the Zenz saltation correlation	m
ϵ	voidage	—
ϵ_{ch}	voidage at choking	—
ϵ_{mf}	voidage at minimum fluidization	—
θ	loading factor used in Konno and Saito correlation—	
μ	gas viscosity	N s/m^2
ρ_b	solids bulk density	kg/m^3
ρ_g	gas density	kg/m^3
ρ_p	particle density	kg/m^3
ψ	sphericity of a solid particle	—
ω	pseudo-velocity used in the Zenz saltation correlation	m/s

12.13 REFERENCES

Arastoopour, H., Modi, M.V., Punwani, D.V., and Talwalkar, A.T. (1979). Paper presented at International Powder and Bulk Solids Conference/Exhibition, Chicago.

Batholomew, R.N., and Casagrande, R.M. (1957). *Ind. Eng. Chem.*, **49**, 428–431.

Berg, C.H.O. (assigned to Union Oil Company of California) (1954a).

Berg, C.H.O. (assigned to Union Oil Company of California) (1954b). US Patent 2,684,870.

Berg, C.H.O. (assigned to Union Oil Company of California) (1954c) U.S. Patent 2, 684, 872.

Berg, C.H.O. (assigned to Union Oil Company of California) (1954d). US Patent 2,684,873.

Bodner, S. (1982). *Proceedings of Pneumatech I, International Conference on Pneumatic Conveying Technology*, Powder Advisory Centre, London.

Campbell, D.L., Martin, H.Z., and Tyson, C.W. (1948). US Patent 2,451,803.

Canning, D.A., and Thompson, A.I. (1980). Paper presented at the Annual Meeting of the American Institute of Mechanical Engineers, San Francisco.

Capes, C.E., and Nakamura, L. (1973). *Can. J. Chem. Eng.*, **51**, 31–38.

Chen, Y.M., Rangachari, S., and Jackson, R. (1984). To be presented in *Ind. Eng. Chem. Fund.*

Culgan, J.M. (1962). Ph. D. Thesis, Georgia Institute of Technology.

Curran, G.P., and Gorin, E. (1968). Prepared for Office of Coal Research, Contract No. 14–01–0001–415, Interim Report No. **3**, Book **6**, Washington, DC.

Davidson, J.F., and Harrison, D. (1963). *Fluidized Particles*, Cambridge University Press.

Institute of Gas Technology (1978). *Coal Conversion Systems Technical Data Book*, Prepared for the US Department of Energy by the Institute of Gas Technology under Contract No. EX–76–C–01–2286.

Doig, I.D., and Roper, G.H. (1963). *Australian Chem. Eng.*, 1, 9–19.

Dries, H.W.A. (1980). In *Fluidization* Eds J.R. Grace and J.M. Matsen, Plenum Press, New York.

Eleftheriades, C.M., and Judd, M.R. (1978). *Powder Technol.*, 21, 217227.

Engineering Equipment Users' Association (1963). *EEUA Handbook*, No. 15, Constable, London.

Findlay, J. (1981). Personal communication.

Flatt, W. (1976). Paper 43, *Proceedings of Pneumotransport* 3, BHRA Fluid Engineering, Bath.

Gajdos, L.J., and Bierl, T.W. (1978). Topical report prepared by Carnegie-Mellon University.

Geldart, D., (1973). *Powder Technol*, 7, 285.

Ginestra, J.C., Rangachari, I., and Jackson, R. (1980). *Powder Technol.*, 27, 69.

Grace, J.R., and Tuot, J. (1979). *Trans. Instn. Chem. Engrs.*, 57, 47.

Hinkle, B.L. (1953). Ph.D. Thesis, Georgia Institute of Technology.

Jones, P.J., and Leung, L.S. (1978). *Ind. Eng. Chem. Proc. Des. & Dev.*, 17, 4.

Jones, P.J., Teo, C.S., and Leung, L.S. (1980). In *Fluidization* (Eds J.R. Grace and J.M. Matsen), p. 469.

Judd, M.R., and Dixon, P.D. (1978). *A.I.Ch.E. Symp. Ser.*, 74 (176), 38.

Judd, M.R., and Rowe, P.N. (1978). In *Fluidization* (Eds J.F. Davidson and D.L. Keairns), Cambridge University Press, p.110.

Jung, R. (1958) *Forschung Ing. Wes.*, 24 (2), 50.

Knowlton, T.M. (1977). Institute of Gas Technology Project IU–4–8 Annual Report for 1977, Sponsored by American Gas Association.

Knowlton, T.M. (1979). *Proccedings of Powder and Bulk Solids Conference/ Exhibition*, Philadelphia, Pennsylvania.

Knowlton, T.M. (1980). *Proceedings of the 6th International Conference on Fluidized-Bed Combustion*, Atlanta, Georgia.

Knowlton, T.M., Aquino, C., Hirsan, I., and Sishtla, C. (1981). *A.I.Ch.E. Symp. Ser.*, 77 (205), 189.

Knowlton, T.M., and Bachovchin, D.M. (1976). In *Fluidization Technology* (Ed. D.L. Keairns), Hemisphere Publications Corporation.

Knowlton, T.M., and Hirsan, I. (1978). Hydrocarbon Processing, 57, 149–156.

Knowlton, T.M., and Hirsan, I. (1980). Paper E3, *Proceedings of Pneumotransport 5*, BHRA Fluid Engineering, London, p. 257.

Knowlton, T.M., Hirsan, I., and Leung, L.S. (1978). In *Fluidization* (Eds J.F. Davidson, D.L Keairns), Cambridge University Press.

Kojabashian, C. (1958). Ph.D. Thesis, Massachusetts Institute of Technology, Cambridge, Massachusetts.

Konno, H., and Saito, S.J. (1969). *Chem. Eng. Japan*, 2, 211–17.

Leung, L.S., and Jones, P.J. (1976). In *Fluidization Technology* (Ed. D.L. Keairns), Vol. II, Hemisphere Publishing Corporation, Washington, DC.

Leung, L.S. (1980). *Proceedings of Pneumotransport*, 5, BHRA Fluid Engineering, London.

Leung, L.S., and Jones, P.J. (1978a). Paper D1, *Proceedings of Pneumotransport 4 Conference*, BHRA Fluid Engineering.

Leung, L.S., and Jones, P.J. (1978b). *Powder Tech.*, 20, 145–160.

Leung, L.S., and Jones, P.J. (1978c). In *Fluidization* (Eds. J.F. Davidson and D.L. Keairns), Cambridge University Press.

Leung, L.S., and Wiles, R.J. (1976). Paper C4–47–58, Presented at Pneumotransport 3 Conference, University of Bath.

Leung, L.S., Wiles, R.J., and Nicklin, D.J. (1971a). *Ind. Eng. Chem. Proc. Des. & Dev., 2, (10)*, 183–189.

Leung, L.S., Wiles, R.J., and Nicklin, D.J. (1971b). Paper presented at Pneumotransport 1 Conference, Churchill College, Cambridge.

Leung, L.S., and Wilson, L.A. (1973). *Powder Technol.,* 7, 343.

Marcus, R.D., Hilbert, J.D., Jr., and Klinzing, G.E. (1984). *Pipelines,* 4, 103–112.

Matsen, J.M., (1973). *Powder Technol.,* 7, 93.

Matsen, J.M. (1976). In *Fluidization Technology* (Ed. D.L. Keairns), Vol II, Hemisphere Publishing Corporation, Washington, DC, p. 135.

Matsen, J.M. (1981). *Proceedings of the 73rd Annual AIChE Meeting,* New Orleans.

Matsumoto, S., Hara, M., Saito, S., and Maeda, S. (1974a). *J, Chem. Eng. Japan,* 7 (6), 425–30,

Matsumoto, S., Harada, S., Saito, S., and Maeda, S. (1974b). *J. Chem. Eng. Japan,* 8, 331.

Mewing, S.F. (1976). B.Eng. Thesis, University of Queensland.

Morikawa, Y., Tsuji, Y., Matsuik, J., and Hani, Y. (1978). *Int. J. Multiphase Flow,* 4, 575.

Patterson, R.C. (1959). *ASME J. Engineering for Power,* **1959**, 43–54.

Punwani, D.V., Modi, M.V., and Tarman, P.B. (1976). Paper presented at the International Powder Bulk Solids Handling and Processing Conference, Chicago.

Richards, P.C., and Wiersma, S. (1973). Paper A1–1–15, Presented at Pneumotransport 2 Conference, Guildford.

Rizk, f. (1973). Doktor-Ingenieurs Dissertation, Technische Hochschule Karlsruhe.

Rose, H.E., and Duckworth, R.A. (1969). *Engineer,* 227, 392–396, 430–433, 478–483.

Sandy, C.W., Daubert, T.E., and Jones, J.H. (1970). *Chem. Eng. Prog. Symp. Ser.,* 66, 105, 133–142.

Sauer, R.A., Chan, I.H., and Knowlton, T.M. (1984). *A.I.Ch.E. Symp. Ser.,* 80 (234), 1–23.

Saxton, A.L., and Worley, A.C. (1970). *Oil and Gas. J.,* **1971**, 92–99.

Schuchart, P. (1968). I. Chem. E. V.D.I./V.Y.G. Symposium on the Engineering of Gas–Solids Reaction, Brighton.

Scott, A.M., Richard, P.D., and Mooij, A. (1976). Paper D10, presented at Pneumotransport 3 Conference, Bath.

Smith, T.N. (1978). *Chemical Eng. Sci.,* 33, 745.

Thomas, D.G. (1962). *A.U.Ch.E.J.,* 8 (3), 363–378.

Tsuji, Y. (1980). Paper H2, *Proceedings of the 5th International Conference on the Pneumatic Transport of Solids in Pipes,* BHRA Fluid Engineering, Cranfield, Bedford, pp. 363–388.

Van Zoonen, D. (1961). *Proceedings of the Symposium on the Interaction between Fluids and Particles,* Inst. Chemical Engineers, London, p.64.

Weidner, G. (1955). *Forschung Ing. Wes.,* 21 (5), 145.

Wen, C.Y., and Galli, A.F. (1971). In *Fluidization* (Eds J.F. Davidson and D. Harrison), Academic Press, London.

Wen, C.Y. and Simons, H.P. (1959). *A.I.Ch.E.J.,* 5 (2), 263–267.

Woebcke, H.N., and Cofield, W.W. (1978). *Coal Processing Technol.,* 3, 79.

Woodcock, C.R., and Mason, J.S. (1976). Paper E1, Pneumotransport 3 Conference, Bath.

Yang, W.C. (1975). *A.I.Ch.E.J.,* 21, 1013–1015.

Yang, W.C. (1976a). Paper presented at Pneumotransport 3 Conference, Bath.

Yang, W.C. (1976b). Paper presented at the International Powder and Bulk Solids Handling and Processing Conference/Exhibition, Chicago.

Yang, W.C. (1982). Paper presented at the 74th A.I.Ch.E. Annual Meeting, Los Angeles, California.

Yerushalmi, J., Turner, D.H., and Squires, A.M. (1976). *Ind. Eng. Chem. Proc. Dev. & Dec.,* **15**, 47–53.

Yousfi, Y., and Gau, G. (1974). *Chem. Eng. Sci.,* **29**, 1939–1946.

Zenz, F.A. (1964). *Ind. Chem. Fund.,* **3**, 65–75.

Zenz, F.A. (1984). Paper presented at Katalistiks Symposium, Vienna, Austria.

Zenz, F.A., and Othmer, D.F. (1960). *Fluidization and Fluid-Particle Systems,* Reinhold, New York.

Zenz, F.A., and Rowe, P.N. (1976). In *Fluidization Technology* (Ed. D.L. Keairns), Vol. II, Hemisphere Publishing Corporation, Washington, DC, p.151.

CHAPTER 13

Instrumentation and Experimental Techniques

J.R. GRACE AND J. BAEYENS

The behaviour of fluidized beds is influenced by many factors — particle properties, gas properties, column geometry, and operating conditions. Two powders which appear to be similar may behave quite differently in the same apparatus while the same powder may fluidize in a different manner in another piece of equipment.

In developing successful fluidization processes, it is essential to minimize scale-up problems by forethought at the experimental stage. While there is no general route to success, there are certain guidelines which should be followed to perform consistent and reliable experiments. We attempt to provide some useful guidelines and to provide typical or useful references in this chapter. While we believe that the guidelines will be useful under most circumstances, we caution the reader that each fluidized bed process has unique elements and special problems which may not be adequately covered by the general treatment given here.

13.1 THE FLUIDIZED POWDER

In order to design a fluid bed system, the engineer should be fully familiar with the characteristics of the powder to be handled. As a first step, one should check the Geldart classification of the powder (see Chapter 3). As a minimum requirement, hydrodynamic equivalence between laboratory powders and process materials requires that they be in the same classification. The extrapolation of results for material in one powder group to another is not advised.

Particle size and density are, however, not the only important parameters.

Variations in particle shape, moisture content, entrained air, porosity, surface configuration, etc., make it impossible to design a standard system to suit all requirements. Some factors which influence design are listed in Table 13.1. For each specific powder, extensive experimentation is preferred to extrapolation, if time and money permit.

Table 13.1 Effect of powder properties on equipment and process design

Powder parameter	Main design influence
Particle size and density	Gas flow requirements, key elements in many correlations, powder group
Aerated specific weight	Hopper volumes, feeder capacities
Size distribution as given by sieve analyses; density distribution	Elutriation, dedusting equipment, seals, segregation, feeding method
Particle shape	Factor only if extreme (e.g. flaky or very angular)
Aeration or deaeration characteristics	Requirements for deaeration, hopper design, standpipe flow
Temperature limitations	Heat transfer surface, insulation needs
Moisture content and/or sensitivity	Feeder design, gas velocity, residence time in bed.
Resistance to attrition	Carryover, distributor design, cyclones
Abrasiveness, corrosiveness	Materials of construction, placement of heat transfer surfaces, distributor design
Agglomeration or cohesive properties	Segregation, use of stirrers, means of withdrawing product, gas flowrate, avoidance of internal surfaces.

13.2 GAS SUPPLY AND METERING

13.2.1 Components

The components of the gas supply system are shown schematically in Fig. 13.1. The blower or compressor must be capable of supplying sufficient gas to the column to give a superficial velocity of at least 1 to 2 m/s and preferably 3 m/s. For studies of circulating beds and fast fluidization, velocities of 5 to 12 m/s should be attainable. In the latter case, the air should be split into a primary and secondary stream whose relative flows are in the range of

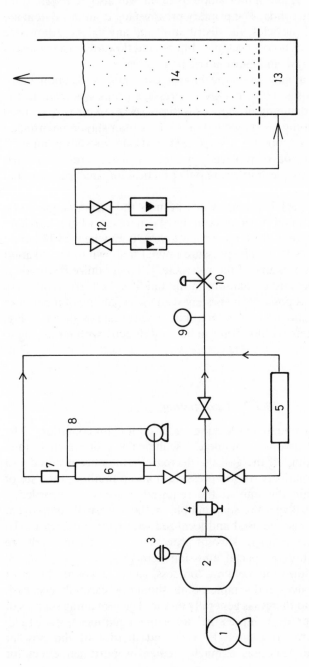

1. Blower
2. Gas pressure
3. Rupture disc
4. Filter
5. Drier
6. Humidification column
7. Mist eliminator

8. Water circulation
9. Humidity indicator
10. Pressure regulator
11. Rotameters
12. Control valves
13. Windbox
14. Fluidized bed

Figure 13.1 Gas supply system for fluidized beds: 1 blower, 2 reservoir, 3 rupture disc, 4 filter, 5 drier, 6 humidification column, 7 mist eliminator, 8 water circulation, 9 humidity indicator, 10 valves, 11 rotameters or orifice meters, 12 control valves, 13 windbox. 14 fluidization column.

approximately 5:1 to 1:3, and minor amounts of air will also be required for aeration of recirculating solids. The primary pressure losses in the system are those across the bed of particles, the distributor, and any valves. Allowance should be made for losses across cyclones, filters, and flow measuring devices as well as pipes, fittings, and noise-abatement devices.

Pressure pulsations arising from the blower can alter fluidization behaviour. For many purposes, elements giving pressure drops upstream of the bed provide sufficient damping of these pulsations. A reservoir may be used to provide further damping, as shown in Fig. 13.1. A filter should be installed to remove oil and water droplets. A surge tank and filter are not required if the blower or compressor is replaced by one or more pressurized gas cylinders. However, these gas bottles are only practical for small columns and limited running times.

Shown next on Fig. 13.1 is a loop which permits the incoming gas to be humidified or dehumidified to a desired level. In cases where humidity control is not necessary (e.g. for large dry particles), it may be possible to do without part or all of this loop. The pressure regulator shown in the diagram is set at the calibration pressure of the rotameters. To minimize fluctuations in float level due to pressure oscillations in the fluidized bed, the calibration level should be as high as possible, consistent with the capability of the blower or compressor. The rotameter valves should be downstream as shown in Fig. 13.1, so that the rotameter calibration can be used directly without having to correct for pressure.

13.2.2 Leak testing

There are two ways to evaluate gas leaking under pilot plant conditions. The pressure drop method involves completely sealing the unit which is then pressurized. After turning off the gas, the decrease in pressure is noted as a function of time. Alternatively, in the pressure hold-up method, the rate of gas addition to maintain the unit at the required pressure is recorded.

To locate points of leakage, the soap bubble method is usually employed. Special gas tracers can also be used and localized with special detectors. To test equipment components for leakage of process gases, other methods are available such as the use of special emission detectors.

Aside from the fluidization column, process valves account for most leakage; safety relief valves and sample points should be carefully checked. Leakage from piping and flanges is generally minor. The remaining sources of leakage are the blower or compressor, dedusting equipment, etc. Leak tightness becomes more critical as the size and density of the powder decrease, since a small leak has a much greater proportional effect for powders of low U_{mf}.

13.3   FLUIDIZATION COLUMNS

13.3.1   Two-dimensional Columns

'Two-dimensional' columns are of rectangular cross-section, the width being considerably greater than the thickness. The fluidized particles are contained in the gap between two flat transparent faces, separated by a distance which is usually in the range 10 to 25 mm. A schematic diagram of a typical set-up appears in Fig. 13.2.

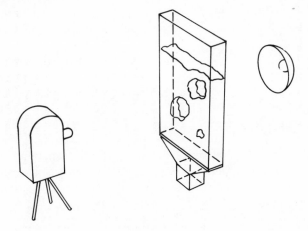

Figure 13.2   Schematic diagram of two-dimensional column.

Dimensions and uses of a number of two-dimensional columns which have been reported in the literature are given in Table 13.2. There is no standard column size, with reported thicknesses as small as 5 mm and as large as 63 mm and heights up to 7.3 m. Table 13.2 shows that a wide variety of experiments have been carried out in two-dimensional columns.

The two-dimensional column has proved to be especially useful:

(a) For studies of bubble properties. Bubbles span the bed thickness and hence are readily viewed with the aid of back lighting.

(b) As an educational tool in helping those unfamiliar with fluidized beds to visualize and understand basic phenomena.

(c) As a means of qualitatively viewing the fluidization characteristics of given powders.

While two-dimensional columns are useful for qualitative purposes, there are important quantitative differences between two- and three-dimensional fluidized beds. These arise from quantitative differences in rise velocities of

Table 13.2   Representative uses of two-dimensional columns in fluidization research

Reference	Column dimensions, cm (width × thickness × height)	Aspect studied
Zenz and Weil (1958)	61 × 5.1 × 370	Particle entrainment
Wace and Burnett (1961)	30.5 × 2.5 × 61	Gas tracers and pressure near bubbles
Harrison and Leung (1962)	45 × 1.8 × 180	Bubble coalescence
Rowe and Partridge (1962)	30.5 × 2.5 × 61	Particle motion caused by bubbles
Rowe, Partridge and Lyall (1964)	30.5 × 0.5 to 0.9 × 46	Cloud formation around bubbles
Hiraki, Yoshida and Kunii (1966)	80 × 6.3 × 300	Bubble size, frequency, and velocity in freely bubbling bed
Lockett and Harrison (1967)	30 × 1 × 80	Voidage variation around bubbles
Pyle and Harrison (1967)	25 × 1 × 120	Velocity of isolated bubbles
Ellis, Partridge and Lloyd (1968)	36 × 1.3 × 122	Bubble size distribution for application to three-dimensional bed
Grace and Harrison (1969)	46 × 1.9 × 122	Bubble velocities in swarm, distribution of gas between phases
Partridge, Lyall, and Crooks (1969)	30 × 1.4 × 30	Particle slip surfaces
Geldart and Cranfield (1972)	61 × 2 × NS	Fluidization behaviour of large particles
	61 × 5 × NS	
Nguyen and Leung (1972)	61 × 0.6 × NS[a]	Bubble formation
Anwer and Pyle (1973)	48 × 1.3 × 125	Gas drift profiles
Garcia, Grace, and Clift (1973)	51 × 1.5 × 120	Bubble through flow
Littman and Homolka (1973)	51 × 1.3 × 145	Pressure field around bubbles
Tuot and Clift (1973)	50 × 1.2 × 120	Heat transfer due to bubble passage
Chavarie and Grace (1975)	56 × 1 × 245	Chemical reactor modelling
Yerushalmi, Turner, and Squires (1976)	51 × 5.1 × 730	High velocity fluidization
Yoshida, Fugii, and Kunii (1976)	24 × 2.0 × 51	High temperature fluidization
Lockwood (1977)	51 × 5.1 × 200	Effect of tube array on bubbling
Nguyen and Grace (1978)	56 × 1 × 245	Transient forces on tubes
Sit and Grace (1978)	56 × 1 × 245	Interphase mass transfer
Varadi and Grace (1978)	31 × 1.6 × 51	Influence of pressure on bubbling
Wen et al. (1978)	30.5 × 1.2 × NS	Dead zones near grid
Jin et al. (1980)	40 × 1.5 × 224	Bubble-induced particle motion
Gbordzoe, Bulani, and Bergougnou (1981)	89 × 1.3 × 500	Effect of mobile bubble breakers
Lignola, Donsi, and Massimilla (1983)	21 × 1.7 × 96	Gas concentration profiles in and near bubbles
Valenzuela and Glicksman (1984b)	43 × 2.5 × NS	Residence time of fines in beds of coarse particles

[a] NS = not specified.

isolated bubbles, different bubble coalescence properties, differences in bubble shapes and wake characteristics, increased jet stability in two-dimensional columns, different mechanisms of solids ejection into the freeboard, and reduced solids mixing. Rowe and Everett (1972) used X-rays to study the influence of the thickness of a rectangular column of 0.30 m width. The thickness was increased in stages from 14 mm (a two-dimensional column) to 0.30 m (a fully three-dimensional column). For the range of conditions studied, there was a significant change in bubble properties for thicknesses less than about 0.10 m, with thickness having little discernible influence once this value was exceeded, The minimum column thickness (Lyall, 1969) is 30 mean particle diameters for spherical particles, but this ratio should be somewhat greater when the particles are angular, sticky, or have a broad size distribution.

Some factors to be considered in designing and building two-dimensional columns are as follows:

(a) Materials of construction. The front and rear faces should be transparent and are generally of plastic or armour glass. Plastic materials are cheaper and easier to work with, but are more subject to scratching, discoloration, and electrostatic effects. The spacers separating the front and rear plates may be of any material, but it is best that the entire inside surface of the column be of uniform roughness and electrical properties. Special construction techniques are required to make two-dimensional columns suitable for studying fluidization at elevated pressures (Subzwari, Clift, and Pyle, 1978; Varadi and Grace, 1978; Kawabata et al., 1981).

(b) Distributor and windbox. The column should be supported in such a way that the windbox–distributor assembly can be removed in order to replace the distributor, repair leaks, or make other modifications. The most popular types of distributor are perforated plates, porous plates, and supported filter paper or cloth. The pressure drop across the distributor should be designed to be approximately 10 to 50 per cent. of that across the bed of particles. In order to improve the uniformity of gas flow across the distributor, the windbox may be partitioned, with the flow to each section of the distributor measured and controlled separately. It is essential to provide an effective seal around the outside of the distributor plate to prevent gas from short-circuiting up the walls. A pressure relief valve may be installed in the windbox to prevent the pressure from exceeding its design maximum. Alternatively, a U-tube manometer can be used for this purpose, with the manometer fluid discharged if the pressure in the windbox is excessive.

(c) Alignment. The column must be perfectly vertical

(d) Reinforcement. The front and rear faces tend to bulge, due to hydrostatic pressure forces inside the fluidized bed. Consequently, it is usually necessary to install horizontal metal supports at intervals to keep the faces parallel. Spacing of these supports depends on the thickness and stiffness of the walls, the width and depth of the bed, and the tolerance level of distortion.

(e) Electrostatic effects. Because of their large surface area, two-dimensional columns are more subject to electrostatic charging effects than normal fluidized beds. In some cases, this may cause a layer of particles to build up on both faces, totally obscuring the phenomena under study. General techniques for reducing electrostatic effects are discussed in Section 13.3.7 below. A technique which is often useful for two-dimensional columns is to cover all internal faces of the column (including the spacers) with a layer of Cellotape or Scotchtape.

(f) Ports. Provision should be made for emptying the bed, usually by installing a port of inside diameter at least 20 mean particle diameters just above the distributor. The layer of solids which cannot be extracted by opening this port and fluidizing the bed can be removed by tipping the entire column or by vacuum cleaning. Pressure taps should be installed just above the distributor and at regular intervals (typically 0.1 to 0.2 m) above the grid. The taps are usually drilled through one of the spacers in order not to interfere with viewing. If ports are required for bubble injection, gas sampling, etc., these are best installed before the column is erected. Injection ports should be at least 40 mm above the distributor. Tubes should not protrude beyond the inner surface of the column.

(g) Opening the column. The column should be constructed in such a way that one of the faces can be removed for modification, adding or replacing surface coatings, cleaning, etc. The column must be leak tested carefully after each opening.

(h) Photography. Sufficient space should be left behind the column for placing floodlights to give even illumination. Tracing paper attached to the outside back surface helps to give an evenly bright backdrop. There should be a clear viewing area in front, extending 8 m or more if possible to allow photography with negligible distortion.

(i) Vibration. Gentle vibration or tapping is sometimes required to initiate fluidization or to prevent bridging effects. Bridging, if not prevented, may cause dead zones or lead to instability of injected bubbles.

13.3.2 Semicylindrical (or Half-) Columns

A column geometry which is intermediate between a two-dimensional column and a cylindrical three-dimensional geometry consists of a cylindrical vessel which has been sliced in half, with a flat plate installed along the diametral plane. A diagram showing this geometry appears in Fig. 13.3. The idea is that one should view, through the flat face, what happens across the diameter of a

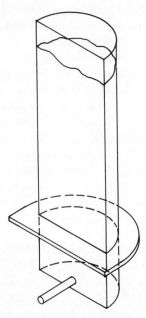

Figure 13.3 Schematic diagram of semi-cylindrical column.

full cylindrical column. In this case, front illumination is required for bubble viewing and photography, since bubbles generally do not span the entire bed thickness.*

While this geometry is very popular and generally reliable in spouted bed studies (Mathur and Epstein, 1974; Whiting and Geldart, 1980; Geldart *et al.*, 1981), its use for fluidized beds has been restricted. Hatate, King, Migita, and Ikari (1985) describe its use for bubble visualization. Gabor (1972) employed this geometry in some heat transfer work. Levy *et al.* (1983) studied particle ejection mechanisms at the bed surface in a half-column. Yang *et al.* (1984) have studied jet penetration and bubble properties in a large (3 m diameter) semicircular vessel with a conical grid. However, Rowe, Mac-Gillivray, and Cheesman (1979) found that flat surfaces alter the behaviour of grid jets as compared with the case where the jet is centred in a fully cylindrical vessel. Hence, there is some doubt about how accurately semicylindrical columns represent the behaviour in fully cylindrical vessels.

13.3.3 Laboratory and Pilot Scale Three-Dimensional Columns

Most laboratory and pilot scale fluidized beds are cylindrical. Square and rectangular cross-sections are also quite common, especially when horizontal

---

*Bubbles sometimes move away from the face, thus negating the purpose of the unit. It is therefore useful to slightly tilt the unit so that bubbles 'run up' the face.

tubes are required. The full cylindrical column has the virtues of simplicity and structural strength. Standard steel, glass, and perspex tubes are readily suited to construction of small-scale columns.

Although early work in the field of fluidization was often performed in smaller diameter columns, it now seems to be generally accepted that there is a minimum diameter of about 0.10 m below which results tend to be misleading. Smaller columns almost always give rise to slugging, while bridging and electrostatic effects may also be severe. In no case should the column diameter be less than 100 times the mean particle diameter. The upper extreme of laboratory column diameter is about 0.6 m. This is also a useful size for pilot plant units, large enough to allow useful data to be obtained and to allow a man to work inside, and yet small enough to avoid excessive fabrication, materials handling, and blower costs. The influence of bed diameter on bubble size and the onset of slug flow are treated in Chapter 4.

The usual scale-up or scale-down procedure for fluidized beds is to keep the height fixed while varying the diameter (and hence the aspect ratio) (Grace, 1974). The distributor of the smaller column should be identical (i.e. a portion of that from the full-scale unit rather than a scaled-down replica). Transparent sections or windows should be included if possible. The windbox-distributor assembly should be removable, and safety valves should be provided in case of excessive pressures. A port should be provided just above the distributor for emptying the solids, and pressure taps should be installed at 0.1 to 0.2 m intervals. The height of fluidization columns varies widely, from approximately 1 m for the smallest laboratory units to 20 m or more for pilot plants and full-scale vessels. Entrainment may be reduced by providing a tapered, expanding freeboard. The angle of taper must be greater than the angle of repose so that solids do not lodge on the sloping section. Space restrictions and the need for accessibility usually dictate that cyclones in small-scale units be external rather than internal. The superficial velocity range over which the cyclone operates effectively can be extended by installing a hinged flapper at the inlet to the cyclone to vary the inlet width and gas velocity. Small diameter multiclones are available commercially. Collected solids should be returned to the bed just above the distributor, with the standpipe to be not less than 50 mm in diameter. Circulating bed columns should be at least 6 m tall, with the solids returned to the main column by a solids seal, e.g. by an L-valve, J-valve, or (slow) fluidization section.

13.3.4   Wall Roughness

Surface roughness may affect fluidization behaviour, especially in the slug flow regime (Ormiston, Mitchell, and Davidson, 1965; Geldart, Hurt, and Wadia, 1978) or for immersed tubes (Yates *et al.*, 1984). Roughening probably leads to a condition approaching 'no slip' at the boundary, whereas

Table 13.3 Details of some representative high pressure fluidization studies

Reference	Column dimensions	Maximum pressure, bars	Aspect studied
May & Russell (1954)	5 cm i.d. × 1.5 m	69	Entrainment, appearance, bed density
Lanneau (1960)	7.6 cm i.d. × 9.1 m	5	Density fluctuations, reaction
Godard and Richardson (1968)	10 cm i.d. × 91 cm	14	Particulate expansion
Horsler and Thompson (1968)	Various	69	Hydrogenation of crude oil, gasification of coal and char
Kavlich and Lee (1970)	16 cm i.d. × 2.7 m	86	Electrothermal gasification of coal char
Lee, Pyrcioch, and Schora (1970)	10 cm i.d.	138	Coal hydrogasification
Mogan, Taylor, and Booth (1971)	.25 cm × 1.7 cm × 29 cm	53	Bed expansion
Botterill and Desai (1972)	11.4 cm i.d.	10	Heat transfer
Katosonov et al. (1974)	11.4 cm i.d.	220	Bed density, bubble size
de Carvalho (1976)	5 cm i.d. × 2.4 m	22	Bubble size, bed expansion, elutriation, mass transfer
Knowlton (1977)	29 cm i.d. × 3.3 m	69	Minimum fluidization, minimum bubbling, appearance
Crowther and Whitehead (1978)	2.7 cm i.d. × 92 cm	69	Minimum fluidization, minimum bubbling, expansion, regimes
de Carvalho, King, and Harrison (1978)	10 cm i.d. × 1.6 m	28	Bed expansion and fluctuation
Staub and Canada (1978)	31 cm × 31 cm × 3.2 m	10	Flow regimes, bed expansion, heat transfer
Subzwari, Clift and Pyle (1978)	46 cm × 1.5 cm × 67 cm	7	Flow distribution, bubble properties
Varadi and Grace (1978)	31 cm × 1.6 cm × 51 cm	22	Minimum bubbling, bubble stability
Borodulya, Ganzha, and Podberezky (1980)	10.5 cm diameter × 45 cm	81	Heat transfer
King and Harrison (1980)	10 cm diameter	25	Bubble and slug properties
Knowlton and Hirsan (1980)	30 cm diameter semi cylindrical	52	Jet penetration
Chan and Knowlton (1984)	29 cm diameter	31	Entrainment
Rowe et al. (1984)	17.5 cm × 12.5 cm × 50 cm	80	Minimum fluidization, bubble properties
Weimer and Quarderer (1985)	13 cm diameter × 2.7 m	82	Dense phase voidage and flow, bubble size

particles can slide or roll freely along smooth walls. As a result, roughened walls can promote arching or bridging of particles.

Since wall effects are unlikely to be significant in large columns, laboratory-scale columns should be smooth walled. Offsets, protruding tubes, recesses, etc., which could act as wall roughness elements, should be minimized.

13.3.5   High Pressure Columns

There has been considerable interest in recent years in fluidized beds operating at elevated pressures because of improvements in the homogeneity of fluidization, increased gas throughputs, and chemical equilibrium considerations. 'High pressure fluidization' generally refers to pressures greater than 5 bar, with the range 10 to 80 bar receiving most attention. Details of columns used in some high pressure studies are given in Table 13.3. We have not included details of pressurized fluidized bed combustion (PFBC) pilot plants which are summarized in a recent report (Grace, Lim and Evans, 1983).

Visual observation of beds operating under high pressure has been achieved by:

(a) Thick-walled or toughened transport materials (de Carvalho, 1976; Subzwari, Clift, and Pyle, 1978; Varadi and Grace, 1978)

(b) X-ray photography (de Carvalho, 1976; Rowe et al, 1984)

(c) Enclosing a thin-walled transparent column in a pressurized steel vessel equipped with sight glasses (Knowlton, 1977; Chan and Knowlton, 1984; Chitester et al., 1984)

Gas may be supplied either via a compressor or from pressurized cylinders. Special care is required in venting the exit air to atmosphere to prevent excessive noise and because particle entrainment increases with increasing pressure. All instruments must be capable of withstanding the design pressure. For example, measuring the pressure drop across the bed may require construction of special manometers because of the high absolute pressures. In addition, special precautions may be required to prevent backflow of solids into the windbox during pressurization of the column. Safety standards must be rigidly adhered to when designing and operating high pressure equipment.

13.3.6   Columns Operating Under Vacuum

There has been little work on fluidization under vacuum. Zabrodsky, Antonishin, and Parnas (1976) report some Russian work on heat transfer under vacuum. Germain and Claudel (1976) describe a set-up which was used to study fluidization at pressures less than 30 torr.

13.3.7 Elimination of Electrostatic Effects

The following conclusions can be drawn from the literature (Loeb, 1958; Harper; Richardson and McLeman, 1960; Soo, Ihring, and El Kouh, 1960; Torobin and Gauvin, 1961; Boland and Geldart, 1971/72):

(a) Particulate systems exhibit charging as a result of particle–particle and particle–wall rubbing. The latter becomes more important for units of large wall area (e.g. two-dimensional beds) and fine particles. Both positive and negative charged particles are present and there is some evidence that the air stream may hold a balance of charge.

(b) During fluidization, strong electrostatic forces are likely to be set up in beds of dielectric particles (see Lewis, Gilliland, and Bauer, 1949; Miller and Logwinuk, 1951; Davies and Robinson, 1960; Ciborowski and Wlodarski, 1962; and Katz and Sears, 1969). The charges may alter the fluidization characteristics leading to increased pressure drops, decreased bed expansion, and a general trend towards aggregative behaviour or even channelling conditions.

To eliminate static electrification effects, different techniques can be used:

(a) The use of humid air (60 to 70 per cent. relative humidity) has been shown to increase the conductivities of insulators. Generated charges in the fluidized bed are thus neutralized or conducted to earth.

(b) Insulating surfaces can be treated with conductive films (Forest, 1953; Gabor, 1967) or with a layer of tape, as discussed above for two-dimensional columns. Electrically conducting plastics and transparent metallic coatings are now available commercially.

(c) Air may be ionized by electrical discharge or by radioactive sources (Henry, 1953; Loeb, 1958). Anti-static sprays available commercially may also be added periodically or continuously to the fluidizing gas upstream from the distributor.

13.4 GAS DISTRIBUTORS FOR LABORATORY-AND PILOT-SCALE COLUMNS

At one time it was common practice to use porous plates (e.g. sintered bronze, sintered plastic, or woven stainless steel) as distributors in laboratory-scale fluidized beds. This practice has declined for the following reasons:

(a) Commercially available sintered plates tend to have variations of thickness or permeability which lead to dead regions and regions of high flow. Improvements can sometimes be made on an *ad hoc* basis e.g. by adding a coating layer of glue to reduce flow through overly active regions, but this is time-consuming and uncertain.

(b) These materials tend to be expensive, fragile, and difficult to machine. Pores may also become blocked by entrained dust or oil.

(c) Results are unlikely to be representative of large-scale fluidized beds since porous plates give rise to a large number of very small bubbles while distributors used in commercial operations (see Chapter 4) tend to give jets and much larger initial bubble sizes.

Drilled or punched plates are often employed in laboratory-scale columns. The pressure drop across a perforated plate distributor, Δp_D, can be calculated from the equation:

$$UA = C_d n_{or} A_{or} \sqrt{\frac{2\Delta p_D}{\rho_g}} \qquad (13.1)$$

where C_d is a discharge coefficient which can usually be assigned the value of 0.6 (Kunii and Levenspiel, 1969). The ratio of Δp_D to the pressure drop across the bed should be in the range of 0.1 to 0.5, with the higher value advisable for materials which are difficult to fluidize, shallow beds ($H/D < 0.5$), and applications which are sensitive to bubble size such as certain chemical reactions. For most purposes, a value of 0.3 is quite sufficient. An equation for calculating the ratio of distributor/bed pressure drop has been proposed by Sathiyamoorthy and Rao (1981). The holes should be uniformly distributed across the plate. The size of hole is governed by a compromise between competing factors:

(a) Small orifices, especially those smaller than about 1 mm in diameter, are difficult and expensive to drill or punch.

(b) Better performance is usually given by a large number of small holes than a small number of large holes having the same free cross-sectional area.

The usual hole diameter adopted in experimental fluidized beds is in the range 1 to 6 mm.

Fakhimi and Harrison (1970) proposed a criterion for keeping all grid holes unblocked and in operation. In small-scale units, hole blockage and backflow of solids can be prevented by stretching a piece of cloth or wire mesh over the perforated plate. The cloth or screen is sometimes put on the underside of the plate. This prevents backflow of solids, but is not as effective in preventing hole blockage.

Other types of distributor which may be useful in the laboratoy include:

(a) Filterpaper, cloth, or screens sandwiched between commercially available punched plates, the plates typically having a free cross-sectional area of about 30 per cent.

(b) A shallow packed bed of heavy particles (e.g. lead shot) supported on a screen or on a plate of high free cross-sectional area. For the advantages of this type of distributor, see Hengl, Hiquily, and Couderc (1977).

Use of a low pressure drop screen as the sole gas distributor is not recommended. Similarly, single-orifice plates should not be used except for spouted beds.

When a small-scale pilot-scale unit is to be the basis for scale-up, the distributor should be a full-size portion of that in a larger unit, not a scaled-down version. For design of grids for large-scale fluidized beds, see Chapter 4 and the recent paper by Geldart and Baeyens (1985).

13.5 HYDRODYNAMIC STUDIES

13.5.1 Photographic Techniques

Ciné photography has been one of the chief techniques which has been used to investigate bubble behaviour and particle motion in fluidized beds. Four different set-ups have been employed:

(a) Two-dimensional columns, generally viewed with the aid of back lighting as shown in Fig. 13.2

(b) X-ray or γ-ray photography of three-dimensional columns (e.g. Orcutt and Carpenter, 1971; Rowe, 1971)

(c) Three-dimensional columns photographed from the side through the wall or semi-cylindrical columns (see Section 13.3.32) photographed using front lighting

(d) Three-dimensional columns photographed from above with the aid of top lighting

Each of these has its limitations. Two-dimensional fluidized beds give quantitative differences from three-dimensional beds, as discussed in Section 13.3.1. For X-ray photography, the maximum bed thickness is of order 0.2 m, and gas flowrates must be modest to allow identification of individual bubbles. Normal photography of three-dimensional beds only yields information about the behaviour at the wall, which is generally not representative of behaviour in the interior. Photography from above requires low gas flowrates in order that the bed surface should not be obscured by particles ejected into the freeboard. Despite these limitations, each of these methods has yielded some useful data within the constraints mentioned (Lyall, 1969).

For bubbling fluidized beds, bubble and maximum particle velocities are of order 1 m/s. To obtain resolution within 1 mm therefore requires a shutter speed of about 10^{-3} s or better. Most ciné photography of bubbling fluidized beds has been at speeds of 24 to 80 frames/s, with frame exposure times ranging from 0.02 to 0.002 s. Hence, there is some inherent ambiguity in defining the position of individual particles or bubble boundaries. Sharper resolution can be obtained with high speed photography. High speed photography is also required to view high velocity beds where superficial velocities exceed 3 m/s.

Table 13.4 Example of probes in fluidization research

Reference	Type	Details	Property measured
Gerald (1951)	Piezoelectric crystal	Diaphragm stretched over ring of 5 cm diameter	Particle impacts
Osberg (1951)	Hot wire	Wire supported on mica sandwiched between steel supports	Bed level
Morse and Ballou (1951)	Capacitance	Three parallel plates	Uniformity of fluidization
Bakker and Heertjes (1958)	Capacitance	Three parallel plates (see Fig. 13.4a)	Local void fraction
Yasui and Johanson (1958)	Light transmission	Two probes, each with 0.3 mm diameter lamp and 4 mm prism	Bubble properties
Dotson (1959)	Capacitance	Coaxial cell, 5 cm diameter × 7.6 cm long with central rod and six grounded outer rods	Uniformity of bed density
Lanneau (1960)	Capacitance	Stainless steel strips 19 mm long and 7.6 cm apart	Local voidage fluctuations
Fukuda, Asaki, and Kondo (1967)	Capacitance	1.5 mm diameter brass tips separated by 24 mm	Bed uniformity, bubble properties
Kunii, Yashida, and Hiraki (1967)	Capacitance	Not specified	Bubble frequency
Whitehead and Young (1967)	Light transmission	196 probes, each with 6 mm spacing between source and photodiode	Bubble properties
Park et al. (1969)	Electroresistivity	1 mm of two tips exposed, 9.5 mm apart	Bubble frequency, volume fraction, pierced length
Behie et al. (1970)	Pitot tube	10 mm o.d. ellipsoidal nosed tube with backflushing	Jet momentum
Cranfield (1972)	Inductance	Signal coil encapsulated in epoxy resin; forms cylinder 2 mm diameter x 4.5 mm long	Bubble properties
Geldart and Kelsey (1972)	Capacitance	Two plates 5 mm apart	Bubble properties
Pattureaux, Vergnes, and Mihe (1973)	Piezoelectric	Sensor 4 mm diameter connected to piezoelectric element	Particle collision frequency and amplitude
Werther and Molerus (1973)	Capacitance	Needle 3 mm long and 0.4 mm diameter protrudes from 1.1 mm diameter tube (see Fig. 13.4b)	Bubble properties
Werther (1974)			
Calderbank, Pereira, and Burgess (1976)	Resistivity	Five prongs, three-dimensional (see Fig. 13.4d)	Bubble properties, mass transfer

Reference	Method	Details	Measurement
Okhi and Shirai (1976)	Light transmission	Fibre optics (see Fig. 13.4c)	Local particle velocity
Rietema and Hoebink (1976)	Capacitance and suction	Capacitance probes as Werther combined with suction capillary alongside	Bubble properties coupled with sampling from bubble phase
Mayhofer and Neuzil (1977)	Capacitance	Live central rod with six longer earthed outer rods	Bed uniformity near grid
Vines (1977)	Thermistor	1 mm diameter bead	Regimes near distributor
Masson and Jottrand (1978)	Light transmission	Two light-emitting diode-photodiode pairs with 5 mm gap, one pair 20 mm above the other	Bubble properties
Dutta and Wen (1979)	Optical	LED and photocell separated by 3 mm or more	Bubble properties
Ishida, Hishiwaki, and Shirai (1980)	Light reflectance transmission	Fibre optics bundles connected to signal processing set-up	Velocity and direction of particle flow; diameter and velocity of bubbles
Okhi, Ishida, and Shirai (1980)	Light reflectance	Array of fibre optics sensors	Gives image showing jets and bubbles
Lau and Whalley (1981)	Thermocouples	Measures changes and differences in temperature	Defluidization of agglomerating solids
Ljungstrom and Lindqvist (1982)	Zirconia electro-chemical cell	Commercial cell made by Bosch mounted on tube	Local instantaneous oxygen concentration
Patrose and Caram (1982)	Optical	15 mW laser, five illuminating fibres of diameter 0.13 mm in 0.85 mm tube and correlator	Local particle velocities
Glass and Mojtahedi (1983)	Fibre optic	Four sensors, one ahead of the others	Bubble properties
Raso, Tirabasso, and Donsi	Impact	Deformable 2.3 mm o.d. element in steel shell; strain gauge	Velocities and concentrations of particles in jet region
Flemmer (1984)	Pneumatic: jet impingement	Three 0.5 mm diameter tubes in water-cooled 1.6 mm sheath	Void detection, bubble properties
Ishida and Hatano (1984)	UV light absorption	0.8 mm diameter aligned fibres with screen cylinder in between	Concentration of tracer gas
Glicksman and McAndrews (1985)	Thermistor	2 Thermistors 2 mm apart below 10 W heater	Downward velocity of particles

When it is required to follow the motion of individual particles, front illumination is required, and it is helpful to have a small fraction (typically 1 to 2 per cent.) of the particles coloured to aid identification. It is possible, for example, to apply a surface coating of paint to spherical particles by rolling them down an incline to which a thin layer of fresh paint has been applied. Coloured glass beads are also available commercially. It is important that the properties of the marked particles be representative of those of the entire assembly of particles. Streak photography and use of successive flashes have not found much application in fluidization research.

Frame-by-frame analysis of bubble properties presents some special problems, especially when the flow due to translation of bubbles is measured (Grace and Clift, 1974). These problems arise from transfer of gas between separate bubbles through the permeable intervening dense phase. Bubble areas may be measured by means of a planimeter, by weighing paper cutouts traced from the projected images, or by visually comparing the areas with a set of standard circles.

High speed television recorders are now becoming available. The advantages of instant playback are considerable, and although expensive at present, video systems are likely to displace ciné photography quite rapidly.

13.5.2 Probes in Fluidized Beds

Because fluidized beds are opaque and behaviour near the walls is often not representative of that in the interior, many workers have inserted probes to obtain local property measurements. A partial listing of these probes appears in Table 13.4, and some sketches showing probe geometries are given in Figure 13.3. For earlier reviews of probes in fluidized beds see Dutta and Wen (1979) and Fitzgerald (1979).

The following characteristics are required of immersed probes (Werther and Molerus, 1973):

(a) Minimal disturbance of the bed

(b) Measurement of local properties

(c) Rapid reponse to transients

(d) Mechanical strength

(e) Mobility, i.e. it should be possible to relocate the probe

(f) Compatibility with the solids and gas

Almost all of the probes employed in fluidized studies have been unique 'homemade' devices, rather than items which can be purchased from suppliers. Many of the probes have distinguished between times when a small measuring volume is devoid of particles and when particles are present. The physical property used to discriminate between dense phase and dilute phase

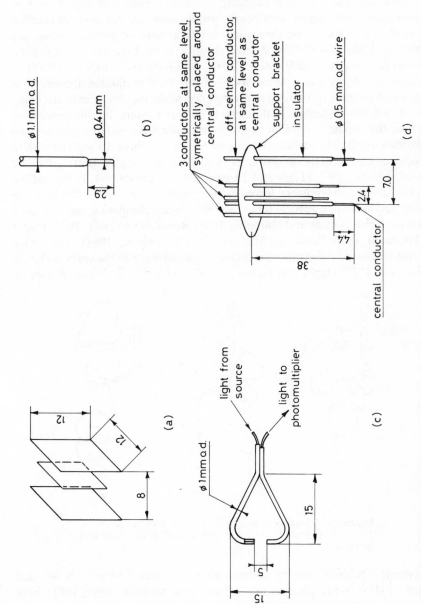

Figure 13.4 Typical probe geometries: (a) early capacitance probe of Morse and Ballou (1951); (b) miniaturized capacitance probe of Werther and Molerus (1973); (c) optical probe of Ohki and Shirai (1976); (d) three-dimensional resistivity probe of Calderbank et al (1976). All dimensions are in millimetres.

has most often been capacitance, light transmission, or electical resistivity. Recording of the probe output gives a direct indication of the uniformity of fluidization and the degree of bubbling. Data processing techniques, such as cross-correlation of signals from nearby probes, can give detailed information on bubble properties. For discussions of treatment of probe signals, see Werther and Molerus (1973), Werther (1974), Park, Lee and Capes (1974), Calderbank, Pereira, and Burgess (1976) and Gunn and Al-Doori (1985).

Probes inevitably cause some local disturbance in fluidized beds. It is therefore important to minimize interference while measurements are being taken. Horizontal cylinders in fluidized beds lead to some bubble generation at or near the equator, to a film of gas on the underside, and to a stagnant cap of particles on the top surface (Glass and Harrison, 1964). In addition, when bubbles impinge directly on obstacles of this nature, bubble splitting may be induced (Cloete, 1967; Hager and Thomson, 1973; Lockwood, 1977) and the object is subject to transient forces (Hosny and Grace, 1984). Bubbles can envelop and adhere to vertical tubes, and this causes elongation and increased velocities of rise (Grace and Harrison, 1968; Rowe and Everett, 1972; Hager and Thomson, 1973; Rowe and Masson, 1980; Yates *et al.*, 1984). Adherence may also be caused if the tube is inclined at a small angle to the vertical (Rowe and Everett, 1972; Hager and Thomson, 1973). If a tube is allowed to vibrate

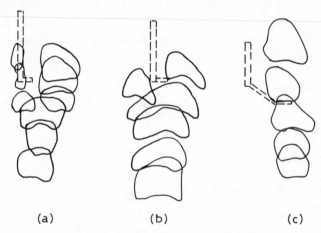

(a)   (b)   (c)

Figure 13.5 Sequences showing interaction of probes and rising gas bubbles, from Geldart and Kelsey (1972). *Reprinted with permission.*

transversely, bubbles can be formed along its entire length (Rowe and Everett, 1972). X-ray photographs (Rowe and Masson, 1980, 1981) have shown that probes can cause bubbles to elongate, to accelerate along the support, to deflect sideways, to flatten prior to reaching the probe, and to split. Small bubbles were found to be more subject to disturbance by the probe than large ones. Probes near the wall of the column, horizontal supports, and multi-prong probes were shown to cause severe interference.

These observations can be used as a guide to designing probes for fluidized beds. Generally speaking, vertical supports are preferable to horizontal supports. The probe itself should be as small and have as few elements as possible. The supporting tube should be suspended from above where possible and at least 30 particle diameters in from the wall of the column. The support should be sufficiently rigid to withstand the buffetting action of the bed with minimal vibration. Geldart and Kelsey (1972) showed the importance of support geometry with respect to whether or not bubble splitting is induced. Their photographs showed that capacitance detector plates, mounted directly on the bottom of an 8 mm diameter suport tube at right angles to the tube, caused bubble splitting and some adherence to the tube. Considerably less interference occurred if the plates were offset from the tube by means of a short, smaller diameter tube inclined at an angle of 45° to the horizontal, as shown in Fig. 13.5(c). Conversely, Rowe and Masson (1981) found a disturbing lack of agreement when comparing bubble heights and velocities measured by X-rays and measured by a light probe. On the other hand, Lord et al. (1982) checked their optical probe data against video records of a two-dimensional bed and found that the probe could be used to provide accurate data. Gunn and Al-Doori (1985) found that proper signal processing and noise discrimination were of critical importance in obtaining accurate data from probes. Caution is clearly required in the use of probe data.

When a probe consists of two parts separated by a gap, e.g. a light source and photodetector or two plates forming a capacitor, the gap width should be sufficient to allow particles to circulate freely through the gap. For rounded particles of narrow size distribution, this requires a gap width of at least 20 to 30 mean particle diameters. Somewhat larger gap widths are required for angular particles or for powders having broad size distributions.

13.5.3 Capacitance Measurements in Fluidized Beds

Many of the probes used in fluidization research have used capacitance measurements to record fluctuations in local bed porosity (see Table 13.4). In addition, a number of workers have avoided probe interference effects by affixing capacitance plates to opposite walls of experimental columns. For two-dimensional beds, resolution has been good enough to allow profiles of local voidage around bubbles (Lockett and Harrison, 1967) and tubes (Fakhimi and Harrison, 1980) to be determined in the manner. For three-dimensional columns, a capacitance signal across the entire column diameter or width allows determination of the arrival of large bubbles, thereby facilitating the measurements of slug or bubble velocities (Angelino, Charzat, and Williams, 1964; Ormiston, Mitchell, and Davidson, 1965). Capacitance elements have also been mounted on vertical and horizontal tubes (Ozkaynak and Chen, 1978; Chandran and Chen, 1982; Fitzgerald, Catipovic, and Jovanovic. 1981) to allow measurements of bubble properties and local voidage which can be related to heat transfer models.

The principle employed in using capacitance variations is simple. The capacitance of solid media differs markedly from that of solids-free gas. The capacitance of a region therefore depends on the concentration of solid particles in that region. Hence, local capacitance variations can be used to provide information on local voidage variations, emulsion density, and on the arrival or departure of bubbles. The probe can be water cooled to allow measurements in high temperature beds (Almstedt and Olsson, 1982), although this requires an increase in probe size. A disadvantage of the capacitance technique is that the measuring volume is never precisely defined. Suitable electrical circuits for monitoring capacitance variations have been presented in a number of the cited references. Werther and Molerus (1973) describe a method of discriminating between signals due to arrival of bubbles and those corresponding to dense phase voidage fluctuations. In their case, electronic counting was found to be better than auto-correlation of signals from nearby probes for measuring bubble velocities. Gunn and Al-Doori (1985) chose a threshold acceptance voltage on the basis of two-dimensional bed ciné films and rejected apparent bubbles with velocities outside a realistic range of 0.05 to 2 m/s.

13.5.4 Other Non-Interfering Diagnostic Techniques

Limited information on the hydrodynamics of fluidized beds can also be obtained by other techniques which are non-interfering. These techniques are also applicable at high bed temperatures and pressures:

(a) By measuring the expanded bed height and height fluctuations visually, some information can be inferred on bubble sizes and whether or not slugging is occurring (King and Harrison, 1980).

(b) Gamma-ray attenuation can be used to measure bed level, density profiles and phase hydrodynamics (Batholomew and Casagrande, 1957; Baumgarten and Pigford, 1960; Saxton and Worley, 1970; Schuurmans, 1980; Weimer, Guyere and Clough, 1985). X-ray absorption can be used in a similar manner (Bierl et al. 1980).

(c) Microwave scattering has been used to provide information on bubble sizes and velocities (Eastlund, Jellison, and Granatstein, 1982).

(d) Pressure fluctuations provide information on hydrodynamics, as discussed further below.

Table 13.5 Experimental methods used in studying solids mixing and motion

Method/principle	Some references	Quantity measured
High speed photography in two-dimensional bed	Rowe *et al.* (1965) Gabor (1967) Baeyens and Geldart (1973) Loew, Schmutter, and Resnick (1979) Tanimoto *et al.* (1980)	Wake, drift, particle velocities
Addition of tagged tracer particles	Geldart and Cranfield (1972) Whitehead, Gartside, and Dent (1976) Bierl *et al.* (1980)	Solids circulation and velocity
X-ray photography	Nienow and Cheesman (1980)	Segregation patterns
Drag force or impact of particles upon piezoelectric crystal	Heertjes, Verloop, and Willems (1970/71) Donsi and Massimilla (1972)	Local velocity
Rotation of small impeller	Botterill and Bessant (1976)	Local velocity
Heat transfer using thermistor beads	Marsheck and Gomezplata (1965) Vines (1977) Wen *et al.* (1978)	Local velocity
Tracking of radioactive particles	Guinn (1956) Kondukov (1964) Littman (1964) Van Velzen *et al.* (1974) Masson, Jottrand and Dang Tran (1978) Lin, Chen, and Chao (1985)	Local velocity
Tracking of radiopill tracer	Handley and Perry (1965) Merry and Davidson (1973) Rao and Venkateswarku (1973)	Local velocity
Transit time of heated particles	Kawanishi and Yamazaki (1984) Valenzuela and Glicksman (1984a)	Local velocity
Transit time of particles viewed by optical fibre light detectors	Schulz-Walz (1974) Okhi, Walawender, and Fan (1977) Okhi *et al.* (1980) Patrose and Caram (1982)	Local velocity
Thermistor probe	Glicksman and McAndrews (1985)	Local velocity
Response curves using magnetic tracer	Woollard and Potter (1968)	RTD $C(t)$ curve
Chemically active tracer	Kamiya (1955) Yagi and Kunii (1961)	RTD $C(t)$ curve
Radioactive tracer	Guinn (1956)	RTD $C(t)$ curve
Fluorescent tracer	Qin and Liu (1982) Morooka, Kayo, and Kato	Pathlines RTD
Ferromagnetic tracer and inductance	Avidan (1980)	RTD, solids circulation

(e) The rate of collapse of the bed surface after interrupting the gas flow to the bed can be used to estimate the voidage of the dense phase (Rietema, 1967; Geldart and Abrahamsen, 1980; and Chapter 2).

(f) Sound emission from the column may allow an experienced operator to identify slugging conditions or stick-slip flow.

(g) Heat flux probes or temperature-sensitive paint can give indications of hot spots and refractory damage.

13.6 SOLIDS MIXING AND SAMPLING

13.6.1 Overall Approaches

Solids mixing can be examined with a view to determination of axial or radial dispersion coefficients, gross solids circulation rates or turnover times, bubble-induced particle motion, or residence time distribution of solids in the system. Experimental methods for determining particle velocities and residence time distribution (RTD) are summarized in Table 13.5.

Due to the limitations of visual observation and the tedious nature of tracer-suction methods, methods have been developed for direct measurement of local particle velocities inside the fluidized bed. Local velocities are then detected by a small obstacle which detects the drag force exerted on it by a flow of particles, by measurement of the rate of heat transfer, by following the path of a tracer, or by statistical correlation of the optically observed movement of particles. The RTD is determined from response curves using magnetic, chemically active, or radioactive tracers.

The various techniques of Table 13.5 have advantages and disadvantages. For a method to be successful, the particle flow should be obstructed as little as possible, and detection should be independent of particle mass and overall bed characteristics such as voidage. Tracer and optical transit methods seem most appropriate, although the latter requires the use of electronic signal processing. The tracking method developed recently by Lin, Chen, and Chao (1985) is also very powerful, but requires multiple detectors and sophisticated and extensive data storage and processing facilities.

13.6.2 Solids Sampling Techniques

Ernst, Loughnane, and Bertrand (1979) describe a simple device for sampling in-bed solids during operation of the Exxon coal combustion 'miniplant'. A closed stainless steel pipe was made to slide within an open-ended close tolerance pipe by means of pneumatic pistons. Particles enter when the inner pipe is slid outwards exposing a small hole near the closed end. These particles are pushed into a storage vessel outside the combustor by the

Table 13.6 Overall approached to gas mixing patterns

Technique	Typical references	Limitations and comments
Stimulus–response	Gilliland and Mason (1952) Yoshida and Kunii (1968) Atimtay and Cakaloz (1978)	Only overall response is obtained and gas contact time history (bubble gas versus emulsion gas) is ignored
Diffusion-type (dispersion)	Overcashier, Todd, and Olney (1959) Baerns, Fetting, and Schugerl (1963) Kobayashi and Arai (1964) Schugerl (1967) Van Deemter (1980) Yang, Huang and Zhao (1984)	Models assume small and random mixing steps. Approach is better for radial than for axial mixing. Most common approach
Two-phase or two-region models	May (1959); Kobayashi and Arai (1964) Kwauk, Wang, Li Chen, and Shen (1986)	It is virtually impossible to evaluate separately gas flow and dispersion in phases and interphase exchange. Some parameters are often assumed with one or more parameter then chosen to fit the experimental data
Backmixing studies	Cankurt and Yerushalmi (1978) Whitehead *et al.* (1980)	Tracer is looked for upstream of injection level. Little information is obtained on the mechanism of mixing or for systems with limited axial dispersion

pressure (imposed and hydrostatic). Sampling is stopped by retracting the inner pipe to cover the sampling port. Similar simple devices have been described by other workers (e.g. Ljungstrom and Lindqvist, 1982). Because of overall circulation patterns in fluidized beds and dead regions near the grid, it is best to provide a sampling system which can sample solids from more than a single locality. While fluidized beds are generally well mixed, it is best to avoid the regions immediately above the distributor and below the bed surface if representative samples are to be withdrawn from a single level. This advice applies also to sampling from defluidized (slumped) beds. Solids captured in the cyclones can conveniently be sampled by means of lock hoppers or Y-junctions installed between the cyclone and return line.

13.7   GAS MIXING AND SAMPLING

13.7.1   Overall Approaches

The flow pattern of gas in a fluidized bed has been studied both by steady/ unsteady state tracer techniques and by calculations based on reactor performances using reactor models. Results have been expressed by means of stimulus–response curves, diffusion-type models, and a variety of more or less sophisticated two-region models. Limitations of these approaches and typical references appear in Table 13.6.

13.7.2 Gas Sampling Techniques

Gas sampling is often required to measure local gas concentrations in chemical reaction, mass transfer, and gas mixing studies. A requirement for sampling is that the sampling rate be sufficiently small, that there is negligible interference with the surrounding bed, and that a layer of solids, held by suction, does not build up on the sample tube. These conditions are considered to be satisfied if the sampling tube is much smaller than the bed dimensions and if the mean gas velocity in the tube is U_{mf} or less. The end of the sampling tube must be covered with a fine mesh screen or porous tip to prevent inflow of particles. Samples may be withdrawn continuously or in pulses. Tubes may be inserted from the side of the bed or from above. For continuous sampling, it is important that pressure fluctuations (associated with the passage of bubbles, for example) be small relative to the overall pressure driving force in order to obtain a constant flowrate and a proper time-mean concentration.

In testing of reactor models, it is desirable to obtain separate measurements of concentration in the bubble and dense phases. For a two-dimensional column or in the wall region of a three-dimensional bed, this can be achieved by manually actuating a solenoid valve (Chavarie and Grace, 1975) or by synchronization of concentration traces with ciné films (Lignola, Donsi, and Massimilla, 1983). For the interior of a three-dimensional column, more sophisticated electronics are required by which opening of a solenoid is controlled by the output from an optical probe or other sensor (see Table 13.4) capable of distinguishing between the two phases, e.g. the combined capacitance suction probe described by Rietema and Hoebink (1976).

To obtain accurate concentration measurements, deposition, chemical reactions and adsorption in the sampling line must be minimized. This may require special materials, heat tracing of the sample line, or quenching of the gases. Use of a carrier gas is sometimes useful. Tubes should be as short as possible. Enlargements, fitting, etc., should be avoided in order to minimize axial dispersion along the sample lines. Since radial gradients are appreciable in many fluidized beds, care must be exercised when a concentration is to be measured which is supposed to represent an average for a given level or an average outlet value.

In-bed concentrations of oxygen at high temperature can also be measured by zirconia electrochemical cells like those used for measured emissions from automobiles (e.g. Ljungstrom and Lindqvist, 1982). However, the signals are difficult to interpret and subject to significant error in the presence of local homogeneous combustion of gaseous components like CO and methane (Lim et al. 1986).

13.8   CHEMICAL REACTION STUDIES IN FLUIDIZED BEDS OF LABORATORY AND PILOT PLANT SCALE

Experimental fluidized bed reactors range from simple tubes fitted with a porous distributor to highly sophisticated pieces of equipment requiring specialized design. Key requirements in planning a reactor are: (a) to make sufficient provision for instrumentation that useful data can be obtained; (b) to foresee possible problems and modifications and make the reactor flexible enough to allow these to be solved or accommodated. It is always more difficult, for example, to add sampling ports after a reactor is operating than to incorporate additional ports at the design stage, especially for pressurized reactors which are subject to special safety codes and inspections. Table 13.7 lists some common problems that arise in fluidized bed reactors and suggests what can be done at the design stage or during operation to solve these problems.

Since their hydrodynamics are not well understood (see Chapter 11), fluidized beds are usually not in themselves suitable for obtaining kinetic information. If heats of reaction are sufficiently small that reaction under isothermal conditions can be achieved, packed beds can be used to generate the required kinetic data. For larger heats of reaction, spinning basket reactors are often useful. The kinetic data obtained must cover not only the full range of temperature to be encountered in the fluidized bed reactor but also the full range of concentrations. Because conversion in the dense phase tends to be much higher than that in the bubble phase, this means obtaining kinetic data under high conversion conditions if reactor models are to be tested.

Methods for measuring gas and solids concentrations by sampling and by in-bed probes are outlined in earlier sections of this chapter. Many reactions carried out in fluidized beds are highly exothermic or involve substances which are potentially explosive. Monitors and alarms are commonly required for in-bed temperature, pressure, flowrates, and concentrations of key components to keep the reactor within safe limits. The integrity of refractory should be checked at regular intervals, and the reactor shell should be investigated for hot spots during operation. Pressure relief valves or rupture discs should be provided and checked regularly. The refractory should be designed to withstand sudden reactor depressurization.

13.9   MEASUREMENT OF ENTRAINMENT AND OF FREEBOARD AND CIRCULATING BED HYDRODYNAMICS

Measurements of entrainment and related phenomena in fluidized beds pose special problems and are subject to large errors. Many workers who have studied entrainment (e.g. Zenz and Weil, 1958) collected solids captured by a

Table 13.7 Some common problem areas which should be considered when designing or operating experimental fluidized bed reactors

Problem	Provision or action to be considered
Solids segregation	Increase gas velocity; ensure uniform distributor; avoid closely spaced internals; return solids from cyclone near grid
Grid blockage	Install screens over holes; ensure high velocity gas entry; provide for removable or accessible grid
Poor gas distribution	Design removable grid; ensure high distributor pressure drop; install screens to avoid hole blockage
Equipment vibration	Provide vibration pads; install surge tank between blower and bed; vary gas flowrate; introduce baffles to break up slugs; furnish viewing ports to observe bed
Erosion	Avoid jet impingement; consider antiabrasion pads; decrease gas velocity
Attrition	Avoid jet impingement; decrease gas velocity
Electrostatic charges	Humidify incoming gas; use antistatic agents; connect entire column to ground
Particle agglomeration	Alter temperature; dry incoming gas; increase gas velocity; avoid dead regions; improve atomization of liquid feed (if any); stir bed contents or add ultrasonic vibration; add inert solids
Solid deposits	Avoid horizontal and shallow surfaces; dry incoming gases; alter temperature and/or gas velocity
Hot spots, overheating, danger of explosion	Prevent dead regions near grids; ensure that adjacent internal surfaces are at least 30 particle diameters apart; increase gas velocity; increase or provide more uniform cooling
Extinguishing of reaction	Change total transfer surface area or driving force; preheat feed gas or solids
Low conversion or unwanted side reactions	Change operating conditions (e.g. temperature, pressure, feed composition, gas velocity); use improved catalyst; provide more uniform distributor; introduce baffles; introduce quenching in freeboard
Excessive carryover	Add cyclone in parallel or series; reduce attrition (as detailed); lower gas velocity or pressure; introduce baffles just above or below bed surface; furnish viewing ports; provide expanding freeboard
Escape of noxious fumes or dust	Test for leaks; operate under negative pressure; enclose column
Blockage of dipleg	Increase dipleg diameter; heat trace dipleg; avoid bends and inclined portions containing dense phase flow
Rupture of tubes	Avoid erosion and vibration (as detailed); choose tube materials with corrosion and erosion in mind, taking account of local concentration fluctuations; allow for thermal expansion; increase tube support
Poor solids feeding	Avoid bends; increase inlet size; dilute feed with inert particles; aerate feed; consider alternate feed method (e.g. screw feeding, pneumatic feeding)
Failure to reach steady state	Avoid attrition, solid deposits, and hot spots (as detailed); preheat feed; improve heat transfer and insulation; provide more reliable feed system for solids and liquid (if any)

single cyclone over a period of time. These particles were then weighed and sieved. This technique has a number of disadvantages:

(a) It assumes perfect collection in the cyclone.

(b) Since fines are not returned to the bed during the period of collection, bed properties change with time.

(c) No information is given on the hold-up of solids in the freeboard.

Merrick and Highley (1972) accounted for essentially all of the particles carried over by adding a second cyclone and sampling the stack gases. Blyakher and Pavlov (1966) allowed entrained particles to be carried over and used the weight change to determine the total entrainment. Changes in bed composition with time may be minimized by ensuring that the mass of solids collected is much less than the total mass of solids in the bed or that the time of sampling is less than the particle cycle time.

Techniques used to determine the hold-up of solids in the freeboard include static pressure profiles (e.g. Nazemi, Bergougnou, and Baker, 1974), attenuation of a laser beam (Horio *et al.*, 1980), and butterfly or disc valves which trap the suspended solids (e.g. Geldart and Pope, 1978). The static pressure method has the disadvantage that frictional losses and particle deceleration, in addition to the hold-up of solids and gas, contribute to the measured pressure drop, making it difficult to determine the hold-up accurately by this method. The trapping valves are only useful if they can be closed very quickly and if they do not disturb the flow in the freeboard when not in service. Static pressure drops are also commonly used to measure suspension densities in circulating beds. A single porous butterfly valve in the return vessel has been used by a number of investigators beginning with Yerushalmi, Turner, and Squires (1976) to determine solids circulation rates in circulating bed systems.

To determine the transport disengaging height experimentally, the flux of solids should be measured. Since the flux tends to vary with radial position as well as with height, both radial and axial profiles are required. Fournol, Bergougnou, and Baker (1873) employed an isokinetic suction probe, where the suction velocity was set equal to the superficial gas velocity to obtain local flux measurements. A difficulty with isokinetic sampling is that bubble arrival at the bed surface is irregular, causing particle ejection in bursts, rather than in a uniform manner. Bubble eruption also causes the gas flow to be highly turbulent, especially in the lower part of the freeboard (Levy and Lockwood, 1980, 1983). Because bubble bursts cause release of both solids and gas, isokinetic sampling requires suction velocities in excess of the superficial gas velocity (Pemberton, 1982). George and Grace (1978) designed a special trapping device to capture the particles ejected by individual bubbles. Soroko, Mikhalev, and Mukhlenov (1969) used photography to determine the transport disengaging height.

Recent workers have realized the importance of turbulence and of separately measuring gas and particle velocities in the freeboard. Hot wire anemometry has been used to measure gas velocities (Horio *et al.*, 1980; Pemberton, 1982); Horio *et al.* used the optical fibre prove of Okhi and Shirai (1976) to measure particle velocities. Laser Doppler anemometry has been employed by Levy and Lockwood (1980, 1983) to measure both gas and particle velocities in the freeboard, as well as to indicate the size of the particles in the measuring volume. These techniques are difficult to apply in fluidization research because of the presence of a broad spectrum of particle sizes and the relatively high hold-up of solids.

13.10 PRESSURE MEASUREMENTS

Time-averaged pressure measurements are useful in determining the minimum fluidization velocity and expanded bed level and in estimating the hold-up of solids in the unit. Pressure taps should be positioned at regular intervals, typically 0.1 to 0.2 m, up the side of the vessel. One tap is usually installed in the windbox below the grid. There should always be a tap just above the grid (Sutherland, 1964). Although the pressure drop across the grid is an indicator of the gas flowrate, it is advisable to measure the gas flowrate by another method (e.g. rotameter or upstream orifice plate) since the distributor holes may become partially blocked. The pressure drop across the distributor plate can be used to indicate whether hole blockage or changes in grid performance have occurred. Whitehead and Dent (1978) have shown that pressure profiles along the distributor plate can be used to deduce information regarding solids circulation patterns there.

For fully fluidized materials, the pressure drop between two taps is very nearly equal to the weight of solids (and gas) held up between the two levels divided by the bed cross-sectional area. Lower measured pressure drops usually indicate chanelling or partial defluidization, some of the weight of the solids being supported by interparticle contact. When absolute pressure is plotted against height, there is a sharp break in slope at the surface of the expanded bed at low gas velocities, and this gives a convenient means of estimating bed depth. Some deviation from a linear profile for the dense bed may occur at the lowest pressure tap, and this has been ascribed (Rowe and Partridge, 1965) to radial gas flow associated with bubble formation. Changes in slope may also indicate solids segregation in the bed (Chiba, Chiba, Nienow and Kobagoshi, 1979).

The pressure measurements described above are generally obtained with simple manometers or Bourdon-type pressure gauges. A gauze or fine screen should be installed flush with the wall of the vessel to prevent particles from entering the pressure taps. For sticky solids, it may be necessary to purge the manometer lines continuously in order to prevent build-up of a deposit over the pressure taps. The linear velocity in the purge line should be 1 to 2 m/s. In some cases, it is more convenient to use a thin tube covered by a screen lowered from the top of the column to measure the pressure at different levels, instead of taps on the wall. For time-mean pressure measurements, it is often useful to damp out pressure fluctuations by having the pressure lines expanded to a tube of larger bore containing fixed particles (e.g. sand), in effect providing a capacitance.

Aside from the information which can be derived from time mean pressure drops, further information may be obtained from pressure fluctuations if one or more taps are connected to a rapid response (e.g. diaphragm type) pressure transducer and high speed recorder. To minimize damping of the signal, the transducer and connecting tubes should have small volumes. Absence of pressure fluctuations may indicate channelling or defluidization. Pressure fluctuations of irregular amplitude and frequency usually indicate normal bubbling, with the mean amplitude becoming larger with increasing bubble size. Large amplitude oscillations of a fairly regular nature normally denote slugging. High frequency, low amplitude fluctuations tend to be associated with the turbulent regime of fluidization (Yerushalmi *et al.*, 1978). Hence, the output from a pressure transducer can be useful as an indicator of the regime of fluidization and of any shifts in bed behaviour or regime (Broadhurst and Becker, 1976; Ho *et al.*, 1983; Svoboda *et al.* 1984). Within the bubbling and slugging regimes, statistical analysis and correlation of pressure fluctuations have been used to infer bubble and slug properties (Sitnai, 1982; Zhang, Walsh and Beér, 1982; Fan, Ho, and Walawender, 1983). Probes can also be prepared which measure local pressure fluctuations, but interpretation of the signals is difficult (Lord *et al.*, 1982).

13.11   HEAT TRANSFER AND TEMPERATURE MEASUREMENT

13.11.1   Gas-to-solids heat transfer and temperature measurements

In both steady and unsteady state experiments, the gas-to-solids heat transfer coefficient is deduced from a heat balance over a differential section of height dl:

$$\begin{pmatrix} \text{Heat into the} \\ \text{section by gas} \end{pmatrix} - \begin{pmatrix} \text{heat out of} \\ \text{section by gas} \end{pmatrix} = \begin{pmatrix} \text{heat transferred} \\ \text{to particles} \end{pmatrix} \quad (13.2)$$

or

$$-C_{pg}U\,\rho_g dT_g = h_{gp}S_B(T_g - T_p)dx \quad (13.3)$$

Solving this differential equation with specific boundary conditions yields the heat transfer coefficient h_{gp}. It is generally assumed that the solids temperature T_p is uniform throughout the bed. This approach also assumes that gas elements associated with the bubble or dense phase are at the same temperature at a given height, i.e. the interphase heat transfer resistance is ignored. Aside from this last assumption, a principal reason for the experimental scatter in reported values of h_{gp} is the uncertain temperature of gas and solids, as measured by thermocouples embedded in the fluidized system. Bare and suction thermocouples have been used. The differences in thermocouple readings have been discussed by various authors. From thermocouple response and use of 'shielded' thermocouples, Baeyens and Goossens (1973) demonstrated that a bare thermocouple reads the solids temperature, although emulsion gas and solids are at the same temperature in a well-mixed bed. Using suction thermocouples (Walton, Olson, and Levenspiel, 1995; Chang and Wen, 1966), a different reading is obtained depending on the pressure drop of the suction probe. A high pressure drop probe yields a value close to T_g, whereas a low pressure drop probe measures a temperature close to the bubble temperature since bubble gas is sampled preferentially.

It is therefore advisable to determine solids and gas temperature using bare microthermocouples and high pressure drop suction thermocouples, respectively. Errors may be caused by radiation or by a build up of solids on the thermocouple. Solids temperatures can be read by other techniques such as thermistors, miniature microelectric devices (Barker, 1967), or heat-sensitive rings (Yates and Walker, 1978). Optical pyrometry can be used at high bed temperatures.

13.11.2 Bed-to-Surface Heat Transfer

Coefficients for wall-to-bed heat transfer are commonly calculated from a heat balance using the area S of the transferring surface and the difference between the wall temperature T_w and bulk bed temperature T_b, both measured by means of thermocouples. Hence:

$$h_{bw} = \frac{Q}{S(T_w - T_b)} \tag{13.4}$$

where Q is the heat transfer rate.

A typical experimental set-up is illustrated in Fig. 13.6. The experiment involves a heater (or cooler) immersed in (or on the containing wall of) the column. Electrical heating can be used instead of circulating a heat transfer fluid inside the heat transfer surface. Axially and radially moveable thermocouples are used to measure bed temperatures; thermocouples are also fixed to the heater or cooler surface. Axial heat losses at the ends of the heat transfer element should be reduced, e.g by guard heaters (Baeyens and Goossens, 1973). As shown in Table 13.8, a variety of heat transfer surfaces has been used including horizontal and vertical tubes, finned and flattened tubes, vertical surfaces, and spheres. Some investigators have measured spatially and time-average heat transfer coefficients, while others have measured time-average local values and still others have devised means of

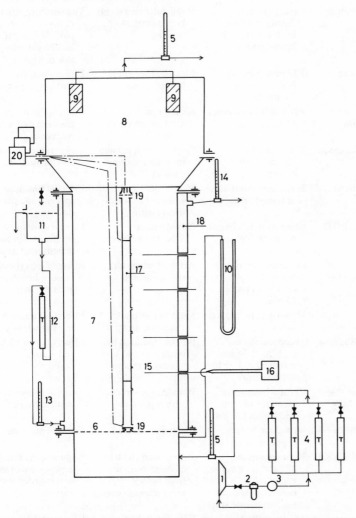

Figure 13.6 Typical experimental set-up for studying bed-to-surface heat transfer: 1 compressor, 2 valve and water trap, 3 pressure gauge, 4 rotameters, 5 thermometer, 6 distributor, 7 fluidization column, 8 expanded freeboard, 9 filters, 10 manometer, 11 constant head tank, 12 water rotameter, 13, 14 thermometers, 15 thermocouples, 16 multi-chanel recorder, 17 heater, 18 cooling jacket, 19 supports for heater, 20 voltage regulators.

Table 13.8   Examples of typical and unusual experimental equipment used to measure bed-to-surface heat transfer coefficients

References	Transfer surface	Details	Aspects studied
Mickley and Trilling (1949)	Vertical 12 mm diameter calrod heater; Externally heated bed also	Single tube on axis of bed, electrical heating	Temperature distributions, effects of particle diameter, bed height, gas-velocity
Jacob and Osberg (1957)	0.13 mm diameter horizontal wire 45 mm long	Electrical heating, bridge circuit	Effect of gas and particle properties
Mickley, Fairbanks and Hawthorn (1961)	Vertical 6.4 mm tube, 610 mm long	Six 100 mm long sections, electrical heating	Influence of height, gas velocity, bubble frequency
Baeyens and Goossens (1973)	10 mm diameter vertical central heater	Single tube; fixed and moveable thermo-couples	Temperature profiles and fluctuations
Bartel and Genetti (1973)	Seven horizontal bare or finned tubes	16 mm tubes with and without fins, electrical heating	Effect of fins, tube spacing, and particle size
Tuot and Clift (1973)	10 mm square vertical platinum film	Guard heater; Wheatstone bridge	Local instantaneous coefficients in two-dimensional column
Andeen, Glicksman, and Bowman (1978)	Flattened horizontal tubes of height 32 mm and width 13 mm	Array of tubes, electrically heated	Effect of superficial velocity bed depth, and tube shape
Canada and McLaughlin (1978)	32 mm o.d. horizontal tubes	5 and 10 row banks, staggered array	Effects of pressure and particle size
Hoebink and Rietema (1978)	30 copper blocks, each 20 × 30 mm arranged in 3 × 10 vertical array	Blocks separated by insulation and kept at same temperature	Effect of heater length and vertical position
Borodulya, Ganzha, and Podberezsky (1980)	Vertical cylinder of diameter 18 mm and length 100 mm	Wooden rod wrapped with copper wire	Effect of pressure and particle diameter
Xavier et al. (1980)	40 × 20 × 10 mm copper block	Electrically heated	Effect of pressure
Abubakar et al. (1980)	30 mm diameter, 277 mm long horizontal tube, isolated longitudinal 5 mm wide strip	Strip heated electrically and insulated from heated remainder of tube	Angular and vertical dependence of local heat transfer coefficient
Botterill, Teoman, and Yuregir (1981)	9.5 mm diameter sphericalorimetric probe	Embedded thermocouple; different surfaces to allow radiation to be evaluated	Effect of particle size and type, bed temperature; contribution of radiation

Table 13.8   (continued)

References	Transfer surface	Details	Aspects studied
Fitzgerald, Catipovic, and Jovanovic (1981)	51 mm diameter tube with five platinum film heating elements around periphery	Resistance heating of elements with simultaneous capacitance and local pressure measurement	Instantaneous local heat transfer coefficients, pressure, and capacitance for horizontal tube
George and Grace (1982)	25 mm o.d. horizontal tubes in arrays	Bed electrically heated; tube cooled by silicone oil	Freeboard transfer, different configurations, and tube spacing
Golan, Lalonde, and Weiner (1982)	100 mm o.d. horizontal tube bundle	Bed heated by propane burning; water cooling	Local, average convective and radiative coefficients measured
Ozkaynak, Chen and Frankenfield (1984)	Radiometer probe of diameter 73 mm	Surface mounted vertically in beds at temperatures to 800°C	Total and radiative heat transfer
Piepers, Wiewiorski, and Rietema (1984)	65 mm vertical tube bundle	One of 19 tubes with 10 electrically heated copper elements	Effect of radial position and height as well as superficial velocity

obtaining instantaneous local values. The listing in Table 13.8 is far from complete. Entries have been chosen to illustrate the wide variety of experimental equipment and to give references where detailed information is provided. Experimental attempts to measure radiation heat transfer in fluidized beds have been reviewed by Ozkaynak, Chen, and Frankenfield (1984).

The methods available for measuring surface temperature and their errors and difficulties are discussed by Baker, Ryder, and Baker (1961) and Walker and Rapier (1965). Thermocouples mounted onto a surface should not perturb the bed or alter the surface roughness. Errors for brazed thermocouples are discussed by Walker and Tapier (1965).

Care should be taken if a direct electrical current is passed through the heat transfer element since pick-up voltages can be of the same magnitude as the thermocouple output. To minimize this effect, electrically insulating cements with high thermal conductivity should be used, e.g. cements based on copper oxide. Care should also be taken to minimize and estimate heat losses in order to obtain accurate surface-to-bed heat transfer coefficients.

Table 13.9 Summary of common instrumentation needs for fluidized
bed equipment

Property	Position	Provision or comments
Gas flowrate	Upstream	Rotameters, orifice plate
Solids feedrate	Inlet	Weigh belt, load cells on hopper
Solids efflux	Outlet	Lock hopper, load cells, filter on cyclone gas outlet
Bed depth	Bed surface	Sight glass, pressure profile, sensor lowered from above, γ-rays
Pressure profile	Bed	Taps at regular intervals connected to manometers
Pressure drop	Grid	Taps just above and below distributor plate
Pressure fluctu-ations	Bed	Tap above distributor connected to transducer and high speed recorder
Temperature	Upstream	Thermocouple in feed line or windbox
Temperature	Bed	Thermocouple in dense bed region (one usually sufficient)
Temperature	Freeboard	Series of thermocouples at regular intervals
Gas composition	Upstream	Sampling from feed stream or windbox to gas chromatograph or other analysers
Gas composition	Bed	Sampling tubes
Gas composition	Freeboard	Sampling tubes
Gas composition	Outlet	Sampling well above bed, e.g. near cyclone inlet
Solids composition	Bed	Grab sampler or exit port(s)
Solids composition	Inlet, outlet	Procedure for collecting representative samples
Heat flux	Tubes, jacket	Thermocouples at inlet and outlet

13.12   INSTRUMENTATION NEEDS

The most common instrumentation needs for fluidized beds are summarized
in Table 13.9. Additional instrumentation may be required for particular
applications.

13.13   NOMENCLATURE

A	cross-sectional area of column
A_{or}	orifice area
$C(t)$	gas concentration response as function of time to pulse stimulus
C_d	discharge coefficient

C_{pg}	specific heat capacity of gas
D	bed diameter
H	expanded bed height
h_{bw}	bed to wall heat transfer coefficient
h_{gp}	gas to particle heat transfer coefficient
n_{or}	number of orifices in the distributor
Q	heat transfer rate
S	surface area of heat transfer surface
S_B	particle surface area per unit bed volume
T_b	bulk bed temperature
T_g	gas temperature
T_p	particle temperature
T_w	wall temperature
U	superficial gas velocity
x	vertical height coordinate
Δp_D	pressure drop across the distributor
ρ_g	gas density

13.14   REFERENCES

Abubakar, M.Y., Bergougnou, M.A., Tarasuk, J.D., and Sullivan, J.L. (1980).'Local heat transfer coefficients around a horizontal tube in a shallow fluidized bed', *J. Powder & Bulk Solids Technol.*, **4**, 11–18.

Almstedt, A.E., and Olsson, E. (1982). 'Measurements of bubble behaviour in a pressurized fluidized bed burning coal using capacitance probes', *Proc. 7th Intern. Fluidized Bed Combustion Conf.*, pp. 89–98.

Andeen, B.R., Glicksman, L.R., and Bowman, R. (1978). 'Heat transfer from flattened horizontal tubes', in *Fluidization* (Eds J.F. Davidson and D.L. Keairns), Cambridge University Press, pp. 345–350.

Angelino, H., Charzat, C., and Williams, R. (1964). 'Evolution of gas bubbles in liquid and fluidized systems', *Chem. Eng. Sci.*, **19**, 289–304.

Anwer, J., and Pyle, D.L. (1973). 'Gas motion around bubbles in fluidized beds', in *La Fluidisation et ses Applications*, Cepadues Editions, Toulouse, France, pp. 240–253.

Atimtay, A., and Cakaloz, T. (1978). 'An investigation of gas mixing in a fluidized bed', *Powder Technol.*, **20**, 1–7.

Avidan, A.A. (1980). 'Bed expansion and solid mixing in high velocity fluidized beds', Ph.D. Thesis, City University of New York.

Avidan, A.A., and Yerushalmi, J. (1982). 'Bed expansion in high velocity fluidization', *Powder Technol.*, **32**, 223–232.

Baerns, M., Fetting, F., and Schugerl, K. (1963). 'Radiale und axiale Gasvermischung in Wirbelschicten', *Chem Ing. Tech.*, **35**, 609.

Baeyens, J., and Geldart, D. (1973). 'Particle mixing in a gas fluidized bed', in *La Fluidisation et ses Applications*, Cepadues Editions, Toulouse, France, pp. 182–195.

452   GAS FLUIDIZATION TECHNOLOGY

Baeyens, J., and Goossens, W.R.A. (1973). 'Some aspects of heat transfer between a vertical wall and a gas fluidized bed', *Powder Technol.*, **8**, 91–96.

Baker, H.D., Ryder, E.A., and Baker, N.H. (1961). *Temperature Measurements in Engineering*, 2nd ed., Wiley, New York.

Bakker, P.J., and Heertjes, P.M. (1958). 'Porosity measurements in a fluidized bed', *Brit. Chem. Eng.*, **3**, 240–246.

Barker, J.J. (1967). 'Microelectronic device for measuring heat transfer coefficients in fluidized beds', *Ind. Eng. Chem. Fund.*, **6**, 139–142.

Bartel, W.J., and Genetti, W.E. (1973). 'Heat transfer from a horizontal bundle of bare and finned tubes in an air fluidized bed', *A.I.Ch.E. Symp. Ser.*, **29**, (128), 85–93.

Bartholomew, R.N., and Cassagrande, R.M. (1957). 'Measuring solids concentration in fluidized systems by gamma-ray absorption', *Ind. Eng. Shem.*, **49**, 428–431.

Baumgarten, P.K., and Pigford, R.L. (1960). 'Density fluctuations in fluidized beds', *A.I.Ch.E.J.*, **6**, 115–123.

Behie, L.A., Bergougnou, M.A., Baker, C.G.J., and Bulani, W. (1970). 'Jet momentum dissipation at a grid of a large gas fluidized bed', *Can. J. Chem. Eng.*, **48**, 158–161.

Bierl, T.W., Gajdos, L.J., McIver, A.E., and McGovern, J.J. (1980). 'Studies in support of recirculating bed reactors for the processing of coal', DOE Rept. FE–2449–11.

Blyakher, I.G., and Pavlov, V.M. (1966). 'Reducing dust entrainment from conical fluidized bed equipment through the use of support grids', *Int. Chem. Eng.*, **6**, 47–50.

Boland, D., and Geldart, D. (1971/72). 'Electrostatic charging in gas fluidized beds', *Powder Technol.*, **5**, 289–297.

Borodulya, V.A., Ganzha, V.G., and Podberezsky, A.I. (1980). 'Heat transfer in a fluidized bed at high pressure' in *Fluidization* (Eds J.R. Grace and J.M. Matsen), Plenum Press, New York, pp. 201–207.

Botterill, J.S.M., and Bessant, D.J. (1976). 'The flow properties of fluidized solids', *Powder Technol.*, **14**, 131–137.

Botterill, J.S.M., and Desai, M. (1972). 'Limiting factor in gas fluidized bed heat transfer', *Powder Technol.*, **6**, 231–238.

Botterill. J.S.M., Teoman, Y., and Yuregir, K.R. (1981). 'Temperature effects on the heat transfer behaviour of gas fluidized beds', *A.I.Ch.E. Symp. Ser.*, **77** (208), 330–340.

Broadhurst, T.E., and Becker, H.A. (1976). 'Measurement and spectral analysis of pressure fluctuations in slugging beds', in *Fluidization Technology* (Ed. D.L. Keairns), Vol. I, Hemisphere, Wasdhington, DC, pp. 63–85.

Calderbank, P.H., Pereira, J., and Burgess, J.M. (1976). 'The physical and mass transfer properties of bubbles in fluidized beds of electrically conducting particles', in *Fluidization Technology* (Ed. D.L. Keairns), Vol. I, Hemisphere, Washington, DC, pp. 115–167.

Canada, G.S., and McLaughlin, M.H. (1978). 'Large particle fluidization and heat transfer at high pressures', *A.I.Ch.E. Symp. Ser.*, **74** (176) 27–37.

Cankurt, N.T., and Yerushalmi, J. (1978). 'Gas backmixing in high velocity fluidized beds', in *Fluidization* (Eds J.F. Davidson and D.L. Keairns), Cambridge University Press, Cambridge, pp. 387–393.

Chan, I.H., and Knowlton, T.M. (1984). 'The effect of pressure on entrainment from bubbling gas-fluidized beds', in *Fluidization* (Eds D. Kunii and R. Toei), Engineering Foundation, New York, pp. 283–290.

Chandran, R., and Chen, J.C. (1982). 'Bed-surface contact dynamics for horizontal tubes in fluidized beds', *A.I.Ch.E.J.*, **28**, 907–914.

Chang, T.M., and Wen, C.Y. (1966). 'Fluid-to-particle heat transfer in air-fluidized beds', *Chem. Eng. Prog. Symp. Ser.*, **62**, (67), 111–117.

Chavarie, C., and Grace, J.R. (1975). 'Performance analysis of a fluidized bed reactor', *Ind. Eng. Chem. Fund.*, **14**, 75–91.

Chiba, S., Chiba, T., Nienow, A.W., and Kobayashi, H. (1979) 'The minimum fluidization velocity, bed expansion and pressure-drop profile of binary particle mixtures', *Powder Technol.*, **22**, 255–269.

Chitester, D.C., Kornosky, R.M., Fan, L–S., and Danko, J.P. (1984). Characteristics of fluidization at high pressure', *Chem. Eng. Sci.*, **39**, 253–261.

Ciborowski, J., and Wlodarski, A. (1962). 'On electrostatic effects in fluidized beds', *Chem. Eng. Sci.*, **17**, 23–32.

Cloete, F.L.D. (1967). *Discussion, Proceedings of International Symposium on Fluidization* (Eds A.A.H. Drinkenburg), Netherlands University Press, Amsterdam, pp. 305–307.

Cranfield, R.R. (1972). 'A probe for bubble detection and measurement in large particle fluidized beds', *Chem. Eng. Sci.*, **27**, 239–245.

Crowther, M.E., and Whitehead, J.C. (1978) 'Fluidization of fine particles at elevated pressures', in *Fluidization* (Eds J.F. Davidson and D.L. Keairns), Cambridge University Press, Cambridge, pp. 65–70.

Davies, G., and Robinson, D.B. (1960). 'A study of aggregative fluidization', *Can. J. Chem. Eng.*, **38**, 175–183.

de Carvalho, J.F.R.G. (1976). 'Fluidization under pressure', Ph.D. Thesis, Cambridge University.

de Carvalho, J.R.F.G., King, D.F., and Harrison, D. (1978). 'Fluidization of fine powders under pressure', in *Fluidization* (Eds J.F. Davidson and D.L. Keairns), Cambridge University Press, pp. 59–64.

Donsi, G., and Massimilla, L. (1972). 'Bubble-free expansion of gas fluidized beds of fine particles', *A.I.Ch.E.J.*, **19**, 1104–1110.

Dotson. J.M. (1959). 'Factors affecting density transients in a fluidized bed', *A.I.Ch.E.J.*, **5**, 169–174.

Dutta, S., and Wen, C.Y., (1979). 'Simple probe for fluidized bed measurements', *Can. J. Chem. Eng.*, **57**, 115–119.

Eastlund, B.J., Jellison, G., and Granatstein, V. (1982). 'Microwave diagnostics of bubbles in fluidized bed combustors', *Proc. 7th Intern. Fluidized Bed Combustion Conf.*, Vol. 1, pp. 99–110.

Ellis, J.E., Partridge, B.A., and Lloyd, D.I. (1968). 'Comparison of predicted with experimental butadiene yields for the oxydehydrogenation of butenes in a gas fluidized bed', *Instn. Chem. Engrs. Symp. Ser.*, **30**, 43–52.

Ernst, M., Loughnane, M.D., and Bertrand, R.R. (1979). 'Instrument methods in fluid bed pilot plant, coal processing technology', *A.I.Ch.E.*, **5**, 166–170.

Fakhimi, S., and Harrison, D. (1970). 'Multi-orifice distributors in fluidized beds: a guide to design', in *Chemeca Proceedings*, Butterworths, Australia, pp. 29–46.

Fakhimi, S., and Harrison, D. (1980). 'The void fraction near a horizontal tube in a fluidized bed', *Trans. Instn Chem. Engrs*, **58**, 125–131.

Fan, L.T., Ho, T–C., and Walawender, W.P. (1983). 'Measurements of the rise velocity of bubbles, slugs and pressure waves in a gas–solid fluidized bed using pressure fluctuation signals', *A.I.Ch.E.J.*, **29**, 33–39.

·Fitzgerald, T.J. (1979). 'Review of instrumentation for fluidized beds', in *Proc. N.S.F. Workshop*, Rensselaer Polytechnic Inst., pp. 292–309.

Fitzgerald, T.J., Catipovic, N.M., and Jovanovic, G.N. (1981). 'Instrumented cylinder for studying heat transfer to immersed tubes in fluidized beds', *Ind. Eng. Chem. Fund.*, **20**, 82–88.

Flemmer, R.L.C. (1984). 'A pneumatic probe to detect gas bubbles in fluidized beds', *Ind. Eng. Chem. Fund.*, **23**, 113–119.

Forrest, J.S. (1953). 'Methods of increasing the electrical conductivity of surfaces', *Brit. J. Appl. Phys.*, **4**, Suppl.2, S37–39.

Fournol, A.B., Bergougnou, M.A., and Baker, C.G.J. (1973). 'Solids entrainment in a large gas fluidized bed', *Can. J. Chem. Eng.*, **51**, 401–404.

Fukuda, M., Asaki, Z., and Kondo, Y. (1967). 'On the nonuniformity in fluidized beds', *Mem. Fac. Eng., Kyoto Univ.*, **29**, 287–305.

Gabor, J.D. (1967). 'Wall effects on fluidized particle movement in a two-dimensional column', in *Proceedings of International Symposium on Fluidization* (Ed. A.A.H. Drinkenburg), Netherlands University Press, Amsterdam, pp. 230–240.

Gabor, J.D. (1972). 'Wall to bed heat transfer in fluidized beds', *A.I.Ch.E.J.*, **18**, 249–250.

Garcia, A., Grace, J.R., and Clift, R. (1973). 'Behaviour of gas bubbles in fluidized beds', *A.I.Ch.E.J.*, **19**, 369–370.

Gbordzoe, E.A.M., Bulani, W., and Bergougnou, M.A. (1981). 'Hydrodynamic study of floating contactors (bubble breakers) in a fluidized bed', *A.I.Ch.E. Symp. Ser.*, **77** (205), 1–7.

Geldart, D., and Abrahamsen, A.R. (1980). 'The effect of fines on the behaviour of gas fluidized beds of small particles', in *Fluidization* (Eds J.R. Grace and J.M. Matsen), Plenum Press, New York, pp. 453–460.

Geldart, D., and Baeyens, J. (1985). 'The design of distributors for gas fluidized beds', *Powder Technol.*, **42**, 67–78.

Geldart, D., and Cranfield, R.R. (1972). 'The gas fluidization of large particles', *Chem. Eng. J.*, **3**, 211–231.

Geldart, D., Hemsworth, A., Sundavadra, R., and Whiting, K.J. (1981). 'A comparison of spouting and jetting in round and half-round fluidized beds', *Can. J. Chem. Eng.*, **59**, 638–639.

Geldart, D., Hurt, J.M., and Wadia, P.H. (1978)). 'Slugging in beds of large particles', *A.I.Ch.E. Symp. Ser.*, **74** (176), 60–66.

Geldart, D., and Kelsey, J.R. (1972). 'The use of capacitance probes in fluidized beds', *Powder Technol.*, **6**, 45–60.

Geldart, D., and Pope, D.J. (1978). 'Elutriation from a multisize fluidized bed, A.I.Ch.E. Meeting, Miami Beach.

George, S.E., and Grace, J.R. (1978). 'Entrainment of particles from aggregative fluidized beds', *A.I.Ch.E. Symp. Ser.*, **74** (176). 67–74.

George, S.E., and Grace, J.R. (1982). 'Heat transfer to horizontal tubes in the freeboard region of a gas fluidized bed', *A.I.Ch.E.J.*, **28**, 759–765.

Gerald, C.F. (1951). 'Measuring uniformity of fluidization', *Chem. Eng. Prog.*, **47**, 483–484.

Germain, B., and Claudel, B. (1976). 'Fluidization at mean pressure less than 30 torr, *Powder Technol.*, **13**, 115–121.

Gilliland, E.R., and Mason, E.A. (1952). 'Gas mixing in beds of fluidized solids, *Ind. Eng. Chem.*, **44**, 218–224.

Glass, D.H., and Harrison, D. (1964). 'Flow pattern near a solid obstacle in a fluidized bed', *Chem. Eng. Sci.*, **19**, 1001–1002.

Glass, D.H., and Mojtahedi, W. (1983). 'Measurement of fluidized bed bubbling properties using a fibre-optic light probe', *Chem. Eng. Res. & Des.*, **61**, 37–44.

Glicksman, L.R., and McAndrews, G., (1985). 'The effect of bed width on the hydrodynamics of large particle fluidized beds', *Powder Technol.*, **42**, 159–167.

Godard, K., and Richardson, J.F. (1968). 'The behaviour of bubble-free fluidized beds', *Instn Chem. Engrs Symp. Ser.*, **30**, 126–135.

Golan, L.P., Lalonde, G.V., and Weiner, S.C. (1982). 'High temperature heat transfer studies in a tube filled bed', *Proc. 6th Intern. Fluidized Bed Combustion Conf.*, **3**, 1173–1184.

Grace, J.R. (1974). 'Fluidization and its application to coal treatment and allied processes', *A.I.Ch.E. Symp. Ser.*, **70** (141), 21–26.

Grace, J.R., and Clift, R. (1974). 'On the two-phase theory of fluidization', *Chem. Eng. Sci.*, **29**, 327–334.

Grace, J.R., and Harrison, D. (1968). 'The effect of internal baffles in fluidized beds: a guide to design', *Instn Chem. Engrs Symp. Ser.*, **27**, 93–100.

Grace, J.R., and Harrison, D. (1969). 'The behaviour of freely bubbling fluidized beds', *Chem. Eng. Sci.*, **24**, 497–500.

Grace, J.R., Lim, C.J., and Evans, R.L. (1983). 'Preparatory Report of pressurized fluidized bed combustion', Report to Government of Canada under Contract 1SU82–00251.

Guinn, V.P. (1956). *Nucleonics.*, **14** (May), 69.

Gunn, D.J. and Al-Doori, H.H. (1985). 'The measurement of bubble flows in fluidized beds by electrical probe', *Int. J. Multiphase Flow*, **11**, 535–551.

Hager, W.R., and Thomson, W.J. (1973). 'Bubble behaviour around immersed tubes in a fluidized bed', *A.I.Ch.E. Symp. Ser.*, **69** (128), 68–77.

Handley, M.F., and Perry, M.G. (1965). *Rheol. Acta, 4*, 225.

Harper, W.R. (1960). *Soc. Chem. Ind. Monograph*, **14**, 115.

Harrison, D., and Leung, L.S. (1962). 'The coalescence of bubbles in fluidized beds', in *Proc. Symp. Interactions Fluids Particles,* Instn Chem. Engrs, London, pp. 127–134.

Hatate, Y., King, D.F., Migita, M., and Ikari, A. (1985). 'Behaviour of bubble in a semi-cylindrical gas-solid fluidized bed', *J. Chem. Eng. Japan.*, **18**, 99–104.

Heertjes, P.M., Verloop, J., and Willems, R. (1970/71). 'The measurment of local mass flow rates and particle velocities in fluid-solids flow, *Powder Technol.*, **4**, 38–40.

Hengl, G., Hiquily, N., and Couderc, J.P. (1977). 'A new distributor for gas fluidization', *Powder Technol.*, **18**, 277–278.

Henry, P.S.H. (1953). 'Electrostatic eliminators in the textile industry', *Brit. J. Appl. Phys.*, **4**, Suppl. 2, S78–83.

Hiraki, I., Yoshida, K., and Kunii, D. (1966). 'Behaviour of bubbles in a two-dimensional fluidized bed', *Chem. Eng. Japan*, **4**, 139–146.

Ho T–C., Yutani, N., Fan, L.T., and Walawender, W.P. (1983). 'The onset of slugging in gas–solid fluidized beds with large particles', *A.I.Ch.E.J.*, **35**, 249–257.

Hoebink, J.H.B.J., and Rietema, K. (1978). 'Wall-to-bed heat transfer in a fluidized bed', in *Fluidization* (Eds J.F. Davidson and D.L. Keairns), Cambridge University Press, pp. 327–332.

Horio, M., Taki, A., Hsieh, Y.S., and Muchi, I. (1980). 'Elutriation and particle transport through the freeboard of a gas–solid fluidized bed', in *Fluidization* (Eds J.R. Grace and J.M. Matsen), Plenum Press, New York, pp. 509–518.

Horsler, A.G., and Thompson, B.H. (1968). 'Fluidization in the development of gas-making processes', *Instn Chem. Engrs Symp. Ser.*, **30**, 58–66.

Hosny, N., and Grace, J.R. (1984). 'Forces on a tube immersed within a fluidized bed', in *Fluidization* (Eds D. Kunii and R. Toei), Engineering Foundation, New York, pp. 111–120.

Ishida, M., and Hatano, H. (1984). 'The behaviour of gas and solid particles in a gas–solid fluidized bed detected by optical fiber probes', in *Fluidization* (Eds D. Kunii and R. Toei), Engineering Foundation, New York.

Ishida, M., Nishiwaki, A., and Shirai, T. (1980). 'Movement of solid particles around bubbles in a three-dimensional fluidized bed at high temperature', in *Fluidization* (Eds J.R. Grace and J.M. Matsen), Plenum Press, New York, pp. 357–364.

Jacob, A., and Osberg, G.L. (1957). 'Effect of gas thermal conductivity on local heat transfer in a fluidized bed', *Can. J. Chem. Eng.*, **35**, 5–9.

Jin, Y., Yu, L.I.Z., and Wang, Z. (1980). 'A study of particle movement in a gas fluidized bed', in *Fluidization* (Eds J.R. Grace and J.M. Matsen), Plenum Press, New York, pp. 365–372.

Kamiya, Y. (1955). *Chem. Eng. Japan*, **19**, 412.

Katosonov, I.V., Menshov, V.N., Zuev, A.A., and Anokhin, V.N. (1974). 'Density of fluidized beds of fine grained materials and its variation under pressure', *Zh. Prikl. Khim.*, **47**, 1861–1866.

Katz, H.H., and Sears, J.T. (1969). 'Electric field phenomena in fluidized and fixed beds', *Can. J. Chem. Eng.*, **47**, 50–53.

Kavlick, V.J., and Lee, R.S. (1970). 'High pressure electrothermal fluidized bed gasification of coal char', *Chem. Eng. Prog. Symp. Ser.*, **66** (105), 145–151.

Kawabata, J., Yumiyama, M., Tazaki, Y., Honma, S., Chiba, T., Sumiya, T., and Endo, K. (1981). 'Characteristics of gas fluidized beds under pressure', *J. Chem. Eng. Japan*, **14**, 85–89.

Kawanishi, K., and Yamazaki, M. (1984). 'Structure of emulsion phase in fluidized bed', in Fluidization (Eds D. Kunii and R. Toei), Engineering Foundation, New York, pp. 37–44.

King, D.F., and Harrison, D. (1980). 'The bubble phase in high-pressure fluidized beds', in *Fluidization* (Eds J.R. Grace and J.M. Matsen), Plenum Press, New York, pp. 101–108.

Knowlton, T.M. (1977). 'High pressure fluidization characteristics of several particulate solids, primarily coal and coal-derived materials', *A.I.Ch.E. Symp. Ser.*, **73** (161), 22–28.

Knowlton, T.M., and Hirsan, J. (1980). 'The effect of pressure on jet penetration in semi-cylindrical gas fluidized beds', in *Fluidization* (Eds J.R. Grace and J.M. Matsen), Plenum Press, New York, pp. 315–324.

Kobayashi, H., and Arai, F. (1964). 4th Symp. Chem. React. Eng., Soc. Chem. Engrs, Osaka.

Kondukov, N.B., Kornilaev, A.N., Skachko, I.M., Akhromenkov, A.A., and Kruglov, A.S. (1964). 'An investigation of the parameters of moving particles in a fluidized bed by a radioisotopic method'. *Int. Chem. Eng.*, **4**, 43–47.

Kunii, D., and Levenspiel, O. (1969). *Fluidization Engineering*, Wiley, New York.

Kunii, D., Yoshida, K., and Hiraki, J. (1967). 'The behaviour of freely bubbling fluidized beds', In *Proc. Intern. Symp. on Fluidization* (Ed. A.A.H. Drinkenburg), Netherlands University Press, Amsterdam, pp. 243–254.

Kwauk, M., Wang. N., Li, Y., Chen, B., and Shen, Z. (1986) 'Fast Fluidization at ICM', In *Circulating Fluidized Bed Technology* (Ed. P. Basu), Pergamon Press, Toronto, pp. 33–62.

Lanneau, K.P. (1960). 'Gas–solid contacting in fluidized beds', *Trans. Instn Chem. Engrs*, **38**, 125–137.

Lau, I.T., and Whalley, B.J.P. (1981). 'A differential thermal probe for anticipation of defluidization of caking coals', *Fuel Proc. Technol.*, **4**, 101–115.

Lee, B.S., Pyrcioch, E.J., and Schora, F.C. (1970). 'Hydrogasification of coal in high-pressure fluidized beds', *Chem. Eng. Prog. Symp. Ser.*, **66** (105), 152–156.

Levy, E.K., Caram, H.S., Dille, J.C., and Edelstein, S. (1983). 'Mechanisms for solids ejection from gas fluidized beds', *A.I.Ch.E.J.*, **29**, 383–388.

Levy, Y., and Lockwood, F.C. (1980). 'Two phase flow measurements in the freeboard of a fluidized bed using laser Doppler anemometry', *Israel J. Technol.*, **18**, 146–151.

Levy, Y., and Lockwood, F.C. (1983). 'Laser Doppler measurements of flow in freeboard of a fluidized bed', *A.I.Ch.E.J.*, **29**, 889–895.

Lewis, W.K., Gilliland, E.R. and Bauer, W.C. (1949). 'Characteristics of fluidized particles', *Ind. Eng. Chem.*, **41**, 1104–1117.

Lignola, R.G., Donsi, G., and Massimilla, L. (1983). 'Mass spectrometric measurements of gas composition profiles associated with bubbles in a two-dimensional bed', *A.I.Ch.E. Symp. Ser.*, **79** (222), 19–26.

Lim, C.J., Ko, G.H., and Grace, J.R. (1986) 'Oxygen probe measurements and their accuracy in coal combustors'. To be published.

Lin, J.S., Chen, M.M., and Chao, B.T. (1985). 'A novel radioactive tracking facility for measurement of solids motion in gas fluidized beds', *A.I.Ch. E.J.*, **31**, 465–473.

Littman, H. (1964). 'Solids mixing in straight and tapered fluidized beds', *A.I.Ch.E.J.*, **10**, 924–929.

Littman, H., and Homolka, G.A.J. (1973). 'The pressure field around a two-dimensional gas bubble in a fluidized bed', *Chem. Eng. Sci.*, **28**, 2231–2243.

Ljungstrom, E., and Lindqvist, O. (1982). 'Measurements of in-bed gas and solids compositions in a combustor operating at pressures up to 20 bar', *Proc. 7th Intern. Fluidized Bed Combustion Conf.*, **1**, 465–472.

Lockett, M.J., and Harrison, D. (1967). 'The distribution of voidage fraction near bubbles rising in gas–fluidized beds', *Proceedings of International Symposium on Fluidization* (Ed. A.A.H. Drinkenburg), Netherlands University Press, Amsterdam, 257–267.

Lockwood, D.N. (1977). 'Effects of heat exchanger tube spacing and arrangement on the quality of fluidization', *Proc. Pachec Conf.*, Denver, pp. 1177–1181.

Loeb, L.B. (1958). *Static Electrification*, Springverlag, Berlin.

Loew, O., Schmutter, B., and Resnick, W. (1979) 'Particle and bubble behaviour and velocities in a large-particle fluidized bed with immersed obstacles', *Powder Technol.*, **22**, 45–57.

Lord, W.K., McAndrews, G., Sakagami, M., Valenzuela, J.A., and Glicksman, L.R. (1982). 'Measurement of bubble properties in fluidized beds', *Proc. 7th Intern. Fluidized Bed Combustion Conf.*, **1**, 76–88.

Lyall, E. (1969). 'The photography of bubbling fluidized beds', *Brit. Chem. Eng.*, **14**, 501–506.

Marsheck, R.M., and Gomezplata, A. (1965). 'Particle flow pattern in a fluidized bed', *A.I.Ch.E.J.*, **11**, 167–173.

Masson, H., and Jottrand, R. (1978). 'Measurement of local bubble properties in a fluidized bed', in *Fluidization* (Eds J.F. Davidson and D.L. Keairns), Cambridge University Press, pp. 1–6.

Masson, H., Jottrand, R., and Dang Tran, K. (1978). Intern. Cong. on Mixing in the Chemical Industries, Mons.

Mathur, K.B., and Epstein, N. (1974). *Spouted Beds*, Academic Press, New York.

May, W.G. (1959). 'Fluidized bed reactor studies', *Chem. Eng. Prog.*, **55** (12), 49–56.

May, W.G., and Russell, F.D. (1954). 'High pressure fluidization', Paper given at North New Jersey A.C.S. meeting.

Mayhofer, B., and Neuzil, L. (1977). 'Influence of grid on fluidized bed inhomogeneity', *Chem. Eng. J.*, **14**, 167–173.

Merrick, D., and Highley, J. (1972). 'Particle size reduction and elutriation in a *fluidized bed processes', A.I.Ch.E. Symp. Ser.*, **70** (137), 366–378.

Merry, J.M.D., and Davidson, J.F. (1973). 'Gulf stream circulation in shallow fluidized beds', *Trans. Instn Chem. Engrs*, **51**, 361–368.

Mickley, H.S., Fairbanks, D.F., and Hawthorn, R.D. (1961). 'The relation between the transfer coefficient and thermal fluctuations in fluidized bed heat transfer', *Chem. Eng. Prog. Symp. Ser.*, **57**, 51–60.

Mickley, H.S. and Trilling, C.A. (1949). 'Heat transfer characteristics of fluidized beds', *Ind. Eng. Chem.*, **41**, 1135–1147.

Miller, C.O., and Logwinuk, A.K. (1951). 'Fluidization studies of solid particles, *Ind. Eng. Chem.*, **43**, 1220–1226.

Mogan, J.P., Taylor, R.W., and Booth, F.L. (1971). 'The value of the exponent in the Richardson and Zaki equation for fine solids fluidized with gases under pressure', *Powder Technol.*, **4**, 286–289.

Morooka, S., Kago, T., and Kato, Y. (1984). 'Flow pattern of solid particles in freeboard of fluidized bed', in *Fluidization* (Eds D. Kunii and R. Toei), Engineering Foundation, New York, pp. 291–298.

Morse, R.D., and Ballou, C.O. (1951). 'The uniformity of fluidization, its measurement and use', *Chem. Eng. Prog.*, **47**, 199–211.

Nazemi, A., Bergougnou, M.A., and Baker, C.G.J. (1974). 'Dilute phase holdup in a large gas fluidized bed', *A.I.Ch.E. Symp. Ser.*, **70** (14), 98–107.

Nguyen, T.H., and Grace, J.R. (1978). 'Forces on objects immersed in fluidized beds', *Powder Technol.*, **19**, 255–264.

Nguyen, X.T., and Leung, L.S. (1972). 'A note on bubble formation at an orifice in a fluidized bed', *Chem. Eng. Sci.*, **27**, 1748–1750.

Nienow, A.W., and Cheesman, D.J. (1980). 'The effect of shape on the mixing and segregation of large particles in a gas–fluidized bed of small ones, in *Fluidization* (Eds J.R. Grace and J.M. Matsen), Plenum Press, New York, pp. 373–380.

Ohki, K., Ishida, M., and Shirai, T. (1980). 'The behaviour of jets and particles near the distributor grid in a three-dimensional fluidized bed', in *Fluidization* (Eds J.R. Grace and J.M. Matsen), Plenum Press, New York, pp. 421–428.

Ohki, K., and Shirai, T. (1976). 'Particle velocity in fluidized beds', in *Fluidization Technology* (Ed. D.L. Keairns), Vol. I, Hemisphere, Washington, DC, pp. 95–110.

Okhi, K., Walawender, W.P., and Fan, L.T. (1977). 'The measurement of local velocity of solid particles', *Powder Technol.*, **18**, 171–178.

Orcutt, J.C., and Carpenter, B.H. (1971). 'Bubble coalescence and the simulation of mass transport and chemical reaction in gas fluidized beds', *Chem. Eng. Sci.*, **26**, 1046–1064.

Ormiston, R.M., Mitchell, F.R.G., and Davidson, J.F. (1965). 'The velocity of slugs in fluidized beds', *Trans. Instn Chem. Engrs*, **43**, 209–216.

Osberg, G.L. (1951). 'Locating fluidized solids bed level in a reactor', *Ind. Eng. Chem.*, **43**, 1871–1873.

Overcashier, R.D., Todd, D.B., and Olney, R.B. (1959). 'Some effects of baffles on a fluidized system', *A.I.Ch.E.J.*, **5**, 54–60.

Ozkaynak, T.F., and Chen, J.C. (1978). 'Average residence times of emulsion and void phases at the surface of heat transfer tubes in fluidized beds', *A.I.Ch.E. Symp. Ser.*, **74** (174), 334–343.

Ozkaynak, T.F., Chen, J.C., and Frankenfield, T.R. (1984). 'An experimental investigation of radiation (Eds D. Kunii and R. Toei), Engineering Foundation, New York, pp. 371–378.

Park, W.H., Kang, W.L., Capes, C.E., and Osberg, G.L. (1969). 'The properties of bubbles in fluidized beds of conducting particles as measured by an electroresistivity probe', *Chem. Eng. Sci.*, **24**, 851–865.

Park, W.H., Lee, N.G., and Capes, C.E. (1974). 'Wall effects in point probe measurements of radial bubble distributions', *Chem. Eng. Sci.*, **29**, 339–344.

Partridge, B.A., Lyall, E., and Crooks, H.E. (1969). 'Particle slip surfaces in bubbling gas fluidized beds', *Powder Technol.*, **2**, 301–305.

Patrose, B., and Caram, H.S. (1982). 'Optical fiber probe transit anemometer for particle velocity measurements in fluidized beds', *A.I.Ch.E.J.*, **28**, 604–609.

Patureaux, T., Vergnes, F., and Mihe, J.P. (1973). 'Construction of piezoelectric transducer for statistical study on impact of particles in fluidized beds', *Powder Technol.*, **8**, 101–105.

Pemberton, S.T. (1982). 'Entrainment from fluidized beds', Ph.D. Thesis, Cambridge University.

Piepers, H.W., Wiewiorski, P., and Rietema, K. (1984). 'Heat transfer on a vertical tube bundle immersed in a 0.70 m fluidized bed', in *Fluidization* (Eds D. Kunii and R. Toei), Engineering Foundation, New York, pp. 339–346.

Pyle, D.L., and Harrison, D. (1967). 'The rising velocity of bubbles in two-dimensional fluidized beds', *Chem. Eng. Sci.*, **22**, 531–535.

Qin, S., and Liu, G. (1982). 'Application of optical fibres to measurement and display of fluidized systems', in *Fluidization Science and Technology* (Eds M. Kwauk and D. Kunii), Science Press, Beijing, pp. 258–266.

Rao, V.L., and Venkateswarku, D. (1973). 'Determination of velocities and flow patterns of particles in mass flow hoppers', *Powder Technol.*, **7**, 263–265.

Raso, G., Tirabasso, G., and Donsi, G. (1983). 'An impact probe for local analysis of gas–solid flows', *Powder Technol.*, **34**, 151–159.

Richardson, J.F., and McLeman, M. (1960). 'Pneumatic conveying: solids velocities and pressure gradients in a one-inch horizontal pipe', *Trans. Instn Chem. Engrs*, **38**, 257–266.

Rietema, K. (1967). 'Application of mechanical stress theory of fluidization', in *Proceedings of International Symposium on Fluidization* (Ed. A.A.H. Drinkenburg), Netherlands University Press, Amsterdam, pp. 154–166.

Rietema, K., and Hoebink, J. (1976). 'Mass transfer from single rising bubbles to the dense phase in three-dimensional fluidized beds', in *Fluidization Technology* (Ed. D.L. Keairns) Vol. I, Hemisphere, Washington, DC, pp. 279–288.

Rowe, P.N. (1971). 'Experimental properties of bubbles', Chapter 4 in *Fluidization* (Eds J.F. Davidson and D. Harrison), Academic Press, London, pp. 121–191.

Rowe, P.N., and Everett, D.J. (1972). 'Fluidized bed bubbles viewed by X-rays', *Trans. Instn Chem. Engrs*, **50**, 42–60.

Rowe, P.N., Foscolo, P.U., Hoffman, A.C., and Yates, J.G. (1984). 'X-ray observation of gas fluidized beds under pressure', in *Fluidization* (Eds D. Kunii and R. Toei), Engineering Foundation, New York, pp. 53–60.

Rowe, P.N., MacGillivray, H.J., and Cheesman, D.J. (1979). 'Gas discharge from an orifice into a gas fluidized bed', *Trans. Instn Chem. Engrs.*, **57**, 194–199.

Rowe, P.N., and Masson, H. (1980). 'Fluidized bed bubbles observed simultaneously by probe and X-rays', *Chem. Eng. Sci.*, **35**, 1443–1447.

Rowe, P.N., and Masson, H. (1981). 'Interactions of bubbles with probes in gas fluidized beds', *Trans. Instn Chem. Engrs*, **59**, 177–185.

Rowe, P.N., and Partridge, B.A. (1962). 'Particle movement caused by bubbles in a fluidized bed', in *Proc. Symp. Interactions Fluids Particles,* Instn Chem. Engrs, London, pp. 135–142.

Rowe, P.N. and Partridge, B.A. (1965). 'Aggregative fluidization', *Chem. Eng. Sci.*,

20, 985.

Rowe, P.N., Partridge, B.A., Cheney, A.G., Henwood, G.A., and Lyall E. (1965). The mechanisms of solids mixing in fluidized beds', *Trans. Instn. Chem Engrs*, **43**, 271–286.

Rowe, P.N., Partridge, B.A. and Lyall, G. (1964). 'Cloud formation around bubbles in fluidized beds', *Chem. Eng. Sci.*, **19**, 973–985.

Sathiyamoorthy, D., and Rao, C.S. (1981). 'The choice of distributor to bed pressure drop in gas fluidized beds'. *Powder Technol.*, **30**, 139–143.

Saxton, A.L., and Worley, A.C. (1970). 'Modern catalytic cracking design', *Oil & Gas J.*, **68** (20). 82–99.

Schugerl, K. (1967). 'Experimental comparison of mixing processes in two- and three-phase fluidized beds' in *Proceedings of International Symposium on Fluidization*, Netherlands University Press, Amsterdam, pp. 782–794.

Schulz-Walz, A. (1974). 'Messing von Schuttgutgeschwindigkeiten am Beispiel der Schwingsiebzentrifuge, *Chem. Ing. Techn.*, **46**, 259.

Schuurmans, H.J.A. (1980). 'Measurements in a commercial catalytic cracking unit'. *Ind. Eng. Chem. proc. Des & Dev.*, **19**, 267–271.

Sit, S.P., and Grace, J.R. (1978). 'Interphase mass transfer in an aggregative fluidized bed', *Chem. Eng. Sci.*, **33**, 1115–1122.

Sitnai, D. (1982). 'Utilization of the pressure differential records from gas fluidized beds with internals for bubble parameters determination', *Chem. Eng. Sci.*, **37**, 1059–1066.

Sokoro, V.E., Mikhalev, M.F., and Mukhlenov, I.P. (1969). 'Calculation of the minimum height of the space above the bed in fluidized bed contact equipment', *Int. Chem. Eng.*, **9**, 280–281.

Soo, S.L., Ihring, H.K., and El Kouh, A.G. (1960). 'Experimental determination of statistical properties of two-phase turbulent motion', *Trans. ASME, J. Basic Eng.*, **82**, 609–621.

Staub, F.W., and Canada, G.S. (1978). 'Effect of tube bank and gas density on flow behaviour and heat transfer in fluidized beds'. In *Fluidization* (Eds J.F. Davidson and D.L. Keairns), Cambridge Univerisity Press, Cambridge, pp. 339–344.

Subzwari, M.P., Clift, R., and Pyle, D.L. (1978). 'Bubbling behaviour of fluidized beds at elevated pressures', in *Fluidization* (Eds J.F. Davidson and D.L. Keairns), Cambridge University Press, Cambridge, pp. 50–54.

Sutherland, J.P. (1964). 'The measurement of pressure drop across a gas fluidized bed', *Chem. Eng. Sci.*, **19**, 839–841.

Svoboda, K., Cermak, J., Hartman, M., Drahos, J., and Selucky, K. (1984). 'Influence of particle size on the pressure fluctuations and slugging in a fluidized bed', *A.I.Ch.E.J.*, **30**, 513–517.

Tanimoto, H., Chiba, S., Chiba, T., and Kobayashi, H. (1980). 'Mechanism of solid segregation in gas fluidized beds', in *Fluidization* (Eds J.R. Grace and J.M. Matsen), Plenum, New York, pp. 381–388.

Torobin, L.B., and Gauvin, W.H. (1961). 'Fundamental aspects of solids-gas flow', *Can. J. Chem. Eng.*, **39**, 113–120.

Tuot, J., and Clift, R. (1973). 'Heat transfer around single bubbles in a two-dimensional fluidized bed', *A.I.Ch.E. Symp. Ser.*, **69**, (128), 78–84.

Valenzuela, J.A., and Glicksman, L.R. (1984a). 'An experimental study of solids mixing in a freely bubbling two-dimensional fluidized bed', *Powder Technol.*, **38**, 63–72.

Valenzuela, J.A., and Glicksman, L.R. (1984b). 'Residence time and dispersion of fine particles in a 2–d fluidized bed of large particles', in *Fluidization* (Eds D. Kunii and R. Toei), Engineering Foundation, New York, pp. 161–168.

Van Deemter, J.J. (1980). 'Mixing patterns in large-scale fluidized beds', in *Fluidization* (Eds J.R. Grace and J.M. Matsen), Plenum Press, New York, pp. 69–89.

Van Velzen, D., Flamm, H.J., Langenkamp, H., and Castle, A. (1974). 'Motion of solids in spouted beds', *Can. J. Chem. Eng.*, **52**, 156–161.

Varadi, T., and Grace, J.R. (1978). 'High pressure fluidization in a two-dimensional gas fluidized bed', in *Fluidization* (Eds J.F. Davidson and D.L. Keairns), Cambridge University Press, Cambridge, pp. 55–58.

Vines, S.N. (1977). 'Gas–solid regimes in the distributor region of a fluidized bed', Ph.D. Thesis, University of Virginia.

Wace, P.F., and Burnett, S.J. (1961). 'Flow pattern in gas–fluidized beds', *Trans. Instn Chem. Engrs*, **39**, 168–174.

Walker, V., and Rapier, A.C. (1965). UKAEA Report. JRG1026 (W).

Walton, J.S., Olson, R.L., and Levenspiel, O. (1952). Gas–solid film coefficients of heat transfer in fluidized coal beds', *Ind. Eng. Chem.*, **44**, 1474–1480.

Weimer, A.W., Gyure, D.C., and Clough, D.E. (1985). 'Application of a gamma-radiation density gauge for determining hydrodynamic properties of fluidized beds', *Powder Technol.*, **44**, 179–194.

Wen, C.Y., Krishnan, R., Khosravi, R., and Dutta, S. (1978). 'Dead zone heights near the grid of fluidized beds', in *Fluidization* (Eds J.F. Davidson and D.L. Keairns), Cambridge University Press, Cambridge, pp. 32–37.

Werther, J. (1974). 'Bubbles in gas fluidized beds', *Trans. Instn Chem. Engrs,* **52**, 149–169.

Werther, J., and Molerus, O. (1973). 'The local structure of gas fluidized beds', *Int. J. Multiphase Flow*, **1**, 103–138.

Whitehead, A.B., and Dent, D.C. (1978). 'Some effects of distributor slope and auxiliary gas injection on the performance of gas–solid fluidized beds', in *Fluidization* (Eds J.F. Davidson and D.L. Keairns), Cambridge University Press, Cambridge, pp. 44–49.

Whitehead, A.B., Gartside, G., and Dent, D.C. (1976). 'Fluidization studies in large gas–solid systems', *Powder Technol.*, **14**, 61–70.

Whitehead, A.B., Potter, O.E., Nguyen, H.V., and Dent, D.C. (1980). 'Gas backmixing in 0.61 m and 1.22 m square fluidized beds', in *Fluidization* (Eds J.R. Grace and J.M. Matsen), Plenum Press, New York, pp. 333–340.

Whitehead, A.B., and Young, A.D. (1967), 'Fluidization performance in large scale equipment', in *Proceedings of International Symposium on Fluidization* (Ed. A.A.H. Drinkenburg), Netherlands University Press, Amsterdam, pp. 294–302.

Whiting, K.J., and Geldart, D. (1980). 'A comparison of cylindrical and semi-cylindrical spouted beds of coarse particles', *Chem. Eng. Sci.*, **35**, 1499–1501.

Woollard, I., and Potter, O.E. (1968). 'Solids mixing in fluidized beds', *A.I.Ch.E.J.*, **14**, 338–391.

Xavier, A.M., King, D.F., Davidson, J.F., and Harrison, D. (1980). 'Surface–bed heat transfer in a fluidized bed at high pressure', in *Fluidization* (Eds J.R. Grace and J.M. Matsen), Plenum Press, New York, pp. 209–216.

Yagi, S., and Kunii, D. (1961). 'Fluidized solids reactors with continuous solids feed, *Chem. Eng. Sci.*, **16**, 364–371.

Yang, G., Huang, Z., and Zhao, L. (1984). 'Radial gas dispersion in a fast fluidized bed', in *Fluidization* (Eds D. Kunii and R. Toei), Engineering Foundation, New York, pp. 145–152.

Yang, W–C., Revay, D., Anderson, R.G., Chelen, E.J., Keairns, D.L., and Cicero, D.C. (1984). 'Fluidization phenomena in a large-scale cold-flow model', in *Fluidization* (Eds D. Kunii and R. Toei), Engineering Foundation, New York, pp.

Yates, J.G., Cheesman, D.J., Mashingaidze, T.A., Howe, C., and Jefferis, G. (1984). 'The effect of vertical rods on bubbles in gas fluidized beds', in *Fluidization* (Eds D. Kunii and R. Toei), Engineering Foundation, New York, pp. 103–110.

Yates, J.G., and Walker, P.R. (1978). 'Particle temperatures in a fluidized bed combustor', in *Fluidization* (Eds J.F. Davidson and D.L. Keairns), Cambridge University Press, Cambridge, pp. 241–245.

Yerushalmi, J., Cankurt, N.T., Geldart, D., and Liss, B. (1978). 'Flow regimes in vertical gas–solid contact systems', *A.I.Ch.E. Symp. Ser.*, **74** (176), 1–13.

Yerushalmi, J., Turner, D.H., and Squires, A.M. (1976). 'The fast fluidized bed', *Ind. Eng. Chem. Proc. Des. & Dev.*, **15**, 47–53.

Yoshida, K., and Kunii, D. (1968). 'Stimulus and response of gas concentration in bubbling fluidized beds', *J. Chem. Eng. Japan*, **1**, 11–16.

Yoshida, K., Fujii, S., and Kunii, D. (1976). 'Characteristics of fluidized beds at high temperatures', in *Fluidization Technology* (Ed. D.L. Keairns), Vol. 1, Hemisphere, Washington, DC, pp. 43–48.

Zabrodsky, S.S., Antonishin, N.V., and Parnas, A.L. (1976). 'On fluidized bed-to-surface heat transfer', *Can. J. Chem. Eng.*, **54**, 52–58.

Zenz, F.A., and Weil, N.A. (1958). 'A theoretical–empirical approach to the mechanisms of particle entrainment from fluidized beds', *A.I.Ch.E.J.*, **4**, 472–479.

Zhang, M.C., Walsh, P.M., and Beér, J.M. (1982). 'Determination of bubble size distributions from pressure fluctuations in a fluidized bed combustor', *7th Intern. Fluidized Bed Combustion Conf.*, **1**, 75–75J.

Index